Investigations in
BIOLOGY

Richard J. Montgomery
Hagerstown Junior College

William D. Elliott
Hagerstown Junior College

D. C. HEATH AND COMPANY
Lexington, Massachusetts   Toronto

Cover: Larry Ulrich/DRK Photo

Published simultaneously in Canada.

Printed in the United States of America.

International Standard Book Number: 0–669–12010–3

10 9 8 7 6 5 4 3 2 1

PREFACE

This laboratory text conforms in content organization with *Biology: Discovering Life* by Joseph S. Levine and Kenneth R. Miller but can be used in conjunction with any biology textbook written for an introductory biology course.

The laboratory activities in this book are designed for professors who believe that laboratory instruction is an essential ingredient in the biology curriculum. In that regard, these activities are meant to complement the lecture, not be secondary to it. During our many years of teaching, we have developed these forty-three activities as they were being used by our students, whose continued feedback allowed us to improve on our initial ideas.

By providing a variety of activities that sample the enormous scope of modern biology and involve students directly in the process of being a practicing biologist, we have tried to convey our excitement about the field of biology. Both the experimental and descriptive sides of biology investigation are included, with experimental procedures, data collection, and data analysis presented at an understandable yet challenging level. Whenever possible, simple homemade equipment is assembled by the students so that they may better understand and appreciate its operation.

We have developed several activities not usually found in introductory biology laboratories. These include:

- Laboratory Activities 4–7, which constitute "The Study of a Freshwater Stream," a tested aquatic field study that can be done in its entirety or as individual activities.
- Laboratory Activity 19, an investigation of the unique relationship between numbers and plant morphology.
- Laboratory Activity 20, which explores the physical properties of water and how these properties relate to the biology of organisms.
- Laboratory Activity 22, on model building and the isolation of DNA.
- Laboratory Activity 42, which introduces students to the field of chronobiology and enables them to measure their own circadian rhythms.

Each activity is divided into sections. The Objectives list the skills and knowledge the students will have acquired when the activity is finished. While not meant to be exhaustive, the Introduction will provide the student with sufficient background information to do an activity without constant reference to a textbook. Questions in the Evaluation will normally require the student to fill in data or record observations. Most importantly, evaluation questions are designed to tie ideas together and to have students understand and think about what they have done and observed. Each evaluation may be easily removed for grading.

Acknowledgments

Many people were helpful during the development of this laboratory text. Our students have always been extremely cooperative about completing new activities and offering suggestions for improvement. Likewise, our teaching colleagues have helped enormously by using these materials in their classes and evaluating them. Over the years, many staff members at Hagerstown Junior College have helped in ways too numerous to mention—we thank them for their efforts. A special note of gratitude goes to our Audio-Visual Department for the photographs used in the activities on the fetal pig dissection and model building of DNA. Eve Walton of the National Cancer Institute in Frederick, Maryland, was extremely helpful in developing the procedures for DNA extraction.

The assistance given to us by the late Mary Le Quesne and the editorial staff of D. C. Heath has been immeasurable. The developmental editing done by Kathi Prancan has made this an infinitely better book. We are grateful to the following manuscript reviewers for their extremely valuable suggestions and comments: Carla Bowan, Los Angeles Mission College; Robert Blystone, Trinity University; Daniel Hornbach, Macalester College; Ron Leavitt, Brigham Young University; Fred McCorkle, Central Michigan University; David Prescott, University of Colorado; Ralph Reiner, College of the Redwoods; and Paul Wright, West Carolina University. We owe much to our wives, who have been helpful, understanding, and patient throughout the whole process. Finally, we thank Hagerstown Junior College for giving us the freedom to teach as we think best for over twenty years.

Richard J. Montgomery
William D. Elliott

This laboratory text was prepared as a guide for your work during the introductory biology course. As a student of biology, one of the most important experiences you will have is to assume the role of a practicing biologist. Although lectures and textbooks will provide you with many valuable experiences, it is in the laboratory that you will come into contact with the frustrations, rewards, and excitement typical of the daily work of a practicing biologist. Our purpose in writing this laboratory text was to help clarify the meaning and methods of biology in the hope that you will have a better understanding of the living world around you.

You can make the most efficient use of your time in the laboratory if you follow these guidelines:

- Read and be sure you understand the safety information below. Sign the Laboratory Safety Contract and submit it to your instructor.
- Read the laboratory activity for the coming week ahead of time. This will save lengthy explanations at the start of the laboratory period concerning procedures.
- Try to organize your data into neatly prepared charts, tables, and graphs.
- Remember that honesty is the cornerstone of science. Simply because your neighbor's data differ from yours does not mean that you are wrong. Keep in mind that scientists live and work by sharing their ideas and the results of their experiments, and you should do the same with the other members of your class.

The experiences you will have during your year of biology can be as rewarding and enriching as you make them. We wish you much success.

Laboratory Safety Procedures

Your work in the biology laboratory will occasionally involve the use of equipment and chemical reagents that have the potential of doing harm if they are not handled properly. In order to make your learning experience in the laboratory meaningful and safe, it is essential that you adhere to the following general safety rules at all times. Specific safety procedures will be given in each laboratory when they are needed; these may also be supplemented by your instructor.

- Wear safety goggles during all activities involving the use of caustic or corrosive chemicals and at any other time as directed by your instructor.
- Contact lenses may represent a hazard under certain circumstances. If this is the case, your instructor will inform you and require you to wear goggles specifically designed for people with contact lenses.
- Do not eat, drink, or smoke in the laboratory.
- Never smell any chemical directly from the container. To smell a chemical, fan the vapors toward you with your hand.
- Never pipette by mouth. Always use an automatic pipetting device.
- Immediately report any chemical or bacteriological spills to your instructor so that proper cleanup procedures may be carried out.
- Learn the location and proper use of all safety equipment: fire blanket, eyewash fountain, first aid kit, fire extinguisher, and fire alarm.
- Avoid wearing loose, baggy clothing in the laboratory. Tie back long hair and roll up long sleeves when working near Bunsen burners.
- Treat all toxic and flammable chemical reagents with extreme care. Follow any specific instructions given by your instructor concerning the chemicals used in a particular laboratory activity.
- Dispose of used chemicals, solid waste, and broken glass in the proper container and according to your instructor's directions.
- Keep your area clean by wiping your work surface after use. Clean all equipment and return it to its proper place in the laboratory after use.
- Notify your instructor immediately of any accident, no matter how trivial it may appear.

Laboratory Safety Contract

I have read, understand, and agree to follow the safety procedures described on page iv as well as any other written or verbal instruction provided by my instructor.

_____

Student Signature

_____

Date

CONTENTS

The Method of Science

OBJECTIVES

At the end of this laboratory activity, you should be able to:

- state the general procedure used in the scientific method.
- define a positive and negative correlation.
- name a type of investigation in biology that would use correlation as a statistical tool.
- state a hypothesis for a particular problem.
- calculate a correlation coefficient.

INTRODUCTION

Lay people frequently have an idealized picture of scientists and the way in which they work. Indeed, students often mistakenly believe that the scientific method used by scientists is too complicated for the average person to understand. This is not the case; most people who attempt to solve a problem in a logical and organized manner are using the scientific method to a certain degree. The scientist has been trained to apply this tool in a much more rigorous and disciplined manner, thus making it a powerful means of solving problems.

Some textbooks provide a list of steps that are said to constitute the scientific method. This gives the impression that all scientists follow these same steps in performing their experiments. Actually, researchers do not always follow an exact sequence of steps as they work toward solving a problem. Nevertheless, a scientist's observation of a complex phenomenon in nature elicits a series of mental activities that eventually lead to experimentation and data collection, and finally to a resolution, or at least to a better understanding of the problem.

One of the most outstanding characteristics of scientific work is the scientist's use of careful observation. Because science deals with material things, a phenomenon has little value in the scientific world unless it can be observed and measured several times by several researchers. Careful observation leads to questions, and these questions eventually lead to hypotheses. You may have heard a hypothesis defined as an "educated guess," but a well-stated hypothesis is more than this. It not only postulates an answer to a question, but also gives researchers a sense of direction in their work. A workable hypothesis must be logical, based on sound observation, and most importantly, testable.

In many cases a hypothesis is tested by doing a **controlled experiment.** An experiment is a situation designed to produce results supporting the hypothesis being tested. Well-designed experiments have an **experimental** group and a **control** group. In the experimental group (or groups), one factor—the experimental **variable**—is altered in some manner. In the control group, the test is performed under the same conditions as in the experimental group except that there are no changes made to the experimental variable. Experiments produce quantitative data and help scientists arrive at conclusions that indicate whether the hypothesis is substantiated. An unsubstantiated hypothesis may be reworked and tried again, or it may be abandoned for a new hypothesis (see Figure 1.1).

Not all hypotheses are easily tested by controlled experiments; some must be tested by making detailed observations of conditions that already exist. For example, evolutionary biologists test hypotheses by collecting and observing fossils. Similarly, an ecologist studying the distribution of vegetation in a certain area can collect data only about existing growth patterns and then draw conclusions based on this information.

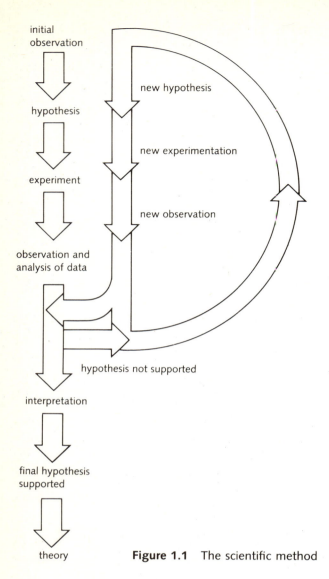

initial
observation

hypothesis

experiment

new hypothesis

new experimentation

new observation

observation and
analysis of data

hypothesis not supported

interpretation

final hypothesis
supported

theory

Figure 1.1 The scientific method

Laboratory Activity 1 is designed to introduce you to the scientific method, data collection, and the statistical analysis of your data. Because this is your first activity in the laboratory, you will not do a controlled experiment involving unfamiliar equipment and complicated procedures, but will collect data that is already available in the laboratory. Before you begin, carefully read the sample experiment, which illustrates how a hypothesis may be tested.

Sample Experiment: The Impact of Eye Coloration on the Fruit Fly Population

Drosophila melanogaster is the scientific name for the common fruit fly that swarms around decaying fruit in the summer. For decades geneticists have used

Figure 1.2 Eye color in *Drosophila melanogaster*

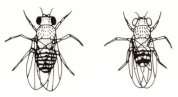

this organism to study the principles of heredity, and much of our knowledge of genetics has come from research with this insect. The fruit fly usually has large red eyes; however, other colors occasionally emerge in a population. In some cases, even white eyes develop (see Figure 1.2).

Among animals, males and females are often attracted to each other by body coloration. A researcher might logically wonder whether eye color influences mate selection in the fruit fly, and if so, how this factor would eventually affect the frequency of different eye color of a given population. This is a reasonable question and leads rather easily to the following statement of a hypothesis.

1. If white eye color has a negative influence on mate selection in the fruit fly, then a population of fruit flies with an equal number of white- and red-eyed male flies would show a decline in the number of white-eyed male flies after several generations of mating.

To test this hypothesis, a scientist would allow a population of 50% red-eyed and 50% white-eyed male flies to mate randomly with red-eyed females. Then the researcher would count the number of white-eyed male offspring and tabulate the data as shown in Table 1.1.

The data in Table 1.1 would produce a graph like the one shown in Figure 1.3.

The graph in this hypothetical experiment reveals that the number of white-eyed male flies declined appreciably during the 20 generations of the experiment. Thus, the hypothesis *appears* to have been substantiated. We will now examine in detail how a scientist would have arrived at this conclusion.

When evaluating data that involve the relationship between two variables (as in this experiment), it is useful to use a statistical tool known as **correlation** to help in the interpretation. This tool determines the strength of the relationship between the two variables, in our case the percentage of white-eyed male flies and the number of generations. Calculating a correlation coefficient allows

Table 1.1 Percentage of white-eyed male flies in a population of *D. melanogaster* in relation to number of generations

Generation number	White-eyed male flies (%)
0	50
2	41
4	29
6	27
8	31
10	28
12	25
14	21
16	14
18	9
20	8

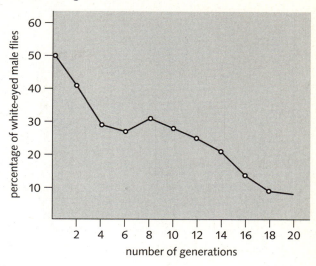

Figure 1.3 Percentage of white-eyed male flies as a function of generation time

us to determine how well we can predict what kind of change will occur in variable *Y* (percentage of white-eyed male flies) as a result of a particular change in variable *X* (number of generations). If the association between these variables is strong, then we can feel confident that a given change in *X* will be associated with a given change in *Y*.

A word of caution is in order: a strong correlation between two variables does not mean that a cause-and-effect relationship exists. There are no set rules for determining whether a cause-and-effect relationship exists; one must rely heavily on

past experience, intuition, and plain common sense about the way in which the natural world works. A good example is the strong correlation that has been established between cigarette smoking and lung cancer. In spite of this correlation, the exact biological link between cigarette smoking and cancer has not been established.

Statisticians have several ways of defining correlations. The one we will use is the Pearson product-moment coefficient of correlation, signified by *r*. Table 1.2 shows the best way to arrange data for calculating the Pearson *r*.

Table 1.2 Pearson *r* calculation

Variable *X*: Number of generations				Variable *Y*: Percentage of white-eyed male flies			
Observation (*N*)	Generation (*X*)	Deviation from mean (d_X)	Deviation squared (d_X)^2	White-eyed flies (%) (*Y*)	Deviation from mean (d_Y)	Deviation squared (d_Y)^2	Product of deviations (d_X)(d_Y)
1	0	−10	100	50	24.36	593.41	−243.60
2	2	−8	64	41	15.36	235.93	−122.88
3	4	−6	36	29	3.36	11.29	−20.16
4	6	−4	16	28	2.36	5.57	−9.44
5	8	−2	4	30	4.36	19.01	−8.72
6	10	0	0	28	2.36	5.57	0
7	12	2	4	25	−0.64	0.41	−1.28
8	14	4	16	20	−5.64	31.81	−22.56
9	16	6	36	14	−11.64	35.49	−69.84
10	18	8	64	9	−16.64	276.89	−133.72
11	20	10	100	8	−17.64	311.17	−176.40
	$\Sigma = 110$ $\overline{X} = 10$		$\Sigma = 440$	$\Sigma = 282$ $\overline{Y} = 25.64$		$\Sigma = 1626.55$	$\Sigma = -808.00$

Figure 1.4 Graphs showing types of correlations

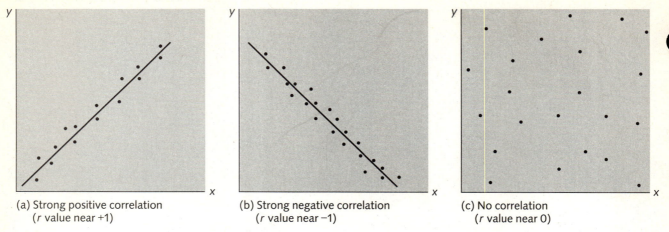

(a) Strong positive correlation
(r value near +1)

(b) Strong negative correlation
(r value near −1)

(c) No correlation
(r value near 0)

Values for r in a correlation lie between -1 and $+1$. The closer a value is to 1, either positive or negative, the greater the association between the two variables. A value close to zero indicates no correlation; a correlation of 1 is perfect. The graphs in Figure 1.4 show three different types of relationships that may occur between X and Y.

When the points in a graph lie close to a straight line and have a positive slope, this means that as X increases, Y also increases. If the inverse is true (that is, if a negative correlation exists), then as X increases, Y decreases. Table 1.2 shows a sample calculation for the Pearson r using the data from the *D. melanogaster* experiment. The following steps explain each calculation.

1. Enter the number of observations (N) in Column 1. This will be used to calculate the average or mean.

2. Enter raw data: generation (X) in Column 2 and percentage of white-eyed flies (Y) in Column 5.

3. Sum each column and calculate the mean for each variable as follows.

$$X = \frac{\Sigma(X)}{N}$$ (Σ is the summation sign.)

$$= \frac{110}{11} = 10$$

$$Y = \frac{\Sigma(Y)}{N}$$ (Rounding to the second decimal place is common practice and will be sufficient for our purposes here.)

$$= \frac{282}{11} = 25.64$$

4. Calculate the deviation of each observation from its mean and enter it in Columns 3 and 6 for generation and percentage of white-eyed flies, respectively.

5. Square the deviations in Columns 3 and 6 and enter the results in Columns 4 and 7, respectively.

6. Total the values in Columns 4 and 7.

7. Multiply the values in Columns 3 and 6 and enter these products in Column 8. Be careful of the sign. Sum the values in Column 8.

8. Calculate the Pearson r by using the following equation.

$$r = \frac{\Sigma[(d_x)(d_y)]}{[\Sigma(d_x)^2][\Sigma(d_y)^2]}$$

9. Substitute the values in Table 1.2.

$$r = \frac{-808}{\sqrt{(440)(1626.55)}}$$

$$= \frac{-808}{\sqrt{71568.2}}$$

$$= \frac{-808}{845.98}$$

$$= -0.96$$

Our correlation is -0.96 and is close enough to 1 to say that a strong negative correlation exists between number of generations and percentage of white-eyed male flies in the population. The negative correlation indicates that the number of white-eyed male flies should decline with each generation.

Table 1.3 Correlations between body measurements of 2400 adult male RAF volunteers, ages 17–38 years (averaged correlations for all ages)

	Standing height	Sitting height	Arm length	Leg length	Thigh length	Abdomen girth	Hip girth	Shoulder girth	Weight
Standing height	—	0.732	0.677	0.864	0.608	0.321	0.490	0.386	0.627
Sitting height	0.732	—	0.421	0.498	0.201	0.173	0.417	0.384	0.548
Arm length	0.677	0.421	—	0.683	0.447	0.266	0.483	0.363	0.466
Leg length	0.864	0.498	0.683	—	0.817	0.312	0.424	0.304	0.556
Thigh length	0.608	0.201	0.447	0.817	—	0.515	0.440	0.234	0.525
Abdomen girth	0.321	0.173	0.266	0.312	0.515	—	0.667	0.526	0.709
Hip girth	0.490	0.417	0.483	0.424	0.440	0.667	—	0.562	0.785
Shoulder girth	0.386	0.384	0.363	0.304	0.234	0.526	0.562	—	0.681
Weight	0.627	0.548	0.466	0.556	0.525	0.709	0.785	0.681	—

From *Annals of Eugenics*, 1947, by Burt and Banks. Reprinted by permission of Cambridge University Press.

MATERIALS

Tape measure (2 m)
Stick (2 m)
Scale

PROCEDURE

During this laboratory activity you will be expected to write a hypothesis and collect and analyze data about the size relationships between various parts of the human body. An example of a simple hypothesis is: the taller a person is, the more he or she weighs. This experiment can be conducted without using complicated equipment, and it illustrates how a biologist goes about collecting data without experimental and control groups. This study was actually done in 1947 using 2400 Royal Air Force personnel, and the correlations are given in Table 1.3.

The list on the following page contains the body measurements you will use in this activity and instructions on how to measure them. You have been observing people for most of your life and you should be able to use this information to hypothesize certain relationships that may exist. Figure 1.5 shows the location of several body parts that may be unfamiliar to you. Study the list carefully and then do the following with a partner.

1. Choose any two measurements that seem reasonable to pair and establish one as the X variable and the other as the Y variable. In Question 1 in the Evaluation, state why you chose to investigate this pair and then write a hypothesis concerning the relationship you would expect to find between these variables.

2. For each student in your laboratory class, measure in centimeters (kilograms for weight) the two variables you have chosen. Record this data in Question 2 in the Evaluation.

Body Measurement

Standing height	Remove shoes and stand against the wall.
Sitting height	Measure from chair seat to top of head.
Arm length	Measure from acromion process to end of middle finger.
Leg length	Measure from top of ilium to bottom of foot while standing.
Thigh length	Measure from top of ilium to top of patella.
Hip girth	Measure around hips at top of ilium.
Shoulder girth	Measure around shoulders at midpoint of deltoid muscle.
Weight	Use scale provided and convert to kilograms (1 kg = 2.205 lb.).

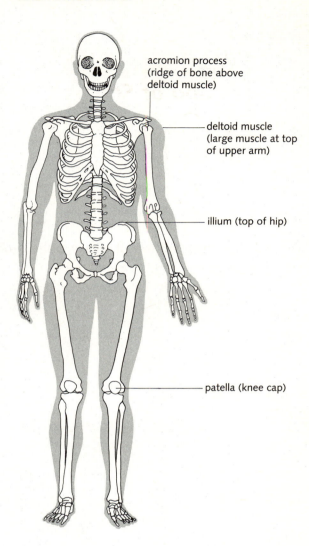

Figure 1.5 Location of several body parts used to obtain measurements

acromion process (ridge of bone above deltoid muscle)

deltoid muscle (large muscle at top of upper arm)

illium (top of hip)

patella (knee cap)

E V A L U A T I O N **1**

The Method of Science

1. (a) State the problem (relationship) you chose to investigate.

   (b) State your hypothesis as it relates to this problem.

2. Record the data in the following chart.

Variable X _____				Variable Y _____			
Subject #	Subject sex	X	Y	Subject #	Subject sex	X	Y

3. Using the sample correlation in the introduction as a guide, complete the following blank table to determine a correlation coefficient for the data you have collected.

Variable X _____				Variable Y _____			
Observation (N)	(X)	Deviation from mean (d_x)	Deviation squared $(d_x)^2$	(Y)	Deviation from mean (d_y)	Deviation squared $(d_y)^2$	Product of deviations $(d_x)(d_y)$
	$\Sigma =$		$\Sigma =$	$\Sigma =$		$\Sigma =$	$\Sigma =$
	$\overline{X} =$			$\overline{Y} =$			

4. Use the graph paper provided to prepare a graph of the data. Remember that variable X is plotted on the horizontal axis and variable Y on the vertical axis. After you have plotted all of the data points, draw a line that best shows the trend of these points. If you need help in preparing your graph, refer to Appendix A. Write the correlation coefficient in the top right corner of the graph.

   In the space below write a statement that shows that you understand the relationship between the calculated correlation coefficient and the line on the graph.

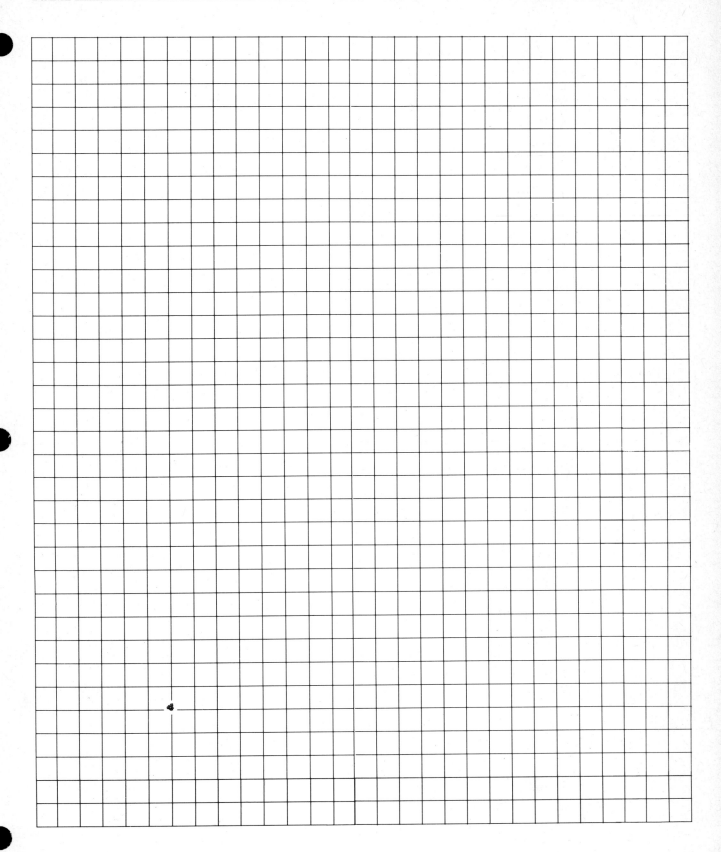

5. Compare your study to the Royal Air Force study and give at least one reason why your results may not be as reliable.

6. Identify each one of the following correlations as positive or negative.

(a) Amount of light shining in eye and size of pupil _____

(b) Rate of heartbeat during exercise _____

(c) Quantity of food in stomach and sensation of hunger _____

(d) Amount of student study time and grade point average _____

The Microscope

OBJECTIVES

At the end of this laboratory activity, you should be able to:

- identify the parts of a microscope and state the function of each.
- indicate the importance of resolution, magnification, and contrast when working with a microscope.
- prepare a specimen for observation under the microscope.
- list the metric units in microscopy and be able to convert from one unit to another.
- estimate the size of an object in the microscope field.
- list in the proper order the steps to follow when focusing on an object in the microscope field.
- demonstrate the proper way to clean a microscope lens.

INTRODUCTION

Magnifying lenses are very inexpensive and easy to obtain nowadays, but this is a relatively recent development. Although a few ancient scholars had some ideas concerning magnification, it was the English scientist Roger Bacon (1214–1294), writing in his *Opus Majus*, who first described the use of lenses as spectacles. A short time later, this idea was used to produce eyeglasses. We probably cannot appreciate the magnitude of this invention. Can you imagine how unproductive millions of people would be if eyeglasses were not available?

The invention of spectacles eventually led to the development of the microscope. Some believe that in 1590, the Dutch eyeglass maker Zacharias Janssen conceived the idea of combining two lenses into an optical device, producing the first compound microscope. Others give credit to Galileo (1564–1642), who modified his famous telescope into a microscope.

During the seventeenth century, the microscope was used to study a world that until this time was unknown to anyone. One of the most famous early microscopists was Robert Hooke (1635–1703), an experimental philosopher known for his broad scientific interest. Hooke became the Curator of Experiments at the Royal Society of London, where he was expected to produce demonstrations weekly, including the examination of a variety of specimens under the microscope. He made many observations, including examinations of the edge of a razor and snow crystals. His description of the microscopic openings in a slice of cork gave us the word "cell" as it is used in biology today.

The early compound microscope had relatively low magnification. Hooke and others knew that magnification would improve if the lens could be made smaller and more spherical. In an attempt to bring this about, a very different kind of "simple" microscope was developed. The most famous design was constructed by Anton van Leeuwenhoek (1632–1723). His microscope consisted of small plates with holes that held a lens. The specimen was placed on glass and moved by a screw device until it was focused under the lens. The entire apparatus was only one by two inches in size. One reason for the great success of van Leeuwenhoek's design was that it allowed light to pass through the specimen.

The compound microscope, using a system of lenses, was further refined by the addition of condensers, lights, mirrors, and diaphragms. These many refinements have produced the kind of microscope you will use in Laboratory Activity 2. This instrument will allow you to view a microscopic world that you may never have seen and will help you to achieve a better understanding of the diversity of life on our planet.

The use of the compound microscope in the laboratory can be a rewarding or frustrating experience, depending on the skill of the user. To keep frustration to a minimum, study this laboratory activity carefully and complete all assigned tasks with diligence and thought. Remember, the microscope is an expensive and valuable instrument; treat it accordingly. Your instructor will tell you when to get your microscope from the storage cabinet.

▶ **CAUTION: Carry the microscope by grasping the arm with one hand and by placing the other hand under the base. Hold it in front of you, chest high.**

MATERIALS

Compound microscope
Lens paper and cleaning fluid

Slide with letter "e"
Prepared slide of bacteria
Immersion oil
Microscope slides
Plastic metric ruler
Cork
Razor blade
Toothpick
Bunsen burner or alcohol lamp
Staining tray
Dropping bottle of crystal violet stain
Water squirt bottle
Onion
Dropping bottle of I_2KI stain
Cover slip

PROCEDURE

The Parts and Operation of the Microscope

Your first task will be to learn the parts of the microscope, where they are located, and how they function. Refer to Figure 2.1 and to your microscope as you learn about each part.

Figure 2.1 A monocular compound light microscope

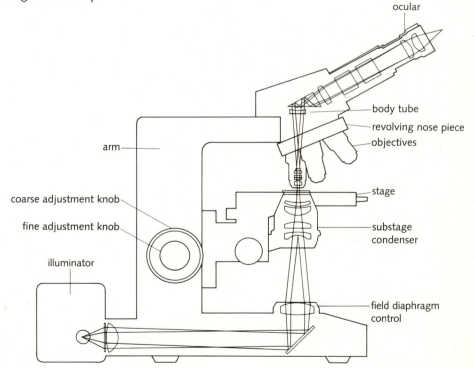

1. The **ocular** (eyepiece) is one of two systems of magnifying lenses in the microscope and is the one closest to your eye when you view a specimen. Its magnification (usually 10X) is stamped on the barrel. The "X" refers to the number of times an object is magnified by a specific lens. What is the magnification of your ocular?

_____

2. The **arm** is used to carry the microscope and support the body tube.

3. The **body tube** is a hollow tube through which light travels from the objective to the ocular. For proper magnification, the body tube must hold the lenses a fixed distance from each other. Usually the tube is bent at an angle to make it more comfortable for you to view an object.

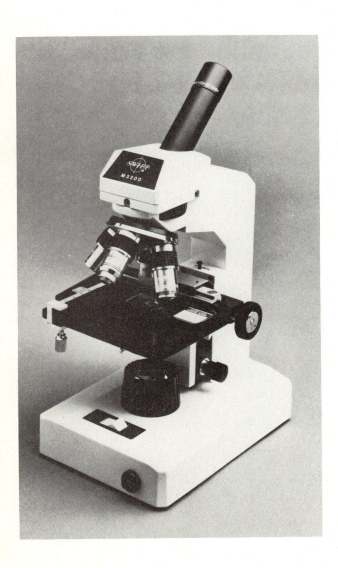

4. The **revolving nosepiece** supports the second lens system of the microscope. The system consists of three or four objectives of various magnifications. The nosepiece can be rotated and each objective positioned over the opening in the stage. The nosepiece must click into place for it to be in proper alignment.

5. The **objectives** are an important part of the optical system of the microscope. Since no single objective can fulfill all the needs of someone using the microscope, there are usually several mounted on the nosepiece. The objectives vary in length, with the shortest lens magnifying the least and the longest one the most. The lowest magnification is usually 4X and is frequently referred to as a **scanning lens.** The next highest magnifications are low power (10X), high power (43X), and oil immersion (97X). These magnifications are not standard; the ones on your microscope may vary slightly.

6. The glass slide used for holding a specimen is mounted on the **stage.** In some cases it is held by two metal clips. More sophisticated microscopes have a calibrated mechanical device that allows the precise movement of the slide.

7. The light source for your microscope may be a **mirror** that reflects light from a lamp. Mirrors often have a concave side for viewing under low power and a convex side for high power. In place of a mirror your microscope may have a built-in **light source** with its own switch and filter.

8. The **diaphragm** is used to control the amount of light entering the lens system. It may be **annular** and consist of a plate with a series of holes of various diameters. Others are of the **iris** type and work by means of a lever that opens and closes the diaphragm.

9. Some microscopes have a **substage condenser system** that is used to focus light on the specimen. The condenser is located below the diaphragm and has its own control knob for moving the lens up or down to regular light intensity.

  In order for you to see a specimen clearly, there must be sufficient contrast among the parts of the specimen. In many specimens, there are opaque parts of pigments that provide sufficient contrast and allow for clear observation. However, proper adjustment of the light intensity will greatly assist you in viewing these objects. If the specimen is translucent, it will be necessary to add a dye or stain to increase contrast. Often these stains are specific for certain cellular structures and allow for easy identification once they are added. Nevertheless, it is important to remember to adjust the light intensity so that you receive the clearest possible image.

10. The **coarse adjustment knob** moves either the body tube or the stage up or down in order to bring the specimen into focus. The gearing mechanism of this adjustment produces a large vertical movement of the stage or body tube with only a partial revolution of the knob. Because of this, the coarse adjustment should only be used with low power (4X, 10X) and *never* with high power (43X) or oil immersion (97X).

11. The **fine adjustment knob** changes the distance between the specimen and the objective very slightly with each turn of the knob. It is used to bring the specimen into sharp focus under low power (10X) and is used for all focusing when using high power (43X) and oil immersion (97X).

12. Microscopes magnify by passing an image through the objectives to the ocular. The **magnification** of the objectives is usually stamped on the side of the lens. The total magnification of the microscope is the product of the magnifications of the ocular and the objective being used. Complete the table below, calculating the total magnification produced by the ocular and each objective for your microscope. Record these total magnifications in the summary table in the Evaluation.

13. **Resolving power** is the capacity of the microscope lens to show detail. Stated simply, it is the lens's ability to distinguish two points that are positioned so close together that they appear as one. The resolving power of the human eye at a distance of 25.4 cm (10 in.) is 0.1 mm (0.004 in.). That is, at a distance greater than 25.4 cm, two dots 0.1 mm apart appear to the naked eye as one dot. On the objective

of your microscope is a number followed by the letter N.A. (numerical aperture). The higher the numerical aperture, the better the resolving power of the lens. For the low power (10X) objective, this value is 0.25 N.A. Record the numerical aperture for each objective lens in your microscope in the summary table in the Evaluation.

Examining a Specimen

Focusing a microscope entails altering the distance between the specimen on the slide and the objective lens. This is accomplished by first using the coarse adjustment knob and then "fine focusing" with the fine adjustment knob. Careful focusing requires consideration of the **working distance** of the objective lens. This is the distance between the lens and the slide when the object is in focus. Some microscopes have the working distance (in millimeters) engraved on the objective. The working distance for the 10X objective is 16 mm.

1. Record the working distance for your other objective lenses in the summary table in the Evaluation.

➤ **CAUTION: The shorter the working distance, the greater the chance the objective will strike the slide and cover slip. If this happens, the slide and specimen may be destroyed and the objective lens damaged.**

Focusing While Using the Low-Power Objectives (4X, 10X)

1. Clean the ocular and objective lenses with lens paper and cleaning fluid.

2. Place a prepared slide of the letter "e" on the stage and center over the stage opening. Use the stage clips to hold the slide in place.

3. Rotate the revolving nosepiece to position the 4X or 10X objective over the slide. It should click into place.

	Scanning lens	Low power	High power	Oil immersion
	_____ X	_____ X	_____ X	_____ X

Ocular

_____ X

4. While looking at the microscope from the side, use the coarse adjustment to move the objective as close as possible to the slide.

5. Look into the ocular and adjust the diaphragm and condenser until you have the maximum amount of light.

6. Slowly turn the coarse adjustment knob, moving the objective away from the slide until the specimen comes into sharp focus. Adjust the light intensity to obtain maximum clarity and sharpen the image with the fine adjustment knob.

7. Draw the letter "e" as it appears:

(a) to the naked eye

(b) in the low power field

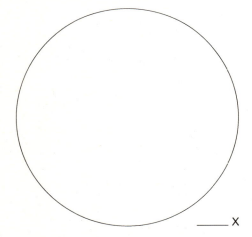

How has its position changed? _____

8. Move the slide forward, backward, to the left, and to the right. How does the image in the microscope appear to move? Record your observations in the table below.

Slide direction	Apparent direction of image
Forward	
Backward	
Right	
Left	

Focusing While Using the High-Power Objective (43X)

1. With the specimen in sharp focus and in the center of the field of view, switch to the high power objective (43X) by turning the revolving nosepiece. Watch the high power objective as you rotate the nosepiece to be sure that it does not touch the cover slip. Good laboratory microscopes are **parfocal,** which means that if the specimen is in focus under low power, it should continue to be in focus when examined under high power.

2. Sharpen the image by using the fine adjustment knob *only.*

3. Indicate what happens to the light intensity and the size of the field of view after moving to high power. Record your observations in the summary chart in the Evaluation.

4. Adjust the diaphragm and/or condenser to obtain the proper illumination for your specimen.

5. Diagram the letter "e" as it appears under high power.

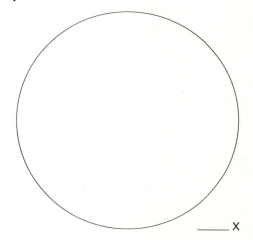

▶ **CAUTION: Never focus the objective toward the slide when looking through the microscope. The objective may hit the slide and damage both the objective lens and slide. This is especially true when you are using high power and oil immersion.**

Focusing While Using the Oil Immersion Lens (97X)

1. Obtain a prepared slide of bacteria and place it on the stage.

2. Focus under low power (10X) on the stained area of the slide. You will not be able to see any detail at this time.

3. Next focus under high power and sharpen the image with the fine adjustment. At this point you should see small specks on the slide. Adjust the light intensity to provide proper contrast.

4. Move the high power objective away from the slide by turning the revolving nosepiece. Halfway between the high power and oil immersion lens, stop the revolving nosepiece and add a drop of immersion oil to the area of the slide you are viewing (see Figure 2.2). Continue moving the revolving nosepiece until the oil immersion lens clicks into place. The lens should

be in contact with the oil. Watch the lens carefully to be sure it does not hit the cover slip or slide when you move it into place.

5. Focus with the fine adjustment knob until you can see individual bacteria on the slide. You may need to adjust the light intensity again. If you have difficulty finding the bacteria, ask your instructor for help. Be patient; using oil immersion is difficult and requires practice.

6. Draw several bacteria in the space provided below. Be sure to indicate the magnification.

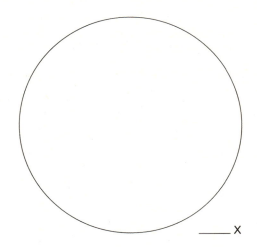

_____ X

Figure 2.2 Procedure for using an oil immersion lens

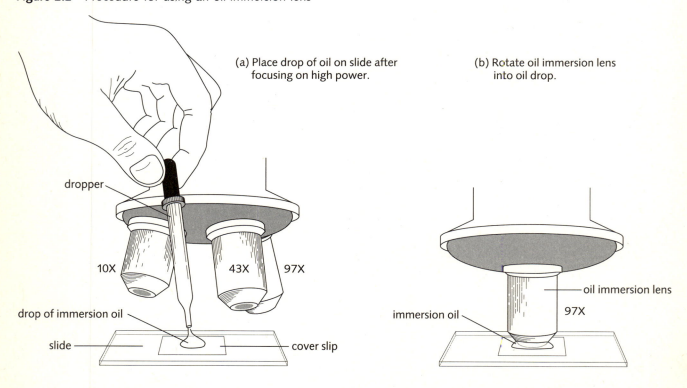

(a) Place drop of oil on slide after focusing on high power.

dropper

10X   43X   97X

drop of immersion oil

slide   cover slip

(b) Rotate oil immersion lens into oil drop.

oil immersion lens
97X

immersion oil

➤ **CAUTION:** **After use, clean the oil immersion lens with lens paper and cleaning fluid. It is important that all oil be removed from the lens.**

Estimating Size in the Field of View

The basic units of linear measurement used in microscopy are derived from the International System, or SI. The basic unit of length is the meter; all smaller units are expressed as a fraction of a meter. Microscopists deal with very small dimensions, so the units they use are very small fractions of a meter. Figure 2.3 shows graphically how the two most common units relate to the millimeter. Table 2.1 summarizes the units most commonly used in microscopic work.

1. Calculate the following values using the data in Table 2.1.

    1.0 mm = _____ μm = _____ nm
    0.5 mm = _____ μm = _____ nm
    2.0 μm = _____ nm = _____ mm

It is possible to estimate the size of an object under the microscope if you know the approximate diameter of the microscope field. Approximate this dimension by doing the following exercise.

1. Place a slide on the stage and then place a clear plastic ruler on the slide with the metric edge along the diameter of the low-power (10X) field. Move the ruler until one of the black millimeter marks is barely visible at the left edge of the field. The next mark should be a little more than halfway across the field (Figure 2.4). The distance from the edge of the first mark to the same edge of the second mark is 1 mm. The remaining distance to the edge of the field is approximately 0.5 mm. Thus, the diameter of the low power field is approximately _____ mm or _____ μm.

Figure 2.3 Ruler showing the relationship between millimeters (mm), micrometers (μm), and nanometers (nm)

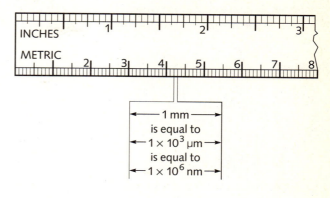

Figure 2.4 Ruler viewed under low power

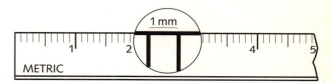

Table 2.1 Units of measurement commonly used in microscopic work

Unit	Symbol	Fraction of meter	Conversion factor
millimeter	mm	0.001 (10^{-3})	
micrometer	μm	0.000001 (10^{-6})	1 μm = 0.001 mm or 1000 μm = 1 mm
nanometer	nm	0.000000001 (10^{-9})	1 nm = 0.001 μm or 1000 nm = 1 μm

2. Although the high power field has greater magnification, the field of view is smaller. You can calculate the diameter of the high-power (43X) and oil immersion (97X) fields using the following inverse relationship:

$$\frac{C}{A} = \frac{B}{D}$$

3. To calculate the diameter of the next higher field of view, make the following substitutions in the above formula and complete the following chart.

A = diameter of field at lower-power magnification

B = magnification of lower-power objective

C = diameter of field at higher-power magnification

D = magnification of higher-power objective

Magnification	Diameter of field
10X (low power)	1500 μm
43X (high power)	_____ μm
97X (oil immersion)	_____ μm

Examination of Selected Cells

As with any newly acquired skill, practice will enable you to perfect your ability to handle the microscope. The following exercises are designed to give you additional experience with the microscope and also to acquaint you with several types of cells.

It seems fitting that you should examine a section of cork to see the cells first examined by Hooke in 1663. Remember, cork is the tough elastic outer tissue of the cork tree (*Quercus suber*) of southern Europe.

1. Slice a very thin wedge of cork with a razor blade. By cutting at an angle, one edge should be thin enough to allow light to pass through.

2. Place on an glass slide; do not add water or a cover slip. Focus on the thinnest edge with low power and look for cell walls enclosing an empty space, the lumen, in which the living cell once existed.

3. Observe under high power. *Do not* attempt to use oil immersion.

4. Make a diagram of several of these cell walls in the Evaluation at the end of this activity.

Of the many specimens examined by van Leeuwenhoek, one was the bacteria that grow on the plaque that forms on our teeth. You can see these same bacteria by completing the following activity.

Preparation of Specimen

1. Wash and dry a microscope slide.

2. With a toothpick, carefully scrape around your teeth at the gum line and then apply these scrapings to the slide.

3. Stir in a circular manner covering an area about the size of a dime until the sample dries.

4. Fix this smear by passing the slide—*smear side up*—through the flame of a Bunsen burner or alcohol lamp three times. As you pass the slide through the flame, count "one and out, two and out, three and out."

5. Place the slide on a staining tray and then flood the smear with crystal violet stain for one minute.

6. Wash off the excess stain with water from a squirt bottle. Dry the slide by gently waving it in the air.

7. Examine the slide under low power, high power, and oil immersion. Refer back to the techniques learned earlier during this activity if necessary. You may see some epithelial cells (Figure 2.5) that were removed from the gums, and you are certain to find several bacteria of the genus *Lactobacillus*, which are common inhabitants of the mouth (Figure 2.6). These bacteria are shaped like straight or curved rods and are approximately 3–5 μm long. Make a diagram of a representative sample of bacteria from your slide in the Evaluation.

Figure 2.5 Squamous epithelial cells

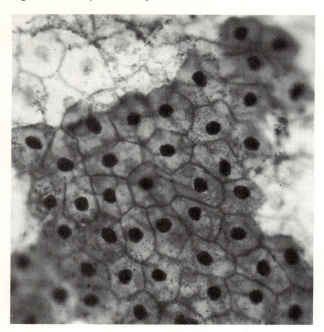

For the final activity, you will prepare a wet mount of onion epidermal cells and attempt to measure their length using the technique you learned in the section on measuring under the microscope.

1. Obtain a fleshy "scale" from the bulb of a sliced onion. Figure 2.7 demonstrates how to peel off the inner epidermal layer from a broken piece of the "scale."

2. Mount this piece of tissue on a microscope slide in a drop of water. Add a drop of iodine (I_2KI) stain and cover with a cover slip.

3. Using low power, observe the rectangular shape of the cells and the darkly stained nucleus. Estimate the length of a cell by lining up a row of cells along the diameter of the field of view. Since you know the diameter of the low power field in micrometers, estimate the length of an onion epidermal cell.

4. Make a diagram of several cells and indicate their lengths in the Evaluation.

Figure 2.6 *Lactobacillus*

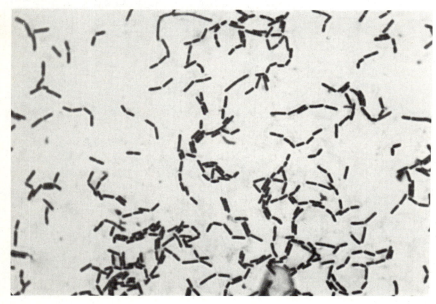

Figure 2.7 Preparation of onion epidermal wet mount

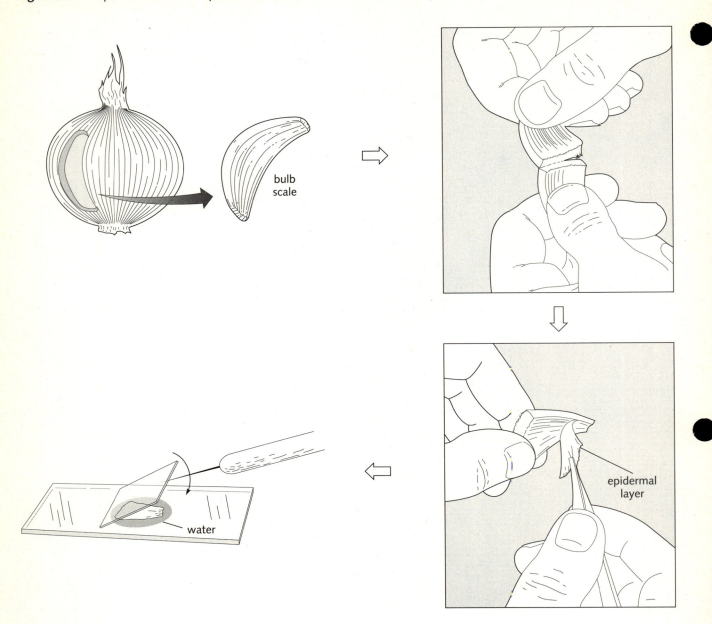

bulb scale

epidermal layer

water

EVALUATION 2

The Microscope

1. The following chart summarizes the most pertinent characteristics of the microscope. Fill in all of the required information to reinforce what you have learned during this activity. This chart will also serve as a valuable reference for future work with the microscope.

	Scanning	Low power	High power	Oil immersion
Objective	4X	10X	43X	97X
Indicate working distance (if available).	mm	mm	mm	mm
Indicate numerical aperture (if available).	_____ N.A.	_____ N.A.	_____ N.A.	_____ N.A.
Indicate total magnification with 10X ocular.	_____ X	_____ X	_____ X	_____ X
Indicate size of field in micrometers.	_____ μm	_____ μm	_____ μm	_____ μm
Shade to show relative brightness of each field.				

2. Diagrams of selected cells

(a) Cork

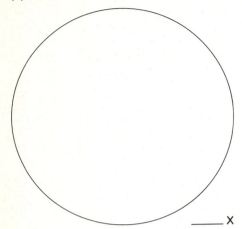

_____ X

(b) Onion epidermal cells

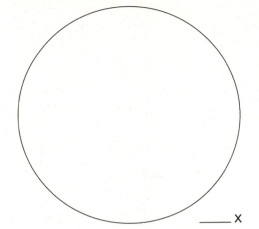

_____ X

(c) Oral bacterial smear

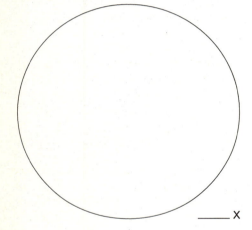

_____ X

Identification of Freshwater Algae

OBJECTIVES

At the end of this laboratory activity, you should be able to:

- understand the construction and use of identification keys.
- use a dichotomous key to identify microscopic algae.

INTRODUCTION

The word *algae* is not considered a formal term in modern classificatioin since the various alga groups are not directly related to each other. Algae include many forms that may consist of a single cell, a filament of cells, a cellular plate, and even some

that are in some ways comparable to vascular plants (see Figure 3.1).

Alga cells contain a variety of chlorophylls, accessory pigments, and storage molecules related to their photosynthetic activities. Because they are photosynthetic and omnipresent in fresh and salt water, they are important to the food chains in many aquatic ecosystems. Algae also produce vast quantities of oxygen needed by aerobic organisms. The number and diversity of algae present in an aquatic ecosystem are good measures of their productivity.

In addition to observing a selection of algae during this activity, you will learn to use a dichotomous key. For many years biologists have used keys to aid in the identification of organisms. Using a key is very much like traveling along a strange road. You come to a fork and read a sign that offers a choice between two directions. You base your

Figure 3.1 Assorted algae

decision on some knowledge of where you are and where you hope to go. In other words, you face a **dichotomous** (dividing into two parts) decision, whether to go to the left or to the right.

In using a key you must have some knowledge of the organism you are "keying out" and you must follow the directions carefully. Figure 3.2 shows a dichotomous key for the vintage identification of the Volkswagen Beetle. It demonstrates the use of a simple dichotomous key. Why not give it a try in the parking lot?

MATERIALS

Compound microscope
Microscope slides
Numbered culture dishes containing pure cultures of selected algae
Cover slips
Pasteur pipettes
Forceps

Figure 3.2 Flowchart for qualitative Beetle analysis

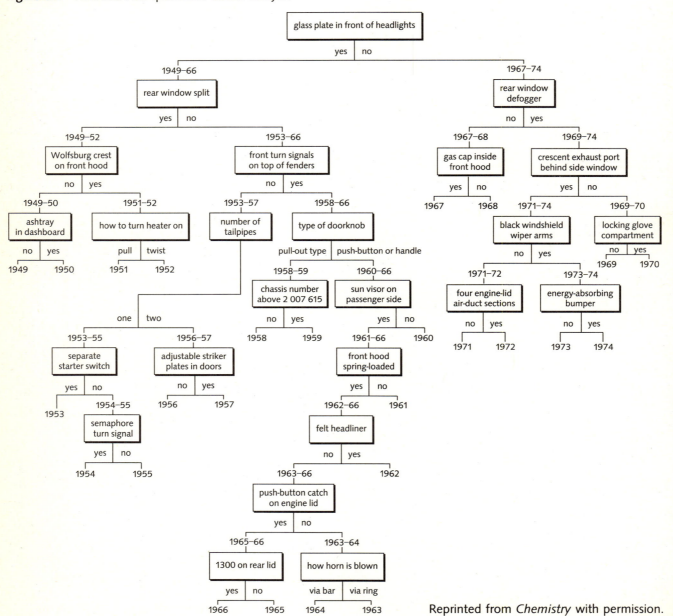

PROCEDURE

In the key you will use to identify each alga, the statements are arranged in pairs. Each pair of statements describes the characteristics of certain algae cells. Prepare your alga sample for observation as described below, and then follow the instructions for identification. Pure cultures of selected algae have been made available for your use, with each type identified by number.

1. Using a pipette or forceps, place a small quantity of culture on a microscope slide and add a cover slip. Each dish will have its own pipette or forcep; do not interchange them.

2. Observe the alga sample carefully, noting the shape and arrangement of the cells.

3. Read statements 1a and 1b. Select the statement that best matches the particular characteristic being described. Note the number that appears at the end of the statement you choose. Move to that number on the key and again compare the statements, deciding which best describes your specimen.

4. Continue this process until you have identified the genus of your alga.

5. After you have identified a particular alga, make a diagram of the organism on the evaluation sheet. List the magnification used, code number, and genus name in the space provided.

Dichotomous Key to Selected Freshwater Algae

1. (a) cells grouped in a colony or end to end in a long filament: 2
   (b) cells single; not grouped in a colony or filament: 3

2. (a) cells joined end to end in a long filament: 4
   (b) cells not grouped in a filamentous colony: 5

3. (a) cells small (less than 25 micrometers); not elongated; spherical and motile: *Chlamydomonas*
   (b) cells single with elongated shape: 6

4. (a) cells rectangular with bright green chloroplasts: 7
   (b) cells lacking chloroplasts and distinct internal structures; blue-green in color: 8

5. (a) colony flat or platelike: 9
   (b) colony spherical: 10

6. (a) cells long; bright green and slightly crescent shaped with round vacuoles at each tip; oval pyrenoid bodies (starch-containing structures) run from tip to clear area in center: *Closterium*
   (b) cells elongated but not crescent shaped; usually motile: 11

7. (a) chloroplasts spiral shaped with pyrenoid bodies: *Spirogyra*
   (b) chloroplasts not spiral shaped: 12

8. (a) filament interrupted by large thick-walled spherical structures (heterocysts): *Anabaena*[1]
   (b) heterocysts absent; may show slightly wavelike movement: *Oscillatoria*[1]

9. (a) colony of four spindle-shaped cells; tip of cells often have spines: *Scendesmus*
   (b) colony flat and somewhat circular; composed of 6–24 polygonal cells: *Pediastrum*

10. (a) colony round with large number of small cells; usually motile, swimming with a rolling motion; bright green in color: *Volvox*
    (b) cluster of several large cells; gelatinous cover may be seen; may or may not be moving with a rolling motion: *Pandorina*

11. (a) cells golden brown; long and narrow; tips rounded or blunted; cell wall etched with fine lines: *Synedra*
    (b) cells green; move by means of flagella; distinct swimming motion: *Euglena*

12. (a) cells nearly filled with chloroplasts, which have a ribbonlike appearance; round sphere (female reproductive structure) larger than cell interspersed along the filament: *Oedogogonium*
    (b) cells contain two star-shaped chloroplasts: *Zygnema*

---

1. Modern classification schemes classify *Anabaena* and *Oscillatoria* in Kingdom Monera, Phylum Cyanobacteria.

EVALUATION 3

Identification of Freshwater Algae

In the spaces provided, diagram the algae you have identified. Label all parts of the alga that allowed you to identify it. Indicate the magnification you used, the code number, and the genus name.

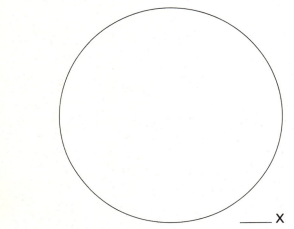
_____ X

\# _____ genus _____

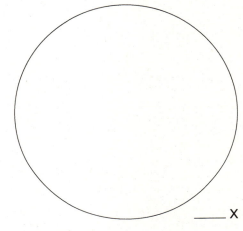
_____ X

\# _____ genus _____

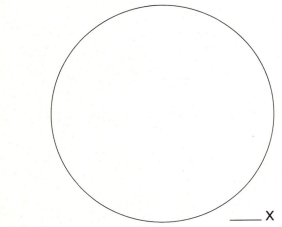
_____ X

\# _____ genus _____

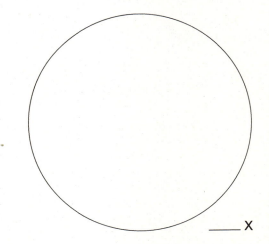
_____ X

\# _____ genus _____

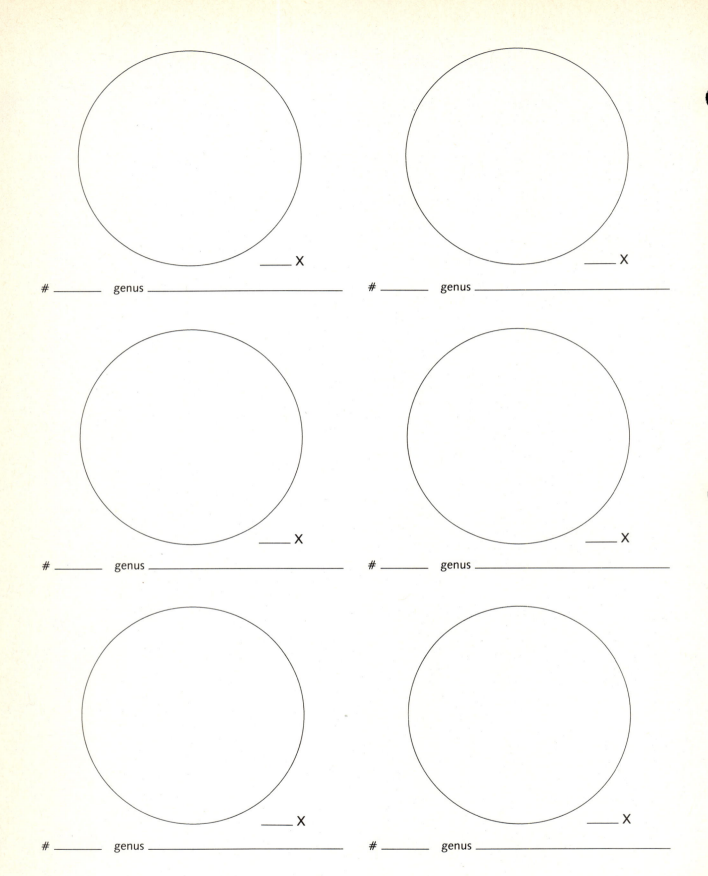

___ X

\# _____ genus _____

\# _____ genus _____

___ X

\# _____ genus _____

\# _____ genus _____

___ X

\# _____ genus _____

\# _____ genus _____

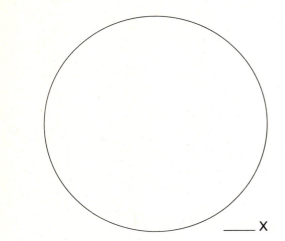

_____ X

_____ genus _____

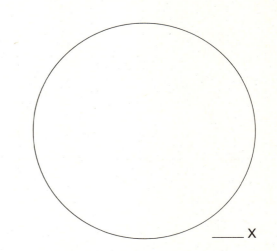

_____ X

_____ genus _____

_____ X

_____ genus _____

The Study of a Freshwater Stream

Students of biology are often intimidated when it is suggested that they enter into any type of field study. We hope to show you by way of the following four laboratory activities that this fear is unfounded. You should remember that it is usually impossible, even for the trained professional, to understand *all* aspects of a particular area being studied. But by taking part in an investigation just as a biologist would, you will find that your study of biology will be more interesting and enriching.

Any number of ecosystems might provide an interesting model to study: a wooded area, a lake, a grassy area of your campus, or even a city lot. Why, then, choose a stream? The main reasons are that streams are usually accessible, are biologically diverse, offer a wide range of physical and chemical parameters that are relatively easy to measure, and are usually pleasant places to work in and around. Finally, humans tend to insult streams in a variety of ways, and this study will give you the opportunity to measure and evaluate several of the abuses that we inflict on these fragile and important ecosystems.

Pursue your study with diligence, intelligence, perseverance, and—above all—sensitivity for your surroundings.

Physical and Biological Overview

OBJECTIVES

At the end of this laboratory activity, you should be able to:

- describe the major physical characteristics of the stream site.
- describe the major biological characteristics of the stream site.
- map the major physical and biological characteristics of the stream site.

INTRODUCTION

Aquatic systems cannot be studied in isolation; they must be viewed in relation to the surrounding area that impinges upon them. For example, as you study your stream site, does the surrounding area show evidence of pollution? Are there a variety of plants and animals present, or do organisms seem scarce? Has man changed the area a great deal or very little from how it was in the past?

The physical characteristics of the stream are most important in determining which organisms live there; thus it is important for you to measure these characteristics as accurately as possible. For example, by determining if your stream is a fast- or slow-running stream, you can usually draw other conclusions as follows.

- Fast-moving streams are usually cold, have good light penetration, and are low in nutrient concentration, organism diversity, and overall productivity.
- Slow-moving streams are warmer, have less light penetration, and are higher in nutrient concentration, organism diversity, and overall productivity.

In addition, the gross biological characteristics of the stream site are significant. The animals and plants in the area adjacent to the stream may alter the environment of the stream itself. For instance, a stream shaded by trees will be cooler during the day, and runoff from a nearby barnyard may increase the nutrients of a stream beyond normal limits.

Your first task will be to find a stream site. Your instructor may assign one or you may be asked to select your own.

➤ CAUTION: **Make certain that the stream you will be studying is no more than 0.5 m deep. This is a major concern to ensure a safe activity, and it is more difficult to study a stream of greater depth.**

MATERIALS

Marked measuring rope or tape measure (20 m)
Meter stick
Watch with second hand
Orange
Celsius thermometer
Camera
U.S. Geological Survey Topographical Map or
  road map

PROCEDURE

Upon arriving at your stream site, you may have trouble knowing where to begin. A good way to gain valuable information is to talk to people who live near the stream. Usually they will be quite interested in your activity and will share information with you. They might be able to tell how the stream got its name, and if they are longtime residents, they may share some additional history of the area with you. Some of this information may be helpful later as you attempt to unravel the ecology of the area.

Locate your site on a map obtained from your instructor. A road map will do, but a U.S. Geological Survey Topographical Map is preferable. After locating your stream site, attempt to determine its source and where it joins with a larger body of water. By knowing your stream's location you can make some assumptions about its general characteristics. For example, as water travels to the sea it slows down, becomes warmer, increases in turbidity, volume, and nutrients, and also shows an increase in the variety of life.

As you do each of the following activities, record your observations in Question 1 of the Evaluation.

1. Look at the land around the site to determine what conditions may influence the quality of the stream. For instance, try to determine if much runoff will occur; if so, what kind of materials will be carried into the stream? Have the adjacent fields been recently fertilized?

2. Do you see any evidence of animal life near the stream; e.g., tracks, nests, or burrows?

3. You should note the large trees that may hang over the stream. Their leaves and branches may fall into the water, and this detritus material will serve as food for organisms in the stream.

4. You may wish to take some photographs to make a permanent record of the general area (see Figure 4.1).

Figure 4.1 Overview of a stream site

Figure 4.2 Hypothetical stream map

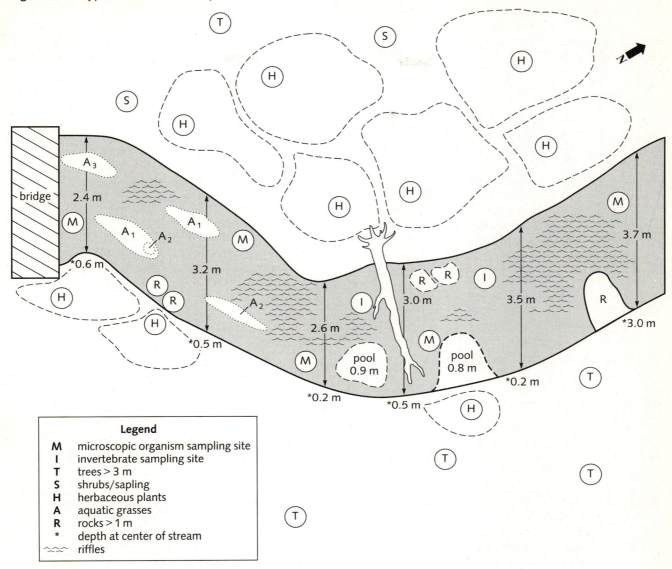

Legend

M microscopic organism sampling site
I invertebrate sampling site
T trees > 3 m
S shrubs/sapling
H herbaceous plants
A aquatic grasses
R rocks > 1 m
* depth at center of stream
~~~~ riffles

Mapping the Stream Site

Select a section of your stream that is about 10–15 m long. Whenever possible your site should contain both **riffles** (small rapids) and **pools** (deep, quiet water). Your first task will be to measure the width and depth of the stream along this section.

1. Use the marked rope or tape measure to determine the length of your section along both shorelines. Mark these points with a stick or rock. Record all of your measurements in Question 2 of the Evaluation.

2. On the graph paper in Question 3, you should begin to prepare a map of your stream site. Draw the contours of the shoreline as well as any major obstructions in or over the water; e.g., large rocks over 0.5 m in diameter or length, fallen trees and branches, litter, bridges, etc. Refer to Figure 4.2 throughout this exercise as you prepare your map.

Figure 4.3 Taking stream measurements

Measuring Stream Velocity and Temperature

One of the most important characteristics of any stream is its **velocity** (distance traveled per unit of time). Velocity is highly variable in different portions of a stream, such as rapids, riffles, pools, or the inner and outer edges of a bend. Try to use the entire distance already established for your stream segment to make this measurement as accurate as possible (see Figure 4.4). If physical obstructions or extremely shallow water prevent this, use the longest segment possible. You can approximate the velocity of your stream segment using the following technique.

1. Throw an orange or stick into the middle of the stream above the first marker and record the time (in seconds) the object takes to float the distance (in meters) from one marker to the next. You may use a stick, but an orange is better since its density, color, and shape enable it to float almost submerged yet still visible. Its spherical shape prevents it from becoming tangled in logs and floating plants.

2. Repeat this activity three times and average the results. Your final average (in meters per second) is the midstream velocity.

Overall stream velocity has been determined to be approximately 0.8 of the average midstream velocity, so you can calculate this velocity using the following formula.

$$\begin{array}{c}\text{Overall stream}\\\text{velocity}\end{array} = \begin{array}{c}\text{Average midstream}\\\text{velocity}\end{array} \times 0.8$$

Limnologists (aquatic biologists) usually classify streams that flow faster than 0.5 m per second as fast streams.

1. Record the stream velocity in Question 2 of the Evaluation.

2. Measure the air temperature and time of measurement during your initial investigation of the stream.

3. Measure the water temperature by placing a thermometer several inches below the surface of the water and holding it there for three minutes.

4. Record your results in Question 2 of the Evaluation.

3. At two-meter intervals along the shoreline, measure across to the opposite shore (see Figure 4.3). Record the width of the stream in Question 2 of the Evaluation. If there is no appreciable change in the width you may extend the length of the intervals to 3 or 4 m.

4. Measure the depth at the center of the stream at each point where you measured across the stream. Record these values on your map. Also record the average depth of your site at the center of the stream in Question 2 of the Evaluation. Measure and record the deepest part of any pool that you find in the stream regardless of its location. If your section of the stream has riffles, indicate their location on the map.

Figure 4.4 Establishing length of stream for measuring stream velocity

Mapping Vegetation

On your map, indicate the location of the major plant types along the stream, using the following coding system.

  T = trees over 3 m in height
  S = shrubs and saplings up to 3 m in height
  H = herbaceous plants, including grasses

1. Record the location of the center of large trees and shrubs and the approximate center of plants that grow over a large area; e.g., grasses and weeds.

2. Use a dotted line to mark the approximate boundary of an area covered by a large growth of plants. Do this for a distance of about 5 m from each shoreline along the entire length of the stream.

3. Locate aquatic vegetation in the same manner. Use the symbol A to mark beds of vegetation and subscripts (A_1, A_2, . . .) to indicate the different species found.

4. Collect samples of each type of plant and bring them to the laboratory for possible identification using plant keys.

EVALUATION 4

Physical and Biological Overview

1. In paragraph form, record your observations of your stream site, including lay of the land, evidence of animal life, type of vegetation, and any evidence of pollution. If you have taken a photograph, attach it to a separate page.

2. Provide the following stream dimensions for your stream site.

    (a) Length _____ m

    (b) Width: maximum _____ m

               minimum _____ m

               average _____ m

    (c) Average depth at center _____ m

    (d) Overall velocity _____ m/sec.

        Slow stream _____ Fast stream _____

    (e) Temperature (in °C)

        Air _____

        Water _____

    (f) Time of day _____

The data you have collected in this activity will be used along with other data you will collect from your site to develop a profile of your stream.

3. Prepare a map of your stream site using Figure 4.4 as a guide.

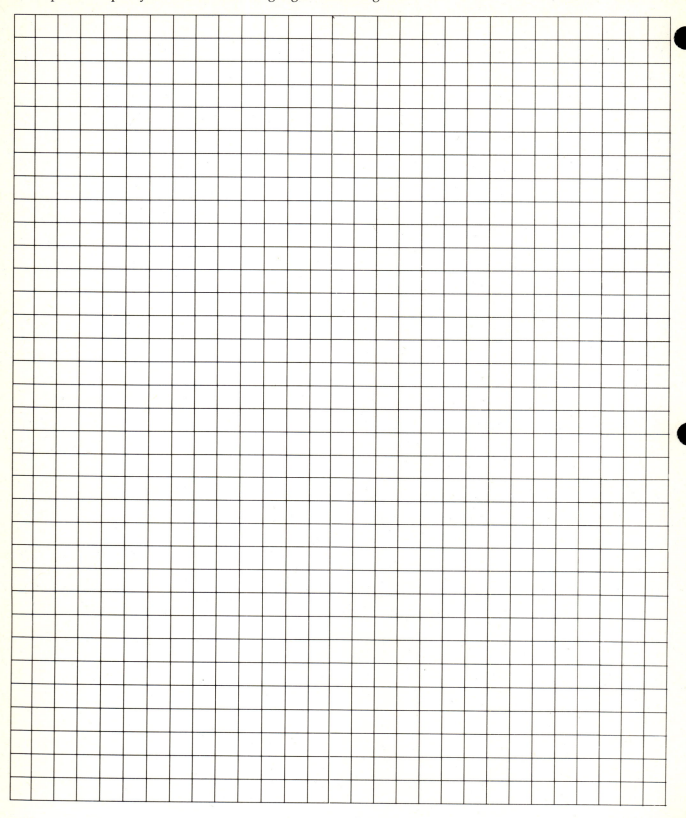

Water Chemistry

OBJECTIVES

At the end of this laboratory activity, you should be able to:

- state the chemical composition of your stream water.
- indicate the acceptable range of concentration for each of the parameters studied.
- describe the biological importance of each parameter studied.
- explain biological oxygen demand.

INTRODUCTION

The ability of water to act as a solvent is well known, and the fact that it can dissolve a variety of substances allows it to support an enormous diversity of organisms. Of the dozens of animal and plant phyla that exist today, only a few have made a successful transition to a terrestrial environment. The rest have remained in their aquatic habitat, where the environment is relatively stable and dissolved nutrients are readily available.

Organisms are composed of chemical elements such as hydrogen, carbon, oxygen, nitrogen, sulfur, and phosphorus. Because they cannot be replenished, these elements are cycled in the environment. This cycling involves not only the food chains of living organisms, but also numerous chemical reactions that occur in the **abiotic** (nonliving) environment. For this reason these cycles are termed **biogeochemical cycles.**

The concentration of minerals and gases dissolved in natural bodies of water is normally very low. For this reason, aquatic biologists measure the concentration of dissolved solids in either **parts per million (ppm)** or **milligrams per liter (mg/L).**

1 ppm = 1 part by weight in 1 million parts by weight

1 mg/L = 1 part by weight in 1 million parts by volume

Because 1 L of water weighs 1000 g, we can use these units interchangeably; that is, 1 mg in 1 L is equivalent to 1 mg in 1000 g or 1,000,000 mg of water. Such small concentrations should not mislead you into thinking they are inconsequential. A difference of only 2 or 3 ppm in the concentration of a particular nutrient may often mean the difference between a healthy or unhealthy stream. Your study will involve selected mineral nutrients and gases. These are described briefly in the following sections.

Carbon Dioxide

A simple example of one biogeochemical cycle is the carbon-oxygen cycle (see Figure 5.1). Carbon dioxide is needed by plants for photosynthesis, and oxygen is needed by all aerobic organisms in cellular respiration. Most carbon dioxide enters the food chain by way of **autotrophs** (green plants) and is converted during photosynthesis to sugar, which is then converted to the other organic compounds needed by plants. These organic compounds are then passed through the food chain, and as they are used by organisms in the process of energy release, carbon dioxide is given off to the atmosphere or water to be reused by green plants. Carbon dioxide is also released as the organic molecules of organisms oxidize during the process of decay. Carbon dioxide is very soluble in water, and aquatic ecosystems provide a vast reservoir for this compound.

When carbon dioxide dissolves in water it may take several forms, such as carbonic acid (H_2CO_3) and the bicarbonate (HCO_3^-) and carbonate ions

Figure 5.1 The carbon-oxygen cycle

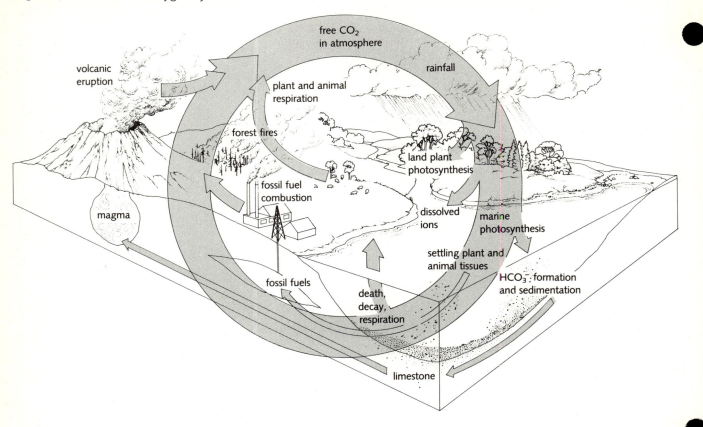

(CO_3^{2-}). These constituents vary in their concentration depending on the acidity of the water and will remain in equilibrium with each other unless some change in pH occurs.

Nitrogen

Another important element that must be cycled is nitrogen (see Figure 5.2). Unlike carbon and oxygen, the reservoir for nitrogen is the atmosphere, where it exists primarily in the molecular or N_2 state. To be cycled it must be converted to compounds such as nitrites (NO_2^-) and nitrates (NO_3^-). The most important of these two ions is nitrate-nitrogen, which is readily absorbed by plants and used to build the many compounds that require nitrogen, such as protein and nucleic acids. After cycling through the food chain, nitrogen is released to the environment through animal wastes and the activity of bacteria during the process of decay. In unpolluted waters nitrate-nitrogen is usually found

in low concentrations, the world average being about 0.3 mg/L (Reid and Wood 1976).[1] However, it still forms an important part of the nutrient base of aquatic communities. Nitrogen, along with phosphorus, is considered one of the most important factors in the growth and productivity of aquatic ecosystems.

Phosphorus

A third cycle that must be considered when evaluating the ecology of a stream is the one involving phosphorus (see Figure 5.3). The element phosphorus travels in combination with oxygen and assumes several chemical forms, including the ions PO_4^{3-}, HPO_4^{2-}, and H_2PO_4. Phosphorus is concentrated in the environment in sediments and rocks

1. G. K. Reid and R. D. Wood, *Ecology of Inland Waters and Estuaries*, 2d ed. (New York: Van Nostrand Reinhold, 1976), p. 485.

Figure 5.2 The nitrogen cycle

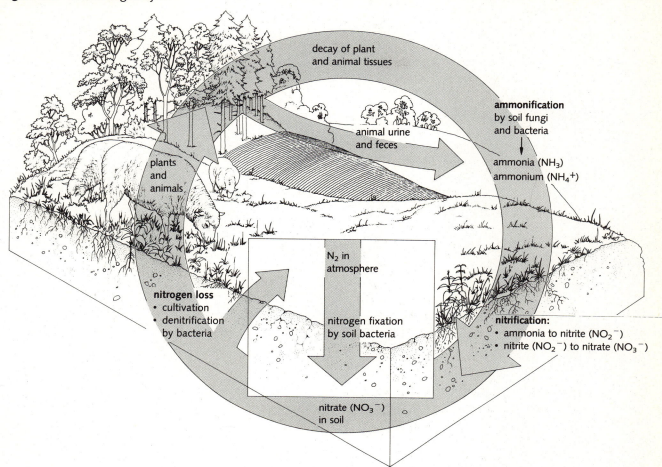

decay of plant
and animal tissues

ammonification
by soil fungi
and bacteria

animal urine
and feces

ammonia (NH₃)
ammonium (NH₄⁺)

plants
and
animals

N₂ in
atmosphere

nitrogen loss
• cultivation
• denitrification
by bacteria

nitrogen fixation
by soil bacteria

nitrification:
• ammonia to nitrite (NO₂⁻)
• nitrite (NO₂⁻) to nitrate (NO₃⁻)

nitrate (NO₃⁻)
in soil

and is relatively insoluble, which means that it is one of the least abundant elements in our waters. A clear, cool lake that is relatively young geologically will often have a concentration of phosphorus that is less than 0.001 mg/L (Eckblad 1978).[2]

Oxygen

The presence of oxygen in an aquatic ecosystem, just as on land, is of considerable importance to the health of the organisms living there. Without dissolved oxygen (DO) aquatic organisms could not carry out their energy-yielding chemical reactions;

because of these reactions, oxygen is constantly being taken out of solution. Oxygen is replenished by the activity of photosynthetic plants or by dissolving at the air-water interface. The solution process is greatly accelerated when water becomes turbulent as it moves through rapids or the riffles of small streams. Temperature also greatly influences the concentration of dissolved oxygen. Figure 5.4 shows how oxygen saturation changes with temperature in salt and fresh water.

Biochemical Oxygen Demand (BOD)

The oxygen of fresh water is constantly being used up by microorganisms as they carry out the process of decay and by the chemical oxidation of compounds that may enter a stream as the result of the

2. James W. Eckblad, *Laboratory Manual of Aquatic Biology* (Dubuque, Iowa: William C. Brown Group, 1978), p. 31.

Figure 5.3 The phosphorus cycle

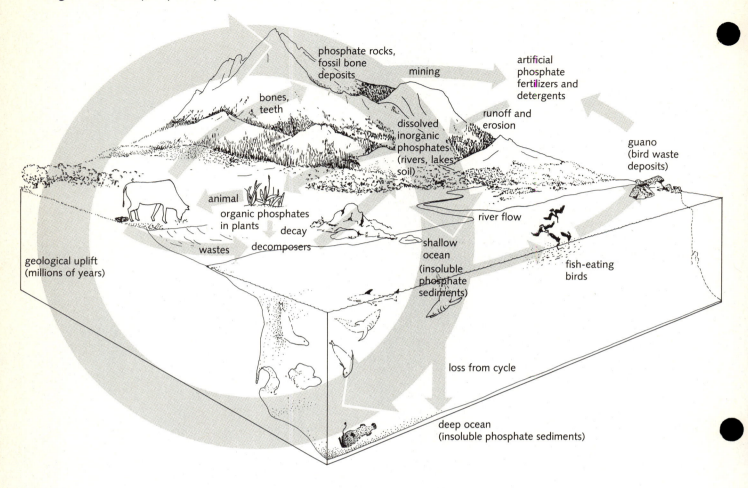

activities of humans. If the demand for oxygen from these sources (as a result of sewage or chemical discharge) is high enough, it may deplete the oxygen concentration of the water to the point where the stream can no longer support a diversity of aquatic life.

pH

It is common to hear aquatic biologists talk about the **pH** or acidity of natural waters. In any given quantity of water some water molecules will dissociate into hydrogen and hydroxyl ions according to the following equation.

$$H_2O \longrightarrow H^+ + OH^-$$

In pure water the number of hydrogen ions (H^+) is equal to the number of hydroxyl ions (OH^-), and we say the water is **neutral.** Any substance that dissolves in water and increases the concentration of hydrogen ions is said to be **acidic.** Likewise, a substance that dissolves in water and increases the concentration of hydroxyl ions is **basic.** The concentration of hydrogen ions in pure water is 0.0000001 g/L. Using scientific notation you may also write this number as 1×10^{-7}. The pH of pure water is 7, which is the negative exponent of the hydrogen ion concentration. The pH scale ranges from 0 to 14; values below 7 are acidic and those above 7 are basic. Most natural waters have a pH range of 7 to 9. Waters outside this range are toxic to aquatic life and cannot be tolerated by organisms for long periods of time. The pH of our natural waters is of particular concern today because of the large quantities of acid rain that are being produced.

The scale in Figure 5.5 summarizes several pertinent facts about pH. It is important to note that a change in one unit of pH represents a tenfold change in the hydrogen ion concentration. For example, a solution with a pH of 4 has ten times the hydrogen ion concentration of a solution with a pH of 5 and 100 times the hydrogen ion concentration of a solution with a pH of 6.

Turbidity

Turbidity is a measure of the amount of suspended solids such as silt, clay, finely divided inorganic matter, and microorganisms. High turbidity reduces light penetration with the resultant loss of photosynthetic activity of the producer population. Heavy concentrations of suspended material may even coat the gills of animals and interfere with their respiratory activity. The amount of turbidity in a stream may vary widely and depends heavily on the amount and kind of runoff from the surrounding area.

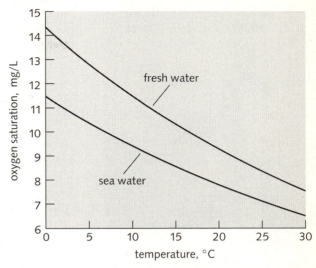

Figure 5.4 Curves for oxygen saturation in salt and fresh water

Reprinted with permission of Macmillan Publishing Company from *Natural Ecosystems*, 2d ed., by W. B. Clapham, Jr., p. 174. Copyright © 1983 by Macmillan Publishing Company.

Figure 5.5 pH scale and tolerance range for selected aquatic organisms

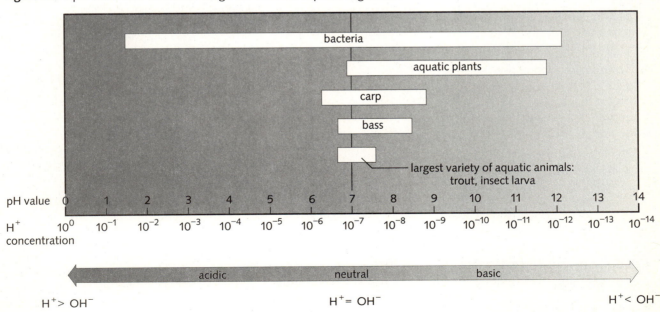

MATERIALS

Commercial water testing kit
Collecting bottle (1 L)
Biochemical oxygen demand (BOD) bottle
Thermos (1 L)
pH paper
pH meter
Oxygen meter

PROCEDURE

During this laboratory activity you will use a commercial laboratory kit to measure the concentration of nitrates, phosphates, carbon dioxide, oxygen, pH (acidity), turbidity, and biochemical oxygen demand (BOD). These kits are usually designed to be used in the field and will have the necessary reagents premeasured. In addition, these test kits usually indicate by a color change the quantity of mineral or gas present in water. The intensity of this color change is measured either with a **colorimeter** (refer to Appendix B) or by comparison with a standard sample. Many kits have the capacity to test for minerals other than those minerals listed in this activity. Perform any tests that your instructor assigns. The following instructions should be supplemented by those specifically given in the water testing kit.

Instructions for Testing Water
Ideally all of your measurements should be done at stream side with fresh samples of water. If circumstances prevent this, use these guidelines for bringing samples to the laboratory for analysis.

1. Thoroughly clean and rinse with distilled water a liter bottle. You want your readings to be as accurate as possible so cleaning and rinsing should not be taken lightly. Even traces of residual chemicals may alter your results.

2. Collect samples on the day the tests are going to be performed. If this is not possible, collect samples as close as possible to your scheduled laboratory period and store them at 4°C for no longer than 24 hours.

Conducting Water Tests

Nitrate, Phosphate

1. Most kits test for nitrates and phosphates using colorimetric procedures. Follow the specific instructions given in your kit.

2. Conduct the phosphate test with great care because only extremely small quantities of phosphate are present in unpolluted water.

3. Record your results in Question 1 in the Evaluation.

pH
The pH of your water sample can be measured in several ways.

1. Some water testing kits will ask you to add certain reagents to your sample to obtain a color change for comparison to a standard.

2. An alternate method is to use litmus paper that changes color depending on the pH of the water. Dip the paper into the water and immediately compare it to the color scale that comes with the kit.

3. The most accurate way to obtain pH readings is to use an electronic pH meter. Specific instructions for using a pH meter are found in Appendix D.

4. Record the pH measurement in Question 1 in the Evaluation.

Oxygen
Since the dissolved oxygen concentration of a stream will vary considerably with temperature, time of day, flow rate, and depth, you should note these conditions when you obtain your sample and, if possible, take several samples for testing. The oxygen test is best done at stream side because the solubility of oxygen in water changes with temperature (see Figure 5.6). If this is not possible, use a thermos bottle to collect your sample, observing the following guidelines.

1. Rinse the thermos several times with stream water so that the temperature of the container approximates that of the stream.

2. Fill the thermos as gently as possible. Avoid undue agitation since this may dissolve more oxygen in the sample than is actually present in the stream.

3. When handling the sample in the laboratory, keep it as still as possible. Work quickly, before any significant temperature change occurs.

There are two ways to measure oxygen concentration. Your instructor will direct you in the method you are to use.

1. Commercial test kits will use a color change when certain reagents are added to water. Follow the specific instructions given in your kit.

Figure 5.6 Student using oxygen meter at stream side

2. An easier method is to use an oxygen meter, which will measure temperature as well as oxygen. Your instructor will have calibrated the meter before you use it. Follow the instructions carefully.

3. Record your results in Question 1 in the Evaluation.

In addition to measuring dissolved oxygen in parts per million, you can also measure it as percent saturation in water at a given temperature. The nomogram (Figure 5.7) allows you to calculate percent saturation as a function of temperature and oxygen concentration.

1. Read the percent saturation using the directions above the nomogram. Record your readings in Question 1 in the Evaluation.

Biochemical Oxygen Demand (BOD)

The standard procedure for measuring the BOD of water is to measure its oxygen concentration and then to incubate the sample in a BOD bottle in the dark at 20°C for five days (Figure 5.8).

At the end of the five days, calculate the BOD value using the following formula.

$$\text{Initial oxygen reading (ppm)} - \text{Final oxygen reading (ppm)} = \text{BOD}$$

Water with a BOD value near 0 or 1 ppm is considered to be of good quality. A value of 3 ppm is considered fairly pure, and a value of 5 ppm or more means the water is of doubtful purity.

1. Record your reading in Question 1 in the Evaluation.

Figure 5.7 The nomogram. Using a straight edge, line up the water temperature on the top scale with the oxygen concentration on the bottom scale. Read the percentage of saturation on the middle scale.

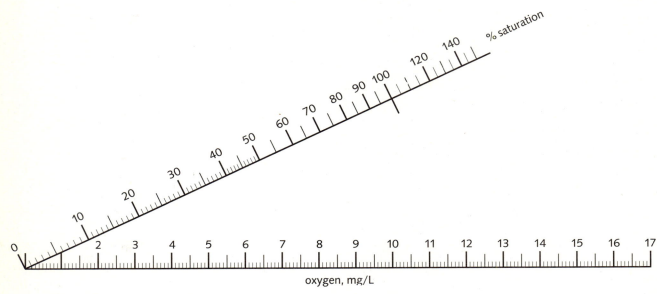

From Welch, P. S., *Limnological Methods* (New York: McGraw-Hill) 1948. Reproduced with permission of McGraw-Hill, Inc.

Carbon Dioxide

Carbon dioxide concentration is usually measured by a process called **titration**. In commercial kits, specific directions will be given. The general procedures are as follows.

1. Add a pH indicator to the water sample and then a second reagent, until a color change occurs. Then convert the mL of reagent used to the concentration of dissolved carbon dioxide in ppm.

2. Record your results in Question 1 in the Evaluation.

Turbidity

Turbidity is expressed as a percentage of light transmittance. There are several ways to measure turbidity; the method you choose will depend on the equipment available in your laboratory. If you use a commercial water testing kit, follow the manufacturer's directions carefully. A spectrophotometer may also be used. Specific instructions for using a spectrophotometer are found in Appendix B.

1. Record your results in Question 1 in the Evaluation.

Figure 5.8 Standard bottle for measuring biochemical oxygen demand

EVALUATION 5

Water Chemistry

1. Complete the following data chart for your stream site.

   Name of stream _____

   Location _____

   Date, time, weather conditions _____

   General comments on stream condition (water level, clear, muddy, etc.)

Chemical/physical parameter	Stream water readings	Normal range*
Nitrate (ppm)		
Phosphate (ppm)		
pH		
Oxygen (ppm)		
% saturation		
BOD (ppm)		
Carbon dioxide (ppm)		
Turbidity (% transmittance)		

*(for your geographical area, if available)

2. Based on the data you have collected, prepare a report on the quality of your stream. The analysis of the data collected for your stream site should be done by researching the literature on **limnology** (aquatic biology) and water quality. The questions that follow are designed to get you off to a good start by focusing your research on the most pertinent aspects of the interaction between the biology of aquatic organisms and water chemistry. They are not meant to be limiting, and you should carry your research as far as you are able in order to understand the stream you are studying. In every case attempt to compare your readings with those that are considered normal for your area. In some cases the normal range for a particular parameter can be obtained from the literature. In other cases, due to the complexity of biological communities, it is difficult to obtain

precise values. This should be a major point of discussion in your evaluation. If your results are outside the normal range, try to explain why this is so and what impact it may have on the quality of life in your stream.

(a) Phosphate. Why is phosphate an important nutrient in aquatic ecosystems? Why do many aquatic biologists consider phosphate to be the chief limiting factor in aquatic productivity? Why are phosphates from fertilizer and detergents considered to be a serious pollutant in aquatic ecosystems?

(b) Nitrate. Why are nitrates a necessary nutrient in aquatic ecosystems? Nitrates are also considered a serious pollutant in streams; why?

(c) pH. Why are extremes of pH dangerous to life in a stream?

(d) Oxygen/carbon dioxide. What is the relationship between these two gases in aquatic ecosystems? Carbon dioxide is considered a part of a major buffering system in aquatic environments. What is a buffer? Why is this buffering system important?

(e) Biochemical oxygen demand. What would a high BOD reading indicate for your stream? How would this affect the stream organisms?

(f) Turbidity. What would be the consequences of continuously muddy conditions in your stream?

Periphyton Diversity

OBJECTIVES

At the end of this laboratory activity, you should be able to:

- make a statement concerning the diversity of the periphyton at your stream site.
- make a statement concerning the general health of your stream site.

INTRODUCTION

An important consideration of any ecosystem is the diversity, or variety, of the species living in it. Even a casual examination of a number of ecosystems will reveal some ecosystems host many species, others only a few. The reasons for this are very complex but, in general, species diversity is a reflection of the number of **niches** present in an ecosystem. A niche is the total life activities of an organism, not just its habitat or home. It follows that if an ecosystem changes, a species's niche may also change. For example, an organism's nutritional needs may not be met, a temperature limit may be reached, or a specialized body part may be rendered a liability instead of an asset. A logical result of these happenings might well be the loss of one or more species from the ecosystem.

Organism diversity thus can serve as an indicator of change in an ecosystem. A significant change in a freshwater ecosystem—such as a reduction of oxygen, an increase of temperature, or an increase of pollutants—may manifest itself in a reduction in the number of species. In a healthy ecosystem, species will normally be distributed in the following manner. **Producers** (green plants) serve as food for **herbivores** (plant-eating animals), which in turn are fed upon by **carnivores** (animals that feed on other animals). In addition, the **decomposers** (bacteria and fungi) digest the dead remains of animals and plants. Major fluctuations in the distribution of organisms within this food chain can create serious problems for ecosystems.

Examining the diversity of certain organisms will yield some indication of the general health of your freshwater ecosystem. The organisms you will study in this activity are the **periphyton,** or **aufwuchs.** The periphyton community develops on a substrate (submerged rocks and plants) over a period of time. Some organisms are firmly attached, while others just feed and rest there. Included in the periphyton are the plaque-forming algae, green filamentous algae, and diatoms (motile algae). Feeding in this crust you will find **protozoans** (heterotrophic protists), **micrometazoans** (microscopic multicellular animals), and insect larvae. Feeding on the remains and wastes of these organisms are the bacteria and fungi. The periphyton forms are excellent for your study because they are composed of both animal and plant species and usually reflect the degree of diversity found in the organisms at large in a freshwater stream. In addition, they are easy to obtain and to study.

➤ **CAUTION: Prior to doing this activity you will need to establish a schedule that will allow you to collect specimens at the appropriate time for your laboratory work.**

MATERIALS

10 notched microscope slides
Triangular file
Protective goggles
String
Sticks or clothes hangers
Paper towel
Map of stream site (from Laboratory Activity 4)
Jar with lid
Cover slips
Pasteur pipette
Compound microscope

PROCEDURE

In moving bodies of water, the free-floating and drifting **plankton** (microorganisms) are constantly being washed away. In some fast-moving headwaters of freshwater streams, these organisms are very scarce and difficult to obtain for study. To overcome this problem you will place microscope slides in your stream and allow the periphyton organisms to collect on them.

In the Laboratory

1. Prepare 10 microscope slides in the following manner. Using a triangular file, score a notch on each side of the slide 1 cm from the end. The cut should be approximately 1–2 mm deep.

▶ **CAUTION: Wear protective goggles while you are filing the microscope slides.**

2. Wrap a string around the slide several times, using the notches to secure it. The attached string should be approximately 1 m in length. See Figure 6.1.

At the Stream Site

1. Once at your stream site, place the slides in the stream, either tying them to an overhanging tree branch or attaching them to a stick pushed into the bank. You can also use a clothes hanger, bent to suit your needs. Make sure your slide is totally submerged in the water but not touching the stream bottom (see Figure 6.2).

2. Record these sampling stations on the map of your stream site that you prepared in Laboratory Activity 4.

3. Allow the slides to remain submerged for 4–5 days.

4. Approximately 1 hour before your scheduled laboratory session, carefully remove the slides from the water. The slides will usually be covered with a greenish-brown film.

5. Cut the string, leaving 6–8 cm attached to the slide. Place a crushed paper towel in the bottom of a small jar and fill it with stream water.

6. Place the slides on their ends on the paper towel and allow them to lean against the side of the jar (see Figure 6.3). Replace lid on jar. With care you should be able to transport these slides with the attached organisms back to the laboratory without harming them.

▶ **CAUTION: Improper transport of your slides will harm your specimens; follow these directions carefully.**

In the Laboratory

Next, you will determine the diversity of the microscopic forms attached to the slides. When you examine your slides you will see precisely the same organisms that are attached to the plants and rocks in your stream. Prepare your slide for study, following these steps.

1. Remove a slide from the jar and wipe one side very carefully with a paper towel.

Figure 6.1 Slide preparation

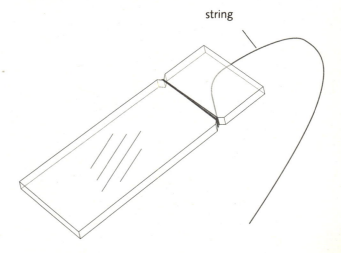

string

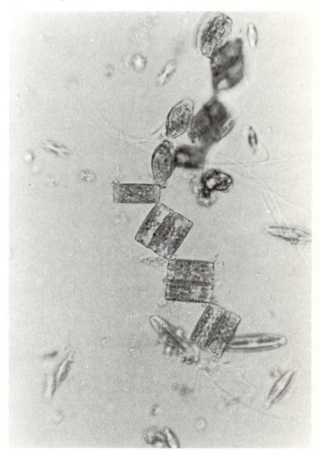

Figure 6.2 Hanging a slide for periphyton collection

2. Add a cover slip to your slide and examine it under the compound microscope. It is important that you keep the area under the cover slip moist throughout the examination. Use a Pasteur pipette to add a drop of stream water at the edge of the cover slip.

In a healthy (unpolluted) stream you should find many representatives of the periphyton community. Figure 6.4 shows a slide of a periphyton

Figure 6.4 Microscopic field showing a periphyton community

Figure 6.3 Jar for collecting slides

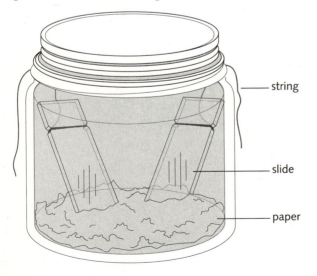

— string

— slide

— paper

community collected from a stream. The population on your slides will vary considerably depending on your locality and the season of the year.

To help you recognize the major groups of organisms on your slides, refer to the sketches in Figure 6.5.

3. Scan the slide under low power to get an overview of your community. In addition to the organisms, you will see considerable extraneous material on the slide. This nonliving material is composed of organic debris and bits of gravel. Living organisms can be recognized because of their uniformity in size and shape.

4. Observe your slide under high power and make an outline diagram of a representative organism of one look-alike group (a group of organisms that look identical to each other) in the field of view. By grouping your organisms in this manner you can conveniently fit them into a taxonomic group without the use of complicated keys. Do not record single organisms. Record your results in Question 2 of the Evaluation.

5. Repeat this process for each look-alike group observed in the field of view.

6. Move to another field of view and observe carefully. Record any *new* look-alike group(s) in the Evaluation. Repeat this process until you no longer find any new organisms in the field of view.

7. Record the relative abundance of each look-alike group in Question 2 of the Evaluation.

8. After observing your individual organisms, scan the slide again and note the presence or absence of blue-green algae. The cells of the blue-green algae are primitive **procaryotic** cells lacking nuclei and chloroplasts. They are usually easy to identify because the chlorophyll is dispersed in the cytoplasm. A large quantity of blue-green algae may indicate that the water has a high level of organic material and low oxygen content. This condition is referred to as **eutrophic** and frequently represents an unhealthy ecosystem. Record your findings in Question 4 of the Evaluation.

This activity has introduced you to the concept of biological diversity. You will have an opportunity to do a more quantitative study of diversity in Laboratory Activity 7.

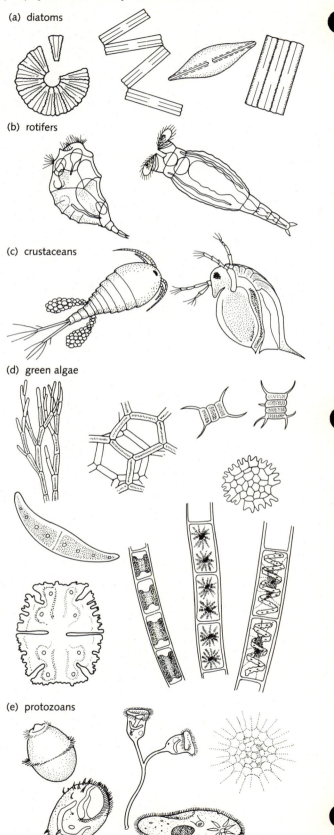

Figure 6.5 Typical members of a freshwater periphyton community (not shown to scale)

(a) diatoms

(b) rotifers

(c) crustaceans

(d) green algae

(e) protozoans

EVALUATION 6

Periphyton Diversity

1. After observing the organisms found attached to the your slide, write a general description of this community.

2. Record your data in the chart below.

Group code	Diagram	Relative abundance		
		Very abundant	Moderately abundant	Sparse
A				
B				

Group code	Diagram	Relative abundance		
		Very abundant	Moderately abundant	Sparse
C				
D				
E				

3. Based on the observations recorded above, what statement can you make concerning the diversity of species in your periphyton community?

4. Were the blue-green algae abundant, moderately abundant, or sparse? Does this information agree with your observations in Question 3?

A Survey of the Macroinvertebrate Population

OBJECTIVES

At the end of this laboratory activity, you should be able to:

- identify the major animal groups in the macroinvertebrate population of your stream site.
- calculate a diversity index and state its importance in evaluating water quality.
- make a statement concerning the quality of your stream site, based on the data collected.

INTRODUCTION

The diversity of life in an aquatic ecosystem varies considerably. Aquatic biologists classify organisms in aquatic environments mostly based on where they live: **neuston** are surface dwellers (water striders, duckweed), **plankton** are distributed through the water mass (microscopic plants and animals such as diatoms and protozoa), **nekton** are the free swimmers (fish, insects), and **benthos** live on or in the bottom substrate.

In this laboratory activity you will concentrate your efforts on studying the benthos. The benthos that you will study are classified as **macrobenthos;** specifically, the **macroinvertebrates** (*macro* = large, *invertebrate* = no backbone). These are defined by biologists as aquatic animals that can be seen with the unaided eye and that are retained by a sieve (U.S. Standard #30) with openings approximately 0.6 mm square. The major taxonomic groups making up the macroinvertebrates are insects and their larval (immature) forms, crustaceans (isopods, scuds), annelids (segmented worms), nematodes

(roundworms), mollusks (clams, mussels, snails), and flatworms (planarians).

Flowing-water ecosystems place great demands on the organisms living there. The major physical parameter that influences their life is the rate of flow of the water. In particularly fast streams, organisms either are washed away or have adapted to life in the fast current. These adaptations assume several forms. For example, many organisms have very flattened bodies (mayflies), others are not only flat but have claws or hooks to hold on to the substrate (riffle beetles, sow bugs), and some even build nests and anchor them firmly to a rock or other substrate (caddis flies). Slower streams have more sediment, and it is possible to find organisms that burrow into the silt or mud, such as nematodes (roundworms), annelids (segmented worms), and certain species of mayfly and dragonfly larvae. Others, such as freshwater shrimp and daphnia, are able to swim in the more slowly moving water and are not dependent upon special adaptations to keep from being washed away.

The macroinvertebrates are very sensitive to stress in the environment and are useful for determining the health of a stream. The diversity of species in the macroinvertebrate population provides a wide range of response to environmental stress, and a change in water quality will readily cause a shift in population density of one or more species. Stream benthos can be classified into three categories according to their tolerance to pollution.

1. Pollution-sensitive (mayflies, stone flies, caddis flies, crayfish)

2. Moderately pollution-tolerant (net-spinning caddis flies, water penny beetles, scuds, hellgrammites, dragonfly nymphs, fingernail clams)

3. Pollution-tolerant (fly larvae, snails, flatworms, leeches, beetles, sow bugs)

Although representatives of all three groups may be present in a stream, environmental stress can be detected by observing an increase in the population of organisms in groups 2 and 3 and a decline in the organisms in group 1.

The idea of biological diversity was introduced in Laboratory Activity 6. In this activity you will extend this concept to include a mathematical calculation of a diversity index and use it as a more formal measure of the stress (if any) your stream is undergoing. The basic premise of a diversity index is that undisturbed biological communities have a large number of species with no individual species predominating. Conversely, in a stressed environment, the number of different species is reduced, and there is a tendency for only a few species to assume a dominant role in the community structure (see Table 7.1).

To calculate a diversity index for your stream site, you will need to place the macroinvertebrates that you collect into look-alike groups as you did in Laboratory Activity 6. Look-alike groups will save you from having to use complicated taxonomic keys to classify organisms. Even though it may sound "unscientific," this classification scheme will save you considerable time and is a reasonably accurate way to classify organisms to the taxonomic level of class, family, genera, and, in some cases, even species.

MATERIALS

Paint brush (2 in. wide)
Spatula
Wide-mouthed jars of isopropyl alcohol (70%) with glycerin
D-ring net or hand screen
Stick or trowel
White enamel pan
Forceps
Suber net
Watch with second hand
Magnifying glass
Dissecting microscope
Calculator with natural log function

PROCEDURE

At the Stream Site

There are several ways to collect benthic organisms, two of which are described below. Figure 7.1 illustrates three different nets you may use to collect these organisms. Your instructor will select the method and device most suitable for you and your partner.

Method I

1. Work through your site, scraping off the organisms from the bottom of the larger rocks with

Figure 7.1 Collecting nets

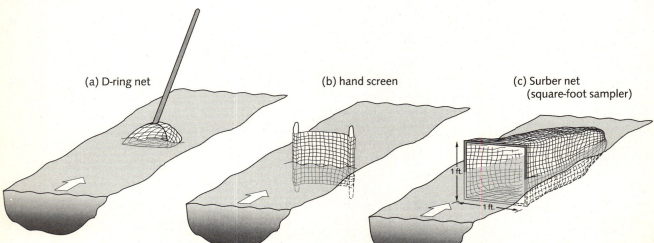

(a) D-ring net

(b) hand screen

(c) Surber net (square-foot sampler)

1 ft.

1 ft.

Table 7.1 Comparison of macroinvertebrate organisms by station in a western Maryland stream*

	Station 1	Station 2	Station 3	Station 4
Class: Tubellaria (flatworms)				
Dugesia sp.	1	2	—	—
Class: Oligochaeta (segmented worms)	1	—	6	1
Class: Hirudinea (leeches)	3	—	4	1
Class: Gastropoda (snails)				
Physa sp.	—	—	—	1
Campaloma sp.	4	3	—	—
Order: Isopoda (sow bugs)				
Asellus sp.	7	2	36	37
Order: Amphipoda (scuds)				
Gammarus sp.	2	—	4	4
Order: Ephemeroptera (mayflies)				
Tricorythodes sp.	4	1	—	—
Baetis sp.	—	—	—	2
Ephemerella sp.	7	6	—	3
Stenonema sp.	1	1	—	6
Stenacron sp.	—	—	—	17
Order: Trichoptera (caddis flies)				
Hydropsyche sp.	14	10	3	11
Cheumatopsyche sp.	46	48	9	9
Orthotrichia sp.	—	3	—	—
Order: Plecoptera (stone flies)				
Acroneuria sp.	—	—	—	1
Order: Megaloptera (hellgrammite)				
Chauliodes sp.	—	—	—	2
Order: Diptera (true flies)				
Family: Chironomidae	17	—	58	1
Chironomus sp.	—	—	3	—
Simulium sp.	1	—	3	—
Antocha sp.	1	—	—	—
Order: Coleoptera (beetles)				
Stenelmis sp.	9	12	—	2
TOTAL ORGANISMS	118	88	126	98

*This table summarizes the data from a stream in western Maryland studied by the authors. Numbers indicate individuals in 3 ft.^{2} (0.28/m^2). Measurements were taken at 4 stations; at that time, one of these stations (3) was subject to considerable stress from waste water effluent. Note that two groups of organisms, *Asellus* (pollution-tolerant crustaceans) and *Chironomidea* (true flies) have rather large populations, while the sensitive species (mayflies, caddis flies, stone flies) are absent or considerably reduced in number.

Figure 7.2 Collecting macroinvertebrates with a D-ring net

a brush or thin spatula. Sample the area within your site for 20–30 minutes. Place the organisms in a wide-mouthed jar of 70% isopropyl alcohol for transport to the laboratory.

2. Return the rocks you disturbed in the stream to their original positions.

3. After removing the organisms from the larger rocks, select several areas (at least three) and use a D-ring net (see Figure 7.1a) to collect organisms from the stream bottom (see Figure 7.2). Hold the net with the open end facing upstream and have your partner stir the bottom of the stream bed with a stick or trowel for about 1 minute.[1]

4. Empty the contents of the net into an enamel pan (Figure 7.3) and remove the organisms with a pair of forceps or spatula. Place all organisms

1. In place of a D-ring net, you may use a hand screen made of two wood or metal rods attached to a piece of nylon net. The screen should be about 2 ft. long and 1 ft. wide (see Figure 7.1b).

in your collecting jar and return the residue to the stream bed.

5. Repeat the process at two more sites.

Method II
This method uses a net called the Suber square foot sampler (Figure 7.1c). The Suber net will allow you to sample the stream bottom quantitatively. Presumably you will collect all the organisms in one square foot of stream bed and you can then express your results as number of organisms per square foot.

1. Hold the net with the open end upstream. The water must be shallow so that it doesn't cover the net.

2. Have your partner lift large rocks within the framed square foot area and brush or scrape off the organisms into the net. Be sure to hold the rocks directly in front of the net opening to avoid losing organisms. Place the rocks outside the frame when you are finished.

Figure 7.3 Sorting organisms in an enamel pan

3. With a stick or trowel, stir the stream bottom within the area of the frame to a uniform depth for several minutes. Keep track of the length of time you sample your site and keep this time constant when sampling the other sites.

4. Shake the contents of the net out onto a white enamel pan and then pick out the organisms from the gravel and sediment.

5. Place the organisms into a wide-mouthed jar of 70% isopropyl alcohol for transport to the laboratory. Return the displaced rocks, sediment, and gravel to the stream.

6. Repeat these steps at two other sites, remembering to spend the same amount of time at each location.

In the Laboratory

1. Spread all of the organisms you have collected onto a white enamel pan and study them carefully. If you are comparing several sites, be sure to do only one site at a time.

2. Study each organism, noting its size, shape, and color. If necessary, use a magnifying glass or dissecting microscope to examine each specimen.

3. After studying each specimen, move it to a section of your tray away from the others. Keep doing this until you have sorted all specimens into look-alike groups (see Figure 7.4).

4. In Question 1 in the Evaluation, make a sketch of a representative organism from each group. Just outline the major features of the body; don't get bogged down in detail. Label the groups A, B, C, and so on.

5. Use the diagrams and information in Table 7.2, "Description of Major Benthic Phyla and Classes," to help you place each organism into a taxonomic group.

Figure 7.4 Look-alike groups sorted into petri dishes

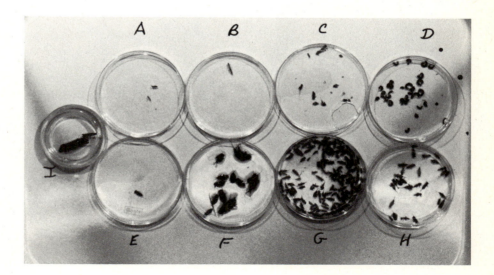

Table 7.2 Description of major benthic phyla and classes

Phylum	Class	Body shape	Color	Locomotion	Habitat	Diagram
Platyhelminthes (flatworms)	Turbellaria (free-living flatworms)	Flat, elongated, unsegmented, 5–10 mm long	Brown, gray-black	Gliding	Under submerged plants and rocks	
	Nematoda (roundworms)	Cylindrical, elongated, unsegmented, less than 1 cm long	Colorless to blackish, often translucent	Constant, rapid, whiplike	In organic material and debris	
Annelida (segmented worms)	Oligochaeta (aquatic earthworms, *Tubifex*)	Cylindrical ringlike segments, about 1–30 mm long	Varies, often red	Crawling	In organic material and debris	
	Hirudinea (leeches)	32 segments, 3–5 grooves per segment, suction disks at anterior and posterior ends	Brown	Crawling	Attached to submerged objects, logs, and plants	
Arthropoda (jointed appendages; hard, chitinous exoskeleton, segmented body)	Crustacea (crayfish, shrimp, sow bugs)	2 pairs of antennae, usually numerous paired appendages, considerable variation in size—from microscopic to 10 cm	Varied	Swimming, crawling	Attached to submerged objects, logs, and rocks	

Phylum	Class	Body shape	Color	Locomotion	Habitat	Diagram
Arthropoda (continued)	Insecta (immature forms)	Segmented body (head, thorax, abdomen), thorax with 3 pairs of jointed legs, 1 pair of antennae on head (see key to aquatic insect larvae)	Brown, black, green	Crawling	Attached to submerged objects, logs, rocks, and plants	
	Arachnida (water mites)	Globular to oval in shape, 8 legs, 0.4–3.0 mm long	Brightly colored	Crawling	Attached to submerged objects, logs, rocks, and plants	
Mollusca (soft body in 1 shell or 2 calcareous [calcium carbonate] shells)	Gastropoda (snails)	Body enclosed in 1 coiled spiral shell, may be cone-shaped, 2–70 mm long	Gray or black shell	"Walking" with foot	Attached to submerged objects, logs, rocks, and plants, or buried in mud	
	Pelecypoda (clams, mussels)	Body enclosed in 2 shells joined by elastic hinge, 2–25 mm long				

6. Identify insect larvae to the order level of classification according to the Dichotomous Key to Aquatic Insect Larvae and the diagrams in Figure 7.5. (You may wish to use more detailed keys, if available, and attempt to assign your specimens to a genera or even a species.)

7. List the groups of organisms by name or letter in Question 2 in the Evaluation. Using the information in the introduction, attempt to indicate whether or not they are tolerant, moderately tolerant, or intolerant to pollution.

Dichotomous Key to Aquatic Insect Larvae[2]

1. (a) larvae with externally developing wings: 2
   (b) larvae with internally developing wings: 5

2. (a) mouth parts adapted for biting: 3
   (b) mouth parts joined in the form of a sucking beak, directed backward beneath the head: *Hemiptera* (water bugs)

3. (a) long, slender tails; labium (lower lip) not longer than head and not hinged on itself: 4
   (b) three tails, either leaflike or as small spinous appendages; extended labium much longer than head and, when at rest, folded on itself like a hinge between forelegs: *Odonata* (dragonflies, damselflies)

Figure 7.5   Representative aquatic insect larvae

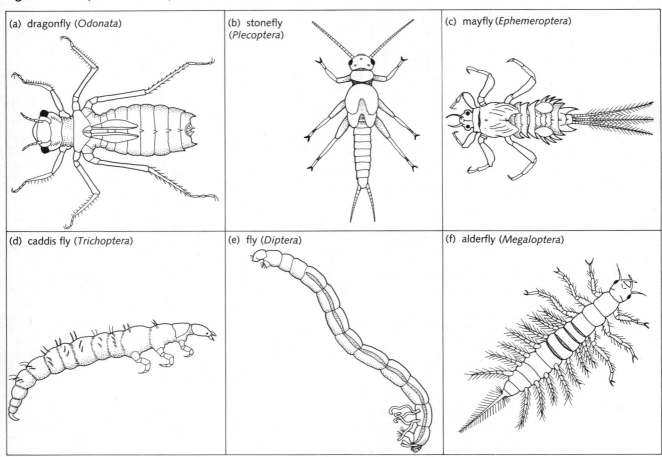

(a) dragonfly (*Odonata*)

(b) stonefly (*Plecoptera*)

(c) mayfly (*Ephemeroptera*)

(d) caddis fly (*Trichoptera*)

(e) fly (*Diptera*)

(f) alderfly (*Megaloptera*)

2. Adapted from Needham, James G. and Needham, Paul R., *A Guide to the Study of Freshwater Biology*, 5th ed. (Oakland: Holden-Day, Inc.) 1962, p. 30. Reproduced with permission of McGraw-Hill, Inc.

4. (a) gills found under plates of thorax (where legs are attached); two tarsal (terminal segment of leg) claws; two tails: *Plecoptera* (stone flies)

   (b) gills usually positioned on sides of abdomen; single tarsal claw; usually three tails: *Ephemeroptera* (mayflies)

5. (a) jointed thoracic legs present: 6

   (b) jointed thoracic legs absent; legless or with abdominal prolegs (short, fleshy limbs): *Diptera* (true flies)

6. (a) long lateral filaments located on abdominal segments: *Megaloptera* (dobsonflies)

   (b) lateral filaments missing; often showing minute gill filaments; cylindrical body generally found in cases attached to rocks: *Trichoptera* (caddis flies)

Calculating the Diversity Index

A formula known as the Shannon-Weaver diversity index allows you to calculate the diversity of your stream site. This measure of diversity produces values from 0 to 4. A diversity index between 0 and 1 indicates water of poor quality; values between 1 and 3 are indicative of intermediate quality water; and values above 3 are typical of good water quality. The Shannon-Weaver index assumes that all organisms have been identified to species. Because you are not an experienced taxonomist, your identifications will probably not be this exact. If you examined the organisms closely when you arranged them into look-alike groups, however, the margin of error will not be significant.

The formula for the Shannon-Weaver index is

$$N = (P_i)(-\ln P_i)$$

where

N = Shannon-Weaver index

$$P_i = \frac{\text{number of organisms in a given species}}{\text{total number of all organisms}}$$

and

$-\ln P_i$ = natural log of P_i

When calculating the index with your own data, follow the steps suggested below, filling in the chart in the Evaluation as you go along. You will need a calculator to determine the natural logs. Study the following sample calculation before attempting your own.

Sample Calculation

A population of organisms collected from a stream bed has the following composition:

Organism A = 25
B = 17
C = 37
D = 5
Total = 84

Organism	Frequency (P_i)	Natural log $(-\ln P_i)$	$(P_i)(-\ln P_i)$
A	$\frac{25}{84} = 0.298$	1.211	0.361
B	$\frac{17}{84} = 0.202$	1.598	0.323
C	$\frac{37}{84} = 0.440$	0.820	0.362
D	$\frac{5}{84} = 0.060$	2.821	0.169
			$N = 1.22$

The Shannon-Weaver index is the sum of the products in the last column. Thus, $N = 1.23$, indicating a stream of intermediate water quality.

Once you feel comfortable with how the formula works, calculate your own diversity index using the data that you have collected from your stream site.

1. Calculate the P_i and $-\ln P_i$ values for each group of organisms in your collection. Record your data in the columns marked "Frequency" and "Natural log" in Question 4 in the Evaluation.

2. Calculate the product $(P_i)(-\ln P_i)$ for each group. Record your data in the last column in Question 4.

3. Add the values in this last column to determine the diversity index (N).

EVALUATION 7

A Survey of the Macroinvertebrate Population

1. In the space provided, prepare a neat line drawing of a representative of each look-alike group.

2. List all of the groups or organisms you collected by letter, general characteristics, and—if possible—name and pollution sensitivity.

Group	Name (optional)	General characteristics	Pollution sensitivity (optional)*

*Tolerant, moderately tolerant, sensitive

3. If you have used a Suber net to collect your organisms, calculate the number of each group per square foot according to the following formula.

$$\frac{\text{Total organisms in group}}{\text{Total number of square feet sampled}} = \text{Organisms/ft.}^2$$

Sample calculation:

$$\frac{327}{10} = 32.7 \text{ organisms/ft.}^2$$

Enter your data in the following table.

Organism	#/ft.²	Organism	#/ft.²

4. Enter your data in the following table and calculate the Shannon-Weaver diversity index for your study area.

Organism	Frequency (P_i)	Natural log (ln P_i)	$(P_i)(-\ln P_i)$

5. Based on the data you have collected, write a statement that indicates the quality of your stream water.

Population Growth of a Microorganism

OBJECTIVES

At the end of this laboratory activity, you should be able to:

- describe the structure of the unicellular alga *Euglena gracilis.*
- list and describe the stages that occur in a sigmoid or S-shaped growth curve.
- prepare a growth curve from data collected with a spectrophotometer or direct counting of cells.
- design a research project and report the results in a scientific paper (optional activity).

INTRODUCTION

Biologists normally recognize two kinds of growth in living organisms. One is the familiar growth that occurs when an individual increases in size. As a **zygote** (recently fertilized egg), you were only several hundred micrometers in diameter. From that very small speck, you have increased your mass many thousands of times, all as a result of adding on new cells. A second kind of growth—the kind we are concerned with in this exercise—is the growth that occurs in populations. In microorganisms, this growth is measured as an increase in the number or total mass of cells rather than as an increase in the size of the individual organism.

If we start a culture with one organism and allow it to divide by cell division, the population will increase in the following manner.

$$1 \rightarrow 2 \rightarrow 4 \rightarrow 8 \rightarrow 16 \rightarrow 32 \rightarrow 64 \rightarrow 128 \ldots$$

We can also express this growth as a geometric progression.

$$1 \rightarrow 2^1 \rightarrow 2^2 \rightarrow 2^3 \rightarrow 2^4 \rightarrow 2^5 \rightarrow 2^6 \rightarrow 2^7 \rightarrow 2^8 \ldots 2^n$$

(2^n refers to the maximum number in the population.)

Under normal conditions this kind of rapid growth does not continue indefinitely. In a laboratory culture such as the one prepared in this activity, not only are nutrients and space limited, but the accumulation of wastes will eventually cause the population death rate to increase and growth to level off. If this course of events is graphed, that is, if we plot population size against time, we obtain the curve shown in Figure 8.1.

This type of curve is referred to as a **sigmoid** or **S-shaped** growth curve. We can divide the line into several segments and describe what is happening in each segment:

Lag phase. The population is not increasing. Cells are metabolizing and increasing in size.

Figure 8.1 Growth curve for a population of microorganisms: (a) Lag phase, (b) Exponential phase, (c) Stationary phase, (d) Decline phase

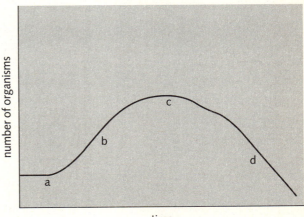

71

Exponential/logarithmic phase. Period of rapid population growth as cells divide at a constant rate.

Stationary phase. Nutrients exhausted and wastes accumulating to toxic levels. Growth levels off as equilibrium is established between the growth and death of cells.

Death (decline) phase. Death rate exceeds growth rate and population drops exponentially. This may take hours, days or months, depending on the type of organisms.

In this laboratory activity you will use a unicellular (one-celled) organism to determine a growth curve. As an optional activity, you may use this same organism to determine how changes in environmental conditions (temperature, light, pollutants, etc.) may effect population growth.

The organism that you will use is a unicellular protist called *Euglena gracilis* (see Figure 8.2). *E. gracilis* has chloroplasts and is photosynthetic, but it also requires dissolved nutrients and is usually found in fresh water that is fairly rich in organic material. In *E. gracilis* a **flagellum** at the anterior end moves in a whiplike manner to propel the organism forward. The base of this flagellum is attached in a flask-shaped opening called a **reservoir**. Adjacent to this reservoir is a small red eyespot. This eyespot is photosensitive and helps to orient the *E. gracilis* toward light. Euglenoids all reproduce by cell division and, when established in a fresh culture medium, will show the sigmoid growth curve described earlier.

MATERIALS

Tubes of stock *Euglena gracilis* culture
Microscope slides and cover slips
Dropping bottle of methyl cellulose
Compound microscope
Tubes of prepared culture medium
Marking pencil or pen
Inoculating loop
Bunsen burner
Sterile pipette (1 mL)
Pipette pump
Test tube rack (or tin can)
Spectrophotometer (optional)
Dropping bottle of ethanol (95%)

Figure 8.2 Diagram of *Euglena gracilis*

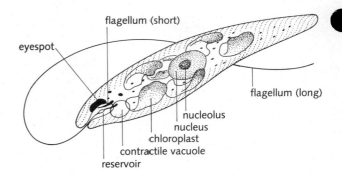

PROCEDURE

Examining Your Organism

Your first activity will be to familiarize yourself with the organism you are growing.

1. Make a wet mount from the culture of *E. gracilis* supplied by your instructor. (If necessary, refer to Laboratory Activity 2, "The Microscope," to review this technique.) If so directed by your instructor, add a drop of methyl cellulose to the drop of culture medium before you add a cover slip. This will slow the swimming motion of the *Euglena* and make observation easier.

2. Examine under the microscope. Look for all of the structures labeled in the diagram. The colorless flagellum may not be visible unless you reduce the light intensity with the diaphragm and/or condenser.

Preparing Cultures

The test tubes of growth medium that you have been given are sterile, which means that they have been placed in an autoclave (see Figure 8.3) and heated in steam to 121°C (252°F) for 15 minutes. This treatment destroys all organisms that may have entered the test tube or medium as it was being prepared. It is important to have a sterile culture medium because you must work with only one species of organism, and contamination by other microorganisms will prevent you from obtaining the desired results. In order to maintain pure cultures throughout your work you must follow **aseptic** (germ-free) **technique.** Your instructor will first demonstrate this procedure, then you will do the following.

Figure 8.3 Autoclave containing test tubes with liquid growth medium

An alternate procedure that will ensure the transfer of an adequate number of organisms involves the use of a sterile pipette. Your instructor will supply you with a sterile 1-mL pipette in a plastic wrapper.

1. Remove the wrapper and place a pipette pump on the wide end of the pipette.

2. Remove the cap from the tube of stock culture and draw up 0.1 mL of culture.

3. Replace the cap and set the tube in the test tube rack. Then remove the cap from the sterile medium and dispense the culture from the pipette into the new medium.

4. Replace the cap on the new culture tube and discard the pipette as directed by your instructor.

Plotting a Growth Curve

You will have to incubate your cultures for a period of time in order to trace the growth of the *E. gracilis* population. The organisms will grow well at room temperature (about 21°C) and should be kept in light for about 12–15 hours each day. Since cell division is rather slow, measuring population growth is not a problem, and in most cases counting may be done once a day. However, it is important to remember that you may have to adjust counting time according to the conditions under which your cultures are growing. Your instructor will help you decide the times for population counts.

1. Obtain two sterile tubes of culture medium and a tube of stock *E. gracilis* culture. Label the culture medium tubes #1 and #2. Heat a wire inoculating loop in a Bunsen burner flame until it glows red. Allow it to cool for 10 seconds.

2. Remove the caps from tube #1 and from the tube of stock culture, holding them between your fingers (watch your instructor).

3. Insert the sterile loop into the stock culture, remove a loop full of culture, and then dip into the medium you wish to inoculate. Note that an inoculation wire with one loop may not transfer a sufficient number of organisms as large as *E. gracilis*, so you may need to use a wire that has three loops (see Figure 8.4).

4. Remove the loop from the inoculated culture, replace the cap, and flame the loop in the burner before you replace it on the tabletop.

5. Repeat this procedure to inoculate tube #2.

Figure 8.4 Inoculating loop for transfer of microorganisms

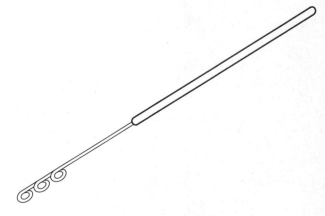

There are two ways to estimate growth in a population of microorganisms. One is to use an instrument called a **spectrophotometer** (a device for measuring the amount of light that passes through a solution) and the other is to make a direct cell count. The procedures for using each method are as follows.

Using the Spectrophotometer to Measure Growth

As your culture of *E. gracilis* grows, the increasing number of cells will make the medium increasingly turbid. This increase in turbidity can be measured with a spectrophotometer as the percentage of light transmittance (optical density) through the culture tube at a given wavelength of light. Specific directions for using a spectrophotometer are given in Appendix E, and you should read these instructions before proceeding any further.

In order to obtain the most accurate readings of population growth, you will have to set the wavelength of light for the maximum absorption for your culture. The following directions will explain how to set your instrument for the proper wavelength. Be sure that you use a stock culture that has been growing for several days and is slightly turbid.

1. Turn on the spectrophotometer and allow it to warm up for at least 5 minutes.

2. Roll the culture tube gently between your hands to evenly distribute the organisms. Wipe the tube clean with a tissue.

3. Set the spectrophotometer to 0% transmittance.

4. Insert a blank tube of sterile medium into the well and adjust the transmittance to 100%. Remove tube from well.

5. Insert the tube of *E. gracilis* culture into the well. Adjust the wavelength knob until you obtain the lowest percentage of transmittance (maximum absorption). Use this wavelength for all future readings of your cultures.

    Wavelength to be used: _____ nm

With the proper calibration of the spectrophotometer you will be able to accurately measure population growth as often as necessary for your cultures. The instructions for taking growth readings are presented below in abbreviated form. If necessary, you should carefully review these instructions in Appendix E to ensure that you are following them correctly.

1. Turn on machine and allow to warm up for 5 minutes.

2. Set wavelength to proper setting.

3. Set transmittance to 0%.

4. Insert test tube with sterile medium (blank tube) into well and set transmittance to 100%. Remove test tube.

5. Insert growing culture of *E. gracilis* into well and measure transmittance. Record your results in the Evaluation.

Direct Cell Count

Direct cell counting has the advantage of allowing you to see the cells you are counting and enables you to ensure that the turbidity you are seeing is a result of the growth of your organism and not some contaminant. The major difficulty with direct counting is getting a uniform sample each time you count. For direct counting do the following.

1. Roll the culture tube between your hands or use a vortex mixer to evenly distribute the organisms.

2. Using aseptic technique, remove two or three loopfuls of culture from the tube and add to a clean glass slide. It is important to use the same size loop(s) each time you make a count.

3. Add a drop of methyl cellulose to slow the organisms' movement and thus make counting easier. Or you may add one drop of 95% ethanol (alcohol) to kill the cells.

4. Add a cover slip.

5. Focus under low power and then high power and count all of the organisms in no fewer than 10 fields of view. Average the count from all fields and record in the appropriate data table.

Measuring the Growth of Your Culture

Now that you have learned the techniques for growing *E. gracilis* cultures and measuring their growth, its time to grow your own cultures and plot a growth curve. Your instructor will provide you with a stock culture of *E. gracilis* to use in establishing your own cultures. You should grow two cultures and average the counts for both tubes. Use the table in Question 1 in the evaluation to record your data. The number of days you make observations will vary depending on the growth of your cultures. You will know the culture is starting to die off when the count levels off for several readings and then starts to drop. At this point you may stop recording data and dispose of your cultures as indicated by your instructor.

Before you start, let's take time to review the overall procedures to be used in establishing a growth curve for *E. gracilis*.

1. Establish two cultures of *E. gracilis* from a stock culture provided by your instructor.

2. Decide on a method of counting—either the spectrophotometer or direct count method.

3. Collect data at the prescribed intervals and record in the Evaluation. Average the counts for both tubes.

4. Prepare a graph showing the growth of your cultures.

Figure 8.5 shows an actual student graph in which percentage of transmittance was used to measure growth. The line labeled "control" shows the typical growth curve, except that readings were not taken during the decline of the culture. In this experiment the effect of various concentrations of mercuric chloride was also being investigated.

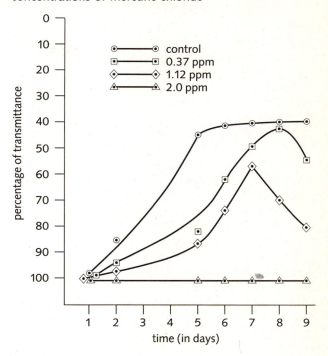

Figure 8.5 Growth of *E. gracilis* in different concentrations of mercuric chloride

OPTIONAL ACTIVITY

The green algae that populate our aquatic environments are known as **autotrophs** (primary producers) and are the base of the many food webs that are found there. Biologists are concerned about changes in various environmental conditions that may disturb the growth of such algae. Altering the environmental conditions through variations in the nutrient content, light (color, intensity), pH, or temperature will allow you to monitor their effects on the growth of *E. gracilis*.

It is also possible to introduce into your cultures various concentrations of potential pollutants, such as pesticides, herbicides, phosphates, or nitrates (see Laboratory Activity 5), and compare the growth of these altered cultures to those that have not been altered. Figure 8.5 shows the growth curves for *E. gracilis* obtained by a student using different concentrations of mercuric chloride. Think carefully about what you might do for a project and be sure to consult with your instructor about the feasibility of your project.

➤ **CAUTION: If you choose to use any type of herbicide or pesticide, be sure to consult with your instructor about its potential toxicity to humans.** *Never* **use such a product until you read and understand the literature supplied by the manufacturer.**

To carry out such an investigation, follow the procedure typical of a good research project. The suggestions below are of a general nature; your instructor may wish to modify them to suit your individual project.

Decide on a research question or problem that can be completed in the time that you have and with the equipment that is available to you. Don't think that your problem must be complicated to be good. Many important questions in biology have been answered with experiments that were elegant in their simplicity. State your problem concisely and clearly and have it checked by your instructor and classmates to see if they understand what you intend to do. Make a statement of a hypothesis at this time. Your hypothesis should guide you in your work, indicating generally how you should set up your experiment and what you expect to be achieved. Refer to Laboratory Activity 1 on the scientific method and review the section on hypothesis writing.

With a few exceptions, the procedures you will follow are similar to the ones you learned in the first part of this activity, these exceptions, however, are very important. First, you must understand that if you are going to alter one factor in the environment of an organism, then you must hold all other factors constant. For example, if you alter the color of light to which the *E. gracilis* is exposed, then temperature, nutrient content, pH, etc. must be the same in all of the cultures. Second, as you learned in Laboratory Activity 1, you must always have a **control** culture (a culture that is not altered in any way) for comparison with the **experimental** cultures (in each of which, one condition is changed). If you don't do this, you will not know what is responsible for the results you obtained.

Know how you are going to collect your data before you start. Will you use the direct count method or a spectrophotometer? Prepare the necessary data tables before you start. Establish a time for reading your cultures and make certain the laboratory will be open. Finally, be sure to have your instructor review your entire project before you begin.

Once you have finished collecting data from your experiment, you will need to present your results. The reporting of results is one of the most important aspects of scientific research and it must be done accurately and completely. Although you are not a scientist, you are assuming the role of a researcher as you do this project and you should report your results as accurately as possible. The following guidelines outline the major sections of your paper.

Format for Paper

Title
Introduction
Procedures
Results
Discussion
Summary and Conclusions

Introduction

The introduction should explain why you undertook this particular project and should make a specific statement of your problem. Review any literature that you researched and list these references at the end of your paper. Use the reference section from an article in a scientific journal as a guide. Finally, state your hypothesis.

Procedures

In this section detail the methods and materials you used to set up your experiment. Explain the manner in which you set up the experimental and control groups and how and when you measured population growth. If your experiment required the use of special media, be sure to include its composition and how it was prepared. In other words, give a detailed description of everything you did to set up your experiment.

Results

This section should contain the data you collected and a description of what happened to your cultures. All of your data should be presented in the form of tables and graphs, modeled after the tables and graphs in this laboratory text. Summarize the information in the tables and graphs, but do not speculate or draw any conclusions at this time.

Discussion

In this section of your paper, summarize the entire experiment and attempt to draw conclusions about your work. Did you answer the question you originally posed when you started the project? Was your hypothesis substantiated? If not, can you suggest why you did not get the expected results? Don't be afraid to offer suggestions for improvement in your methods for use in future projects. Discuss unsolved problems that may have arisen during your project that will provide other students with areas they might explore in the future.

Summary and Conclusions

The summary should be short and include what was done in the experiment, the reasoning behind the experiment, and the results and conclusions.

EVALUATION 8

Population Growth of a Microorganism

1. Average the readings for the two cultures each time you measure growth and record this average in the appropriate column in the following table. Since you will complete either the "direct count" or "percent transmittance" column but not both, put an X through the column not being used.

| Date & time | Number of organisms | | Date & time | Number of organisms | |
	Direct count	Percent transmittance		Direct count	Percent transmittance

2. Graph the above data on the graph paper on the following page. Create an X- and Y-axis, labeling the X-axis with the intervals at which the readings were taken and the Y-axis with the percent transmittance or number of organisms, depending on which method of growth determination you used. Once the data is plotted, identify each segment of your growth curve according to the description in the introduction to this laboratory activity. Appendix B describes the procedure used to make a graph. Refer to it before making your graph for this activity.

3. Optional Activity.   If you did the optional activity involving the effect of altering an environmental condition on the growth of *E. gracilis*, prepare a paper according to the format outlined.

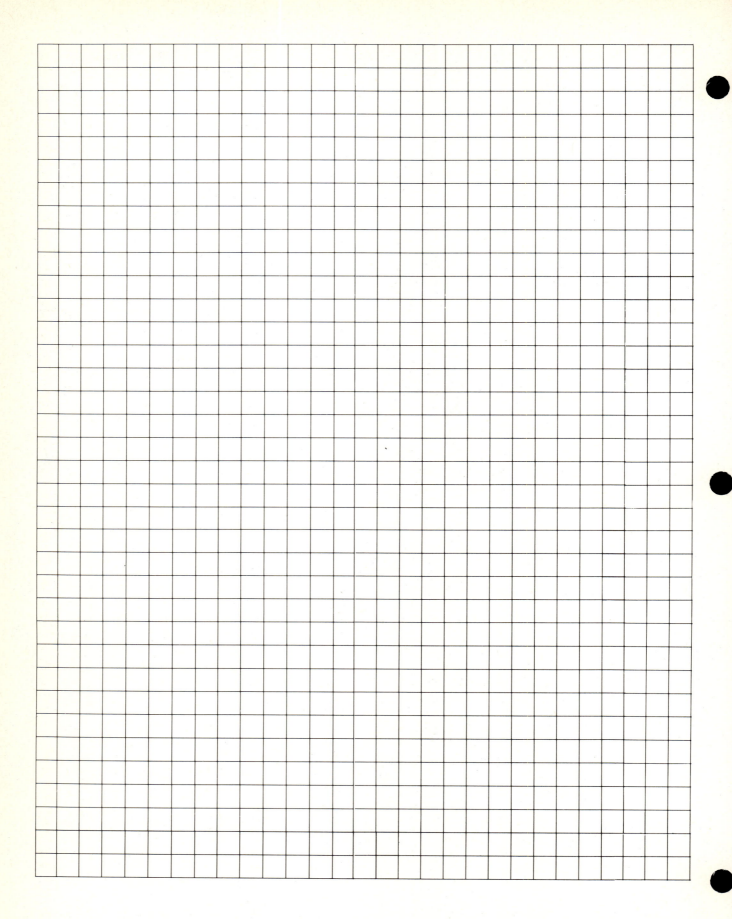

Waste-Water Ecology

OBJECTIVES

At the end of this laboratory activity, you should be able to:

- list and describe the most abundant species of microorganisms found in waste water.
- construct a pyramid of trophic levels of the organisms found in the waste water.

INTRODUCTION

Much has been said recently about water shortages and water contamination in certain areas of America. Some of this shortage has been caused by population growth and the subsequent increased demand for manufactured products. One solution to the water problem is to purify the water quickly after it has been polluted by domestic or industrial use (for example, bathing, washing steel, cooling machinery, or processing food).

Contaminated stream water is purified by physical, chemical, and biological processes that occur in nature and are referred to collectively as self-purification. The physical reactions involved are sedimentation of suspended solids and oxidation of additional solids by sunlight and reaeration. The chemical and biological reactions involved in this process are very complex and involve many organisms of a complicated food chain.

Some of the organisms in this food chain will feed directly on any organic solids present and, in the process, will produce molecules that will then serve as an energy source for other organisms. In this manner complex organic molecules can be decomposed into stable inorganic salts such as phosphates, sulfates, and nitrates. Algae use these in-organic salts and, in return, produce oxygen by way of photosynthesis. This oxygen helps in the aeration of the water.

Bacteria in the water will feed upon certain organic material present, and in turn these bacteria will serve as food for certain **protozoans** (heterotrophic protists). These protozoans can be eaten by other protozoans and still larger **metazoans** (animals composed of many cells).

When the population of this country was not as dense as it is today, human and manufacturing wastes were expelled into nearby streams and the food chain described above purified the stream over time. This natural system of purification has been overloaded; waste water today needs to be purified *before* it returns to a stream.

The modern waste-water treatment plant (see Figure 9.1) simply attempts to duplicate natural stream purification in a more efficient manner within the confines of a physical plant. The operation of a waste-water treatment plant acts to control and enhance the activities and populations of certain freshwater microorganisms.

One very important bacteria involved in the treatment of waste water is *Zooglea ramigera*. This bacteria multiplies rapidly in the presence of the organic material upon which it feeds and then agglutinates into a gelatinous mass called **floc.** The bacteria present in this floc serve as food for free-swimming protozoans. Later, as the population of bacteria increases, stalked **ciliates** (protozoans with cilia) increase in number. In turn, these protozoans serve as food for **nematodes** (roundworms) and **rotifers** (tiny multicellular animals with cilia around their heads). Larger metazoans feed upon these nematodes and rotifers.

The concepts discussed above are ecological ones, and this activity allows you to study the organisms involved in waste-water ecology.

Figure 9.1 A typical sewage plant designed to treat waste water. The waste water used in this activity was taken from a tank similar to the one labeled in the upper right corner.

primary treatment unit

MATERIALS

Waste-water sample in capped bottle
Extralong Pasteur pipettes
Latex gloves
Disinfectant
Microscope slides
Cover slips
Compound microscope

PROCEDURE

In order to examine the waste-water organisms, you will need to prepare a wet-mount slide of the waste water. Use a long Pasteur pipette that will enable you to reach to the bottom of the bottle containing the waste water. Handle the bottle *very carefully* so as not to disturb the organic sediment at the bottom. Most of the organisms will be present in the interface between the sediment on the bottom and the clear fluid on top.

▶ **CAUTION: You are about to handle untreated waste water. Wear latex gloves and wash your hands carefully when finished. Tabletops should also be wiped with a disinfectant.**

1. Squeeze the bulb of the Pasteur pipette *before* putting it into the fluid. Lower the Pasteur pipette into the interface just above the sediment and carefully pull in a small amount of waste water.

2. Remove the Pasteur pipette carefully and prepare a wet-mount slide.

3. Observe the slide under low and high power to become familiar with the host of exciting organisms you will find. It may take several minutes for the organisms to become evident. Be patient.

4. Isolate and study each of the species that are abundant. Make an outline sketch of a representative organism and record its size, shape, movement, and any outstanding features in Question 1 of the Evaluation.

5. Compare your description with the descriptions given in Table 9.1. You will probably be able to identify most of the organisms that you find.

Table 9.1 Typical microorganisms found in waste water

Taxonomic group	Scientific name	Characteristics/behavior	Diagram
Bacteria, members of Kingdom Monera	*Zooglea ramigera*	Floc producer; degrades organic material; aerobic	
Ciliates (Ciliophora), members of Kingdom Protista	*Spathidium sp.*	Vase-shaped; slitlike mouth; feeds on bacteria	
	Paramecium sp.	Slipper-shaped; covered with cilia; oral groove present; feeds on specific bacteria species	
	Didinium sp.	Barrel-shaped; several girdles of cilia; feeds on other ciliates, including paramecia	
	Stentor sp.	Elaborate ciliated food-catching peristome; feeds on bacteria, algae, protozoans	
	Aspidisca sp.	Associated with floc; many modified cilia for crawling; feeds on bacteria	

Table 9.1 Typical microorganisms found in waste water (continued)

Taxonomic group	Scientific name	Characteristics/behavior	Diagram
	Vorticella sp.	Stalked ciliate; attached to floc; feeds on bacteria	
	Opercularia sp.	Stalked ciliate; colonial with more than one body arising from a single stalk; feeds on bacteria	
	Acineta sp.	Rigid, motionless spines used to immobilize prey; feeds on other ciliates	
Phylum Nematoda (roundworms)	Many species	Slender; cylindrical; whiplike motion; feed on protozoans and bacteria	
Phylum Rotifera (wheel animals)	Many species	Larger microscopic multicellular animals; rows of cilia about the head; feed on bacteria, protozoans, other rotifers	

EVALUATION 9

Waste-Water Ecology

1. Make an outline sketch of the microorganisms found in the waste water.
   Under each sketch record its approximate size, shape, movement, and any
   outstanding characteristics.

(a)

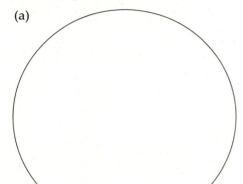

_____ X

Size (μm): _____

Shape: _____

Movement: _____

Outstanding characteristics: _____

(b)

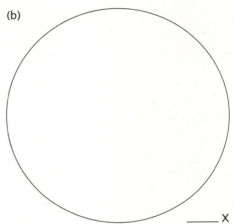

_____ X

Size (μm): _____

Shape: _____

Movement: _____

Outstanding characteristics: _____

(c)

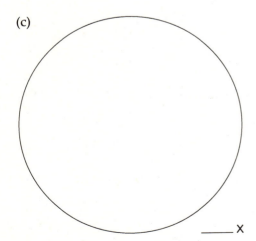

_____ X

Size (μm): _____

Shape: _____

Movement: _____

Outstanding characteristics: _____

(d)

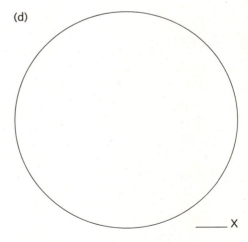

_____ X

Size (μm): _____

Shape: _____

Movement: _____

Outstanding characteristics: _____

(e)

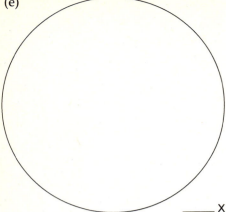

_____ X

Size (μm): _____

Shape: _____

Movement: _____

Outstanding characteristics: _____

(f)

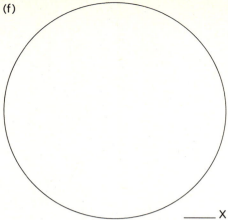

_____ X

Size (μm): _____

Shape: _____

Movement: _____

Outstanding characteristics: _____

(g)

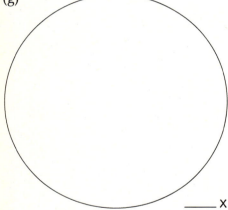

_____ X

Size (μm): _____

Shape: _____

Movement: _____

Outstanding characteristics: _____

(h)

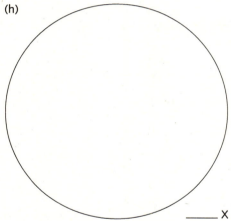

_____ X

Size (μm): _____

Shape: _____

Movement: _____

Outstanding characteristics: _____

(i)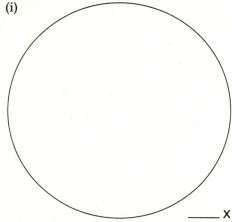

_____ X

Size (μm): _____

Shape: _____

Movement: _____

Outstanding characteristics: _____

(j)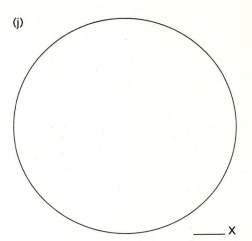

_____ X

Size (μm): _____

Shape: _____

Movement: _____

Outstanding characteristics: _____

2. Identify as many of the organisms in Question 1 as possible. Using the information in the introduction and Table 9.1, determine which organisms feed on which other organisms. Ecologists often illustrate these relationships by using a pyramid of **trophic** (feeding) levels. The floc and some algae compose the base of the pyramid. All other organisms in this ecosystem must live off of this base directly or indirectly. Place the number or letter of each of your sketched organisms on the appropriate level of the pyramid.

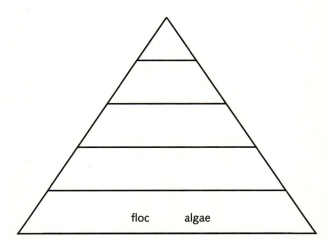

floc   algae

Cell Division: Mitosis

OBJECTIVES

At the end of this laboratory activity, you should be able to:

- identify the various stages of mitosis (prophase, metaphase, anaphase, and telophase).
- diagram the various stages of mitosis.

INTRODUCTION

One of the fundamental characteristics of life is reproduction. Reproduction at the cellular level begins with mitosis, where the duplicated genetic material in the nucleus is separated into two identical nuclei. Later, during **cytokinesis,** the cytoplasm is divided equally. This completes the formation of two daughter cells, each with its own nucleus (see Figure 10.1).

Obviously, mitosis is critical to single-celled organisms (such as algae, fungi, and protozoans) where it serves as a basis for asexual reproduction, but it also plays an important role in diploid multicellular organisms (such as humans). In multicellular oganisms the fertilized egg or zygote duplicates mitotically, and the resulting cells form the foundation for subsequent growth and development.

It is remarkable and fortunate for the student of biology that there is just a single kind of mitosis. Upon learning it, you have all the information needed to begin to understand the basis for normal growth and development in both animals and plants.

Some mature tissues lose their ability to reproduce while others maintain this ability. Mitosis does not occur in all cells, thus it is not as easy as it would first seem to obtain and study mitotically active cells. In the materials chosen for use in this laboratory activity, cells at different stages of mitosis are relatively easy to find.

Biologists identify the stages of mitosis—**prophase, metaphase, anaphase, telophase**—in an attempt to organize artificially the multitude of changes that occur during this process. Mitosis is a continuous process, and the stages are artificial terms used to describe a portion of this continuum. In other words, the stages of mitosis can no more describe the whole process than four photographs of you at birth, at adolescence, as a young adult, and at old age can describe your life.

MATERIALS

Prepared slide of *Allium sp.*, longitudinal section
Compound microscope
Allium sp. or *Zebrina pendula* root tips stored in 70% ethanol
1 molar hydrochloric acid (HCl)
Watch glass
Microscope slides
Cover slips
Dropping bottle of acetocarmine stain
Forceps
Rusty razor blade
Bunsen burner
Paper towel
Dropping bottle of 0.5% toluidine blue O stain
Allium sp. root tips grown in 0.05% colchicine, stored in 70% ethanol

Figure 10.1   Stages of animal mitosis

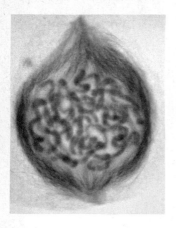
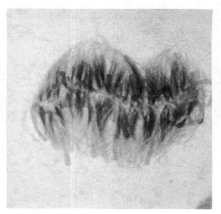
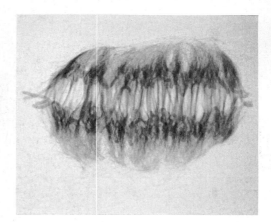
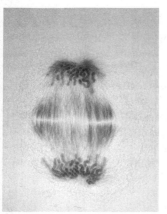

PROCEDURE

The tip of a fast-growing plant root is an excellent source of mitotically active cells. As these fast-growing roots push their way through the soil, the tip is protected by a cap of cells. Just behind this cap, mitosis is occurring in the cells of the **meristematic** region. Beyond this region is a section of root where the cells are growing by **elongation** of the individual cells.

1. Obtain a prepared slide of a longitudinal section of the onion root (*Allium sp.*). This slide was prepared by slicing a root into very thin sections, staining the tissue, and then sealing a section permanently under a cover slip.

2. Observe the root section under low power (100X) so that you can see the complete outline of the root.

3. Examine the tissue under high power (430X) and find the root cap, the meristematic region, and the region of cell elongation.

4. Draw an outline of the longitudinal section of the onion in Question 1 in the Evaluation and label the cap, the meristematic region, and the region of elongation.

5. Examine the meristematic region under high power (430X). The cells of the root tip are capable of mitosis and appear smaller than the cells in the region of elongation. Locate cells at all the different stages of mitosis. Use the photographs in Figure 10.2 to help you identify these stages.

6. Diagram the cells of the meristematic region in prophase, metaphase, anaphase, and telophase in Question 2 in the Evaluation.

Figure 10.2   Stages of plant mitosis

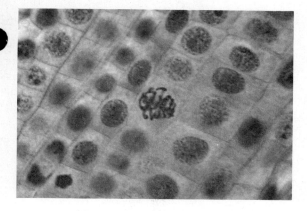

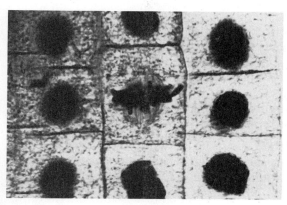

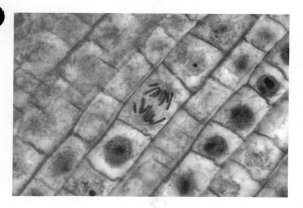

Your instructor will supply you with root tips from either onion (*Allium sp.*) or zebra plant (*Zebrina pendula*) (see Figure 10.3). These roots grow very rapidly, producing 1–2 cm of growth in 3–4 days. The terminal centimeter of root tip was cut off and preserved in 70% ethanol. Your instructor may wish for you to work with either or both of these plants.

If using *Allium sp.* (onion) root tip, follow these instructions to prepare a squash.

1. Add several drops of hydrochloric acid to a watch glass.

2. Cut off the terminal 5 mm of the onion root tip provided by your instructor and place it in the acid. Leave the root tissue in the acid for 2 minutes. This acid will soften the root tissue and cause the cells to separate easily.

3. Wash, rinse, and dry a microscope slide carefully. Add a drop of acetocarmine stain to it. This stain has an affinity for the genetic material in the chromosomes, coloring it purple-red.

4. Place the softened root tip in the acetocarmine stain with forceps.

5. Using a rusty razor blade, cut the tip into tiny pieces. The iron from the rusty razor blade enhances the staining reaction.

▶ **CAUTION: Be very careful not to leave any tissue on the razor blade, it may be the very section containing the meristematic tissue.**

6. Cover the stain and root tissue with a cover slip and heat gently over the open flame of a Bunsen burner. Be very careful not to allow the stain to evaporate during this heating.

Figure 10.3   *Allium sp.* root tip

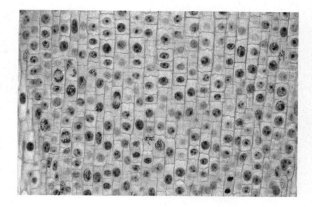

7. Invert the slide on a paper towel and push downward with your thumb over the cover slip (see Figure 10.4).

8. Examining under low and high power, count at least 100 cells at some stage of mitosis. You will probably have to pool your results with several students to obtain this number. Record your data in Question 3 in the Evaluation.

You can prepare a squash of *Zebrina pendula* (zebra plant) root tissue with a slight modification of the procedure listed for the *Allium sp.* root tip.

1. Place several drops of hydrochloric acid in a watch glass.

2. Cut off the terminal 5 mm section of the *Z. pendula* root tip provided by your instructor and place it in the acid for 2 minutes.

3. Prepare a clean slide and add a drop of 0.5% toluidine blue O stain.

4. Place the softened root tip in the stain on the microscope slide with forceps and use a rusty razor blade to cut the tissue into the smallest sections possible.

5. Heat the slide gently over the open flame of a Bunsen burner.

6. Invert the slide and squash as before.

7. Examine under low and high power for the stages of mitosis. Count at least 100 cells at some stage of mitosis. Pool your data with other students if you cannot count 100 cells. Record the number of cells in each stage in Question 3 in the Evaluation.

Figure 10.4   Squash technique

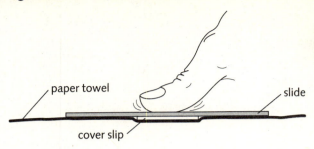

Your instructor has grown some onion (*Allium sp.*) roots in a 0.05% solution of colchicine. Colchicine is an alkaloid substance extracted from the autumn crocus (*Colchicum autumnole*), which acts as a mitotic poison. It interrupts the normal mitotic process by interfering with the formation of the mitotic spindle.

1. Obtain a root tip grown in colchicine and prepare another squash slide; stain the tip with acetocarmine as you did before.

2. Observe under low and high power, counting the cells in each stage of mitosis. Record your data in Question 5 of the Evaluation. Once again you may collect data from your classmates if your individual count is limited.

EVALUATION 10

Cell Division: Mitosis

1. Make an outline drawing of a longitudinal section of an onion root tip and label the appropriate areas: root cap, meristematic region, and the region of elongation. Draw several cells in each of these regions to show their relative size.

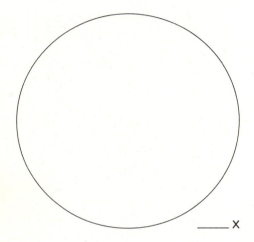

_____ X

2. Using the same longitudinal section of an onion root tip, diagram a cell in each phase of mitosis (prophase, metaphase, anaphase, and telophase) in the space below.

Prophase

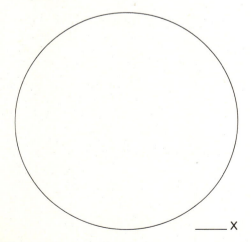

_____ X

Metaphase

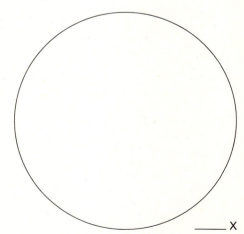

_____ X

Anaphase

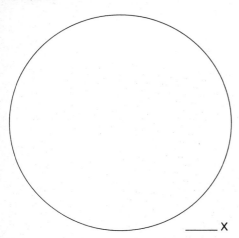

_____ X

Telophase

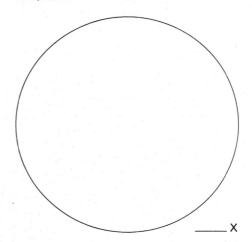

_____ X

3. From either the onion or the zebra plant, record the number and the percentage of cells that you (and your classmates) found in prophase, metaphase, anaphase, and telophase.

	Number	Percent
Prophase		
Metaphase		
Anaphase		
Telophase		

4. Why did you not get an equal number of each?

5. (a) Repeat Question 3 above, recording the data you collected observing the root tip grown in colchicine.

	Number	Percent
Prophase		
Metaphase		
Anaphase		
Telophase		

(b) What major difference did you observe?

(c) Can you explain this phenomenon?

Inheritance in Zea mays

OBJECTIVES

At the end of this laboratory activity, you should be able to:

- show how the F_2 progeny of corn illustrates the laws established by Mendel.
- use the chi square analysis to determine how well observed data fits with theoretical data (goodness of fit).

INTRODUCTION

At about the time that Charles Darwin was writing *The Origin of Species,* an Austrian monk named Gregor Mendel was beginning a series of experiments that would provide the scientific community with a new understanding of the basic mechanisms of heredity. In 1865 Mendel presented a paper on the results of his garden pea hybridization experiments to the Natural History Society of Brunn, Czechoslovakia. Although this paper was a milestone in establishing a new science, it was largely ignored by scientists of the time. It was not until 1900 that three biologists, seeking confirmation for their own experiments in heredity, rediscovered Mendel's paper. Thirty-five years after his research was conducted, Mendel's work was finally recognized for the enormous contribution it made to the field of genetics.

Since then, many plants and animals have been used to study the inheritance of traits, and corn (*Zea mays*) and the fruit fly (*Drosophila melanogaster*) rank high on the list of those that have been most useful. In the next two laboratory activities you will use both of these organisms to illustrate several fundamental concepts in genetics.

In America, corn breeding began in the 1880s but did not make much progress until after the rediscovery of Mendel's paper in 1900. Early hybridizers crossed varieties of corn in an effort to increase the protein and oil content of the endosperm (portion of seed containing nutrients) as well as the total yield of corn per acre. During the years of crossbreeding corn, many mutants have been found that are useful in demonstrating basic genetic principles.

Corn lends itself well to crossbreeding in that the **staminate** (male) flower is well separated from the **pistillate** (female) flower (see Figure 11.1). The staminate flower is commonly referred to as the **tassel** and the **silk** is the elongated style of the pistil. This arrangement of flower parts allows for selective pollination, because the breeder can take pollen from one plant and transfer it to another selected plant to produce the desired cross. Self-pollination is prevented by removing and/or covering the staminate or pistillate flower of the plants involved in the cross (see Figure 11.2). After pollination and fertilization, the base of the pistil will develop into the **seed.** The seed contains the endosperm and the embryo plant. Both the embryo plant and endosperm are surrounded by an **aleurone** layer (outer covering of the endosperm) and the **pericarp** (seed coat) (see Figure 11.3). This entire package, which forms the corn seed, represents the next generation in the life cycle of the corn plant.

One of the mutants frequently used for genetic studies in corn occurs in the pigmentation of the aleurone layer. A second mutant is in the gene that

Figure 11.1 Corn plants
a. Staminate flower tassels

b. Pistillate flower silks

Figure 11.2 Corn plants with tassels covered to prevent self-pollination

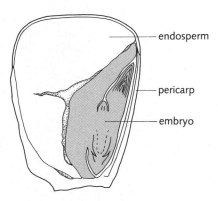

endosperm

pericarp

embryo

Figure 11.3 Corn (*Zea mays*) seed

Figure 11.4 Corn kernels displaying the following characteristics: purple/smooth, purple/wrinkled, yellow/smooth, yellow/wrinkled

controls the amount of starch and sugar in the endosperm and results in seed texture that is either smooth or wrinkled (see Figure 11.4). During this laboratory activity you will work with these two characteristics. The crosses will already have been made and it will be your responsibility to analyze the data that you collect.

Mendel's Laws—A Review

Before you begin it will help to review several of the terms used by geneticists and Mendel's laws of heredity. This review is not comprehensive; you should read carefully the chapters in your textbook that relate to Mendelian genetics.

- **phenotype** the physical appearance of an organism
- **genotype** the genetic make up of an organism
- **allele** one of two or more forms of the same gene
- **locus** a specific position on a chromosome where a gene is located
- **heterozygote** an organism having two different alleles at the same locus on homologous chromosomes
- **homozygote** an organism having identical alleles at the same locus on homologous chromosomes
- **homologous chromosomes** chromosomes having genes for the same kinds of characteristics that pair during meiosis

Mendel's **Law of Dominance** states that when two alleles are present, the one that is expressed as the phenotype is the dominant allele. His **Law of Segregation** states that when gametes are formed, the two alleles of each pair separate from one another and each gamete receives one allele. Table 11.1 illustrates Mendel's first two laws.

Mendel's third law, the **Law of Independent Assortment,** states that traits located on different chromosomes will be inherited independently of each other. For example, when Mendel crossed two pea plants with seeds of different colors (yellow and green) and different textures (round and wrinkled), he found that one trait had no influence on how the other trait appeared in the next generation. When Mendel crossed pea plants heterozygous for these two traits, he obtained offspring that were yellow/round, yellow/wrinkled, green/round, and green/wrinkled in a ratio of 9:3:3:1, indicating that the genes for seed color and texture were located on separate pairs of chromosomes and were inherited independently of each other. You may wish to review an example of this classic cross involving two traits in the chapter of your textbook on Mendelian genetics.

Table 11.1 Mendel's first two laws. The following cross involves pigment production in the aleurone layer of corn seeds. In this cross, Y is the allele for purple aleurone and y is the allele for yellow aleurone.

	Cross			Comments
Parents	YY (purple)	×	yy (yellow)	Parents have two alleles for each trait.
Gametes	Y		y	Law of Segregation: each gamete carries one allele.
Cross	Y	×	y	Parental generation is crossed.
F_1 generation		Yy (all purple)		Law of Dominance: first generation (F_1) offspring show dominant pigment.
F_1 cross	Yy	×	Yy	Heterozygous parents are crossed.
Gametes	Y y		Y y	Law of Segregation: each gamete carries one allele.
F_2 generation		Gametes		Second generation: ratio of 3 purple to 1 yellow.

		Gametes	
		Y	y
Gametes	Y	YY (purple)	Yy (purple)
	y	Yy (purple)	yy (yellow)

The Analysis of Data in Genetic Crosses

When making crosses in genetics it is usually the practice to predict or hypothesize the expected ratio of phenotypes in the offspring. Normally the observed ratio will not be identical to the expected one; that is, it will deviate to a certain extent from what the experimenter had predicted. Such deviations may be due to chance or an incorrect hypothesis. There is also the possibility that certain circumstances or conditions during the experiment may alter the ratios beyond what would be expected based on chance alone. Geneticists deal with this problem of deviation using a statistical analysis known as **chi square** (see Appendix C). You will learn to apply this useful tool during the following exercises.

MATERIALS

F_2 corn showing aleurone and endosperm
  mutants
Tape

PROCEDURE

The Laws of Dominance and Segregation

In the introduction we discussed the trait concerning seed texture in corn. Seeds with a high concentration of starch are smooth, while those with a high concentration of sugar are wrinkled. Experiments with corn have shown that the allele for smooth kernels is dominant (W) to the allele for wrinkled kernels (w).

1. In the Evaluation there is a table similar to Table 11.1, which was used to describe the inheritance of aleurone color. Complete this table for the smooth/wrinkled trait.

   For the next activity, your instructor will supply you with an ear of corn whose F_2 kernels are from a cross between two parents heterozygous for the alleles for aleurone pigmentation (Yy × Yy). Remember that each kernel represents an individual offspring from the F_1 cross. The expected ratio of dominant to recessive phenotypes in this type of cross is 3 purple to 1 yellow.

1. Obtain an ear of corn from your instructor and examine the kernels carefully, noting the differences in color. Some of the seeds will be mottled; these should be counted as purple.

2. Mark the end of a row of kernels with a piece of tape. Count the number of purple and yellow kernels in this row and record your data in Table 11.2. These are the observed phenotypes.

3. Apply a second piece of tape to the next row and repeat the process. Continue counting rows until you have counted all of the kernels.

Table 11.2 Observed phenotypes

Purple	Yellow
Total _____	Total _____

4. When you have finished counting all of the kernels, complete Question 2 in the Evaluation relating to this cross.

The Law of Independent Assortment

When Mendel crossed two parents that were heterozygotic for each of two different traits, he obtained the classic 9:3:3:1 ratio of phenotypes. Using the traits we have been studying in this activity, color and texture, such a cross would be:

YyWw × YyWw

A cross between two parents with this genetic makeup would produce four different phenotypes in the offspring in a ratio of 9:3:3:1. For the next activity, obtain an ear of corn from your instructor that shows the four different phenotypes: purple/smooth, purple/wrinkled, yellow/smooth, and yellow/wrinkled. Remember, this corn is the result of a cross between two heterozygous parents, and the kernels are the F_2 generation.

1. Mark one row of kernels with a piece of tape as you did in the previous activity.

2. Count the four phenotypes in that row and record your data in Table 11.3.

3. When you have finished counting all of the rows, turn to the Evaluation and answer Question 3.

Table 11.3 Observed phenotypes

Purple/smooth	Purple/wrinkled	Yellow/smooth	Yellow/wrinkled
Total _____	Total _____	Total _____	Total _____

EVALUATION 11

Inheritance in Zea mays

1. The following table is similar to Table 11.1, which showed the inheritance of aleurone color in corn. Complete this table for the smooth/wrinkled trait.

	Cross			Comments
Parents	WW (smooth)	×	ww (wrinkled)	Parents have two alleles for each trait.
Gametes	◯		◯	Law of Segregation: each gamete carries one allele.
Cross	◯	×	◯	Parental generation is crossed.
F_1 generation				Law of Dominance: first generation (F_1) offspring show dominant trait.
F_1 cross		×		Heterozygous parents are crossed.
Gametes	◯◯		◯◯	Law of Segregation: each gamete carries one allele.
F_2 generation		Gametes		Second generation: ratio of 3 smooth to 1 wrinkled.

Gametes

2. As indicated in Table 11.1 and in Question 1 above, the F_2 generation should show a 3:1 ratio of dominant to recessive phenotypes. You can test how well your observed count fits with the expected ratio by using the chi square test. This statistical test is explained in Appendix C. You should study this example before completing the following questions.

   (a) Record your data for the cross involving aleurone pigment in the column labeled "observed number" in the chart below. To calculate the "expected number," multiply the total number of seeds counted by the fraction of seeds expected for that particular phenotype. For example, if a total of 460 seeds were counted and you expected a 3:1 ratio, then $\frac{3}{4}$, or 75%, should be purple and $\frac{1}{4}$, or 25%, yellow. The calculation would be:

$$\frac{3}{4} \times 460 \text{ (total seeds)} = 345$$

345 expected to be purple
115 expected to be yellow

(b) Using Appendix C as a guide, calculate the chi square value for this data in the following table.

Phenotype	Observed number O	Expected number E	Deviation O − E	Deviation squared (O − E)²	$\frac{(O - E)^2}{E}$
Purple					
Yellow					

$$\text{Totals} \quad \sum \frac{(O - E)^2}{E} =$$

$$\chi^2 =$$

(c) Using the chi square table in Appendix C, interpret the chi square value you have just obtained. Do your actual results agree with the expected ratio? _____ If not, offer an explanation for this lack of agreement.

3. Answer the following questions for the cross involving the Law of Independent Assortment.

(a) Calculate the expected number of seeds for each phenotype just as you did in Question 2. The following information will assist your calculations.

Phenotype	Expected ratio	Expected number
Purple/smooth	9/16	_____
Purple/wrinkled	3/16	_____
Yellow/smooth	3/16	_____
Yellow/wrinkled	1/16	_____

(b) Complete the following table to calculate the chi square value.

Phenotype	Observed number O	Expected number E	Deviation O − E	Deviation squared $(O - E)^2$	$\dfrac{(O - E)^2}{E}$
Purple/smooth					
Purple/wrinkled					
Yellow/smooth					
Yellow/wrinkled					
TOTALS				$\sum \dfrac{(O - E)^2}{E} =$	
				$\chi^2 =$	

(c) Do the results of your actual count agree with the expected ratio?
_____ If the deviation is greater than that expected by chance, can
you account for this?

Sex-Linked Inheritance
in Drosophila melanogaster

OBJECTIVES

At the end of this laboratory activity, you should be able to:

- handle, determine sex, and culture *Drosophila melanogaster*, and use these skills to set up genetic experiments.
- explain how a sex-linked gene is passed from a parent to the first and second filial generations.

INTRODUCTION

Thomas Hunt Morgan was an American biologist who became interested in Gregor Mendel's laws and saw the need to continue Mendel's work. In order to pursue the study of genetics, Morgan wanted to use an organism better suited to further research than the pea. The pea plant had many characteristics that made it an excellent choice for Mendel, but it had a few major disadvantages for the extensive experiments Morgan wished to undertake, mainly a long reproductive cycle and the need for constant attention. In 1910 Morgan learned of the work being done by Dr. W. E. Castle, who had been using the fruit fly (*Drosophila melanogaster*) for inbreeding experiments. The flies needed very little attention after they were supplied with a sugar-rich medium. Moreover, they took up little laboratory space and had a life cycle of only 10 days at 25°C. Morgan and his students began to use the fruit fly in their genetic experiments. Some of Morgan's students at Columbia University who participated in this research, including Alfred Sturtevant, Calvin Bridges, and Hermann Muller, were destined to become leaders in this new science of genetics.

Morgan and his students collected evidence that became the foundation of the chromosome theory of heredity. In addition, they developed procedures to actually map genes on specific chromosomes and discovered the existence of sex-linked genes. The fruit fly cross you will be doing in this activity is one of the most famous experiments conducted by Morgan and his group. Morgan recognized that the genetic results of this cross were consistent with the movement of the X and Y chromosomes during meiosis. This discovery was key to the development of the chromosome theory.

Sex-linked genes are located on the X chromosome and can be passed on to a male offspring only by the female parent. Since males have only one X chromosome, if they receive a mutant recessive gene from their mother, they will manifest the trait. Sex-linked genes behave in the same manner in humans as they do in the fruit fly. At this time it would be helpful to review the material on sex-linked genes in your textbook.

The Life Cycle of *Drosophila melanogaster*

Drosophila melanogaster is an insect that has a complete metamorphosis, meaning that it goes through four readily identifiable stages in its life cycle (see Figure 12.1). The adult female fly lays its eggs in the soft medium. The eggs are white, very small, and have two pronglike projections extending from one end. In about two days the eggs hatch and the larvae appear as wormlike creatures that work their way through the medium, eating constantly. During this stage they molt twice. There are, therefore, three larval stages separated by two molts; these periods of larval growth are called **instars.** The total larval period lasts about 8 days. The larvae finally

Figure 12.1 Life cycle of *Drosophila melanogaster*

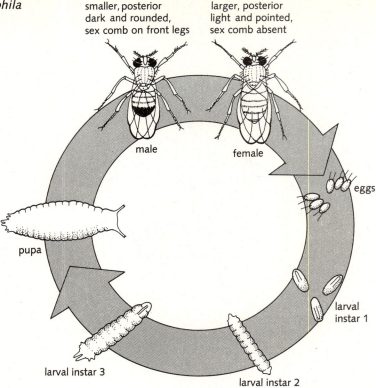

smaller, posterior dark and rounded, sex comb on front legs

larger, posterior light and pointed, sex comb absent

male

female

eggs

larval instar 1

larval instar 2

larval instar 3

pupa

climb onto the side of the vial or onto paper inserted into the medium and form **pupae** (the pupa is immotile, nonfeeding and protected by a case). During the pupal stage of development the larval tissues reorganize to form adult flies. The pupal cases darken as they mature, and in about 4 days the adults hatch. Most of the flies hatch during the early morning (approximately 6:00 A.M.) and are capable of mating within 8–10 hours. Thus any female left in a vial of hatching flies longer than this period of time will probably have mated. In order to obtain virgin female flies, you must remove the females and isolate them in a new vial of medium within 8–10 hours.

MATERIALS

Instant *Drosophila melanogaster* medium
Dissecting microscope
Etherizer
Ethyl ether
Drosophila melanogaster: wild type and white-eyed mutants
3 × 5 white card
Re-etherizer

Camel-hair brush
Culture vials with plugs
Fly morgue

PROCEDURE

Culturing and Handling Fruit Flies

Your instructor will supply you with a culture of fruit flies. These flies will be growing in a prepared medium and will show all four stages of the life cycle. Your first task will be to examine the culture and identify the various stages of development.

1. Using a dissecting microscope, examine your culture. The larva, pupa, and adult stages will be easily seen but the eggs may be difficult to find. Look carefully. The larvae will be crawling in the medium and on the sides of the vial. The pupae attach to the sides of the vial and to any mesh or paper inserted in the medium.

   In order to sex flies and make crosses, the adults must be anesthetized with ether. To anesthetize the flies you will use a special vial-type container called an etherizer (see Figure 12.2).

Figure 12.2 Anesthetizing and handling fruit flies

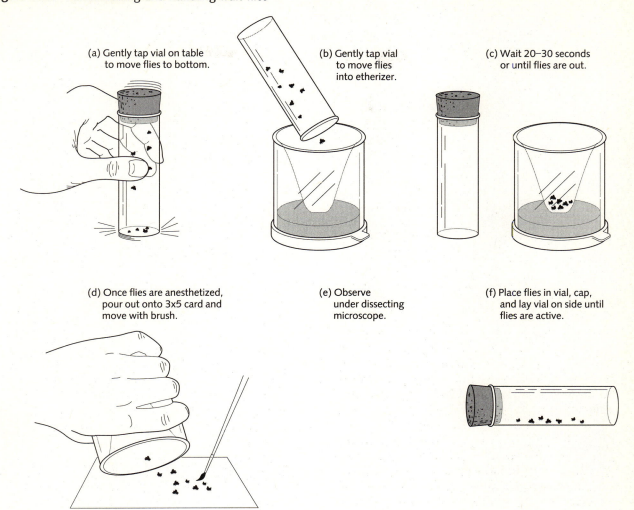

(a) Gently tap vial on table to move flies to bottom.

(b) Gently tap vial to move flies into etherizer.

(c) Wait 20–30 seconds or until flies are out.

(d) Once flies are anesthetized, pour out onto 3x5 card and move with brush.

(e) Observe under dissecting microscope.

(f) Place flies in vial, cap, and lay vial on side until flies are active.

➤ **CAUTION: Ether is explosive! Absolutely do not have any open flames in the laboratory when it is being used.**

1. Obtain an etherizer and remove the large cap from one end and add 5–6 drops of ether to the foam pad. Close it immediately!

2. Take a culture of flies and firmly but gently tap the bottom of the vial on the tabletop so that the flies drop to the bottom.

3. Quickly remove the small plastic insert from the etherizer and the foam plug from the vial of flies.

4. Place the etherizer over the top of the vial and invert the vial and etherizer together. Gently tap the vial to drop the flies into the funnel portion of the etherizer.

5. Quickly replace the insert into the etherizer. The ether will enter the funnel-like chamber and anesthetize the flies.

6. When the last fly has stopped moving, tap the flies onto a 3 × 5 card.

7. Flies may be reanesthetized while you are examining them by using the re-etherizer shown in Figure 12.3. Use it by adding several drops of ether to the cotton attached to the underside of the lid.

➤ **CAUTION: It is important that you do not overanesthetize your flies as this will sterilize or kill them. Overanesthetized flies will hold their wings vertically rather than against their bodies.**

Figure 12.3 Re-etherizer

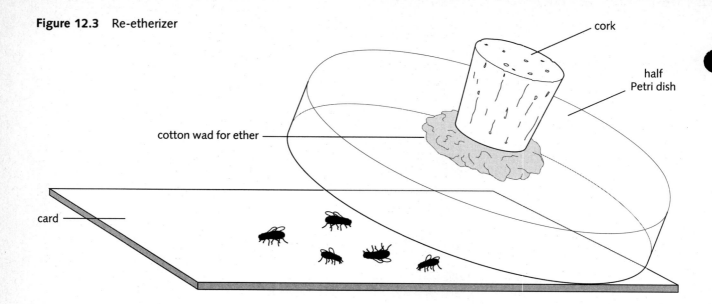

cork

half Petri dish

cotton wad for ether

card

8. Place the card with the anesthetized flies under the dissecting microscope.

9. Using a camel-hair brush, gently sort the flies into males and females. Use Figure 12.1 as a guide.

10. Check your work with your instructor to make sure you have sexed your flies correctly.

11. When you are finished, place the vial on its side and return the flies by gently pushing them inside. Allow the vial to remain on its side until the flies have recovered.

▶ CAUTION: Keep the vial on its side until the anesthetized flies have recovered. Flies that fall into the medium while anesthetized will usually die.

Performing a Cross to Demonstrate Sex-Linked Inheritance

Your first task in making this cross will be to obtain virgin females. Review the life cycle of the fruit fly in the introduction, which explains how you can obtain virgin females.

Make your cross by completing the following steps.

1. Obtain three male white-eyed flies that have been anesthetized and place them in a vial (on its side) containing culture medium.

2. Place three anesthetized virgin female red-eyed (wild type) flies into the vial.

3. Label the vial with the following information: cross, date, laboratory section, and your name. Your cultures should be grown at 20–22°C.

4. Remove the parents after 4–5 days, when larvae are seen in the medium. This is done by anesthetizing the parents and placing them in the fly morgue (a small container of cooking oil).

5. Examine your culture daily after 10–14 days from the date of the cross. Flies that hatch will be the first filial (F_1) generation. Remove flies daily and examine each of the flies to determine its sex and eye color. Record your data in Question 1 in the Evaluation. Flies of this first generation should be collected for 4–5 days.

6. Obtain 3 males and 3 females (they need not be virgins) from the F_1 generation and place them in a vial containing fresh medium. This is the F_1 cross. Label as you did before.

7. Remove the F_1 parents after 4–5 days.

8. When the F_2 (second generation) begins to hatch, collect and count each individual as to sex and eye color. Continue counting flies for 4–5 days. Record your data in Question 1 in the Evaluation.

The techniques learned in this activity can be used to make other crosses in *D. melanogaster*. Your instructor may want you to complete additional crosses involving other mutant strains.

EVALUATION 12

Sex-Linked Inheritance in Drosophila melanogaster

1. Record your results of the white eye–red eye cross in the chart below.

   (a) Parental Cross

   Date parents mated _____   Date parents removed _____

   Date F_1 adults appeared _____

   Female phenotype _____   Number _____

   Male phenotype _____   Number _____

   (b) First Filial Cross

   F_1 Cross: _____ × _____

   Date adults mated _____   Date adults removed _____

   Date F_2 adults appeared _____

   Female phenotype _____   Number _____

   Male phenotype _____   Number _____

2. Use the following Punnett square to illustrate the F_1 cross. You may need
   to review the section in your textbook concerning sex-linked inheritance.
   Let X^r symbolize the allele for white eyes and X^R symbolize the allele for
   red eyes.

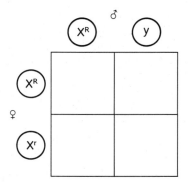

   (a) What is the expected ratio of red- to white-eyed flies?

   Red _____ : White _____

(b) How is the sex of the fly related to eye color?

(c) Morgan and his group obtained 3470 red-eyed flies and 782 white-eyed ones. Is this the ratio you would have predicted?

(d) Which group of flies is deficient in number? Did you get similar results?

(e) Do you have any idea why this deviation might occur?

Human Inheritance: Pedigree Analysis

OBJECTIVES

At the end of this laboratory activity, you should be able to:

- predict from pedigree charts whether a particular trait is controlled by an autosomal dominant, an autosomal recessive, or a sex-linked gene.
- use the product law to determine the probability that a marriage will produce children with a certain trait.
- after knowing the pattern of inheritance in a family pedigree, suggest the most likely genotype for the individuals concerned.

INTRODUCTION

The study of heredity is relatively easy when using organisms such as corn and fruit flies. These organisms can be crossed under well-controlled conditions in laboratories or in the field. They also produce large numbers of progeny, allowing for relatively easy and valid statistical analysis of data. No serious ethical concerns exist, and matings can be designed by the experimenter to match the questions being considered. However, this is not the case when studying inheritance in humans. Matings cannot be controlled, and the number of individuals is limited because the geneticist normally deals with small family units. These difficulties often force geneticists to study human heredity using a **pedigree,** or family history. A pedigree is a convenient and systematic way of following a particular trait as it passes from one generation to another. Pedigrees are frequently used in genetic counseling when a couple is seeking advice about

the possibility that a particular trait will show up in their children. Although it may be possible to determine only the *probability* of having a child with a certain trait, this information may be extremely useful in helping a couple decide whether or not to have children.

The hypothetical pedigree in Figure 13.1 traces the passage of a trait through three generations of a family and shows several of the symbols used in pedigree construction. Familiarize yourself with the notation as you will use it in analyzing pedigrees in this activity.

Figure 13.1 A hypothetical pedigree

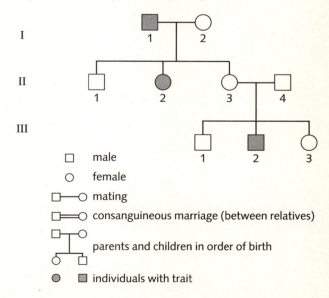

The generations in a pedigree are numbered from top to bottom with Roman numerals. All members of each generation are numbered from left to right with Arabic numerals, regardless of their relationship to other members of that generation.

During this laboratory activity you will study several pedigrees in order to determine the genotype of the individuals involved and the probability that a particular trait will appear in the children of a certain marriage.

PROCEDURE

In order to demonstrate how a pedigree can be analyzed, we will use a family history that shows the inheritance of the albino trait through three generations (see Figure 13.2). To make the analysis a little easier, we will assume that mutations have not occurred unless specifically indicated, that **penetrance** (the degree to which a trait is expressed) is complete, and that anyone marrying into a family (unless otherwise indicated) does not carry the trait.

The first task is to determine the type of inheritance involved. To do this you need to examine the children who have the trait and then look at their parents and, if possible, their grandparents. In this case, the parents of individuals II-3 and III-3 do not have the trait but pass it on to their children, so this rules out the possibility of a dominant

gene. We can also assume that the trait is not sex-linked because the fathers of II-3 and III-3 should have the trait in order for it to appear in their daughters. Based on the limited evidence in this pedigree, then, it is reasonable to assume that the trait is controlled by an **autosomal recessive** gene, meaning that it is neither sex-linked nor dominant. Since you know that the trait is autosomal recessive, you can now determine the genotypes of most of the individuals in the pedigree.

1. In Figure 13.2, write the genotype for each individual on the blank lines under the symbols for male and female, using the letters A for normal pigmentation and a for albinism. When a person is normally pigmented, and you cannot determine whether they are homozygous (AA) or heterozygous (Aa), use the symbol A __.

The final step in our pedigree analysis is to determine the probability that a certain marriage will produce children with the trait in question. To do this we must turn to a law of probability that relates to the simultaneous occurrence of independent events. This **product law** may be stated as follows.

> If A_1 and A_2 are independent events, the probability (P) that they will occur together is the product of their individual probabilities.

For example, if you roll one die, the probability of obtaining a four is $\frac{1}{6}$. This same probability exists for a second die. The probability of obtaining two fours when rolling two dice simultaneously, is

$$\frac{1}{6} \times \frac{1}{6} = \frac{1}{36}.$$

Figure 13.2 A hypothetical pedigree showing inheritance of the albino trait

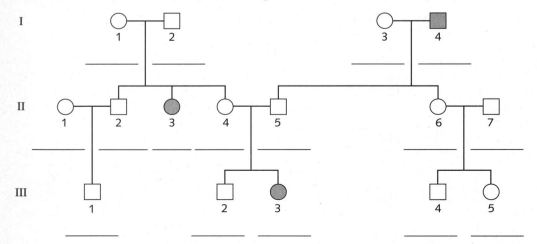

Let's assume that individuals III-1 and III-5 wish to marry and that they want to know their chances of having a child with the albino trait. We can determine this probability by answering the following question: What is the probability that each intended parent is a carrier for the gene?

In the case of intended father (III-1) the probability that his father (II-2) is carrying the allele is $\frac{2}{3}$. We assume that his mother (II-1) is homozygous dominant. This means that the son (III-1) has a probability of $\frac{1}{2}$ of carrying the trait, since his father (II-2) may be heterozygous. The Punnett squares in Figure 13.3 summarize these possibilities.

In the case of the mother of III-5, we are certain that she (II-6) is a carrier since her father (I-4) has the trait. Again we assume the father (II-7) is homozygous dominant. The daughter (III-5), then, has a probability of $\frac{1}{2}$ for carrying the allele. The Punnett square in Figure 13.4 shows this cross.

The final step is to use the product law to determine the probability that an albino child will result from this marriage. You must remember to include in the calculation the probability that an albino child will be produced if both parents are heterozygotes. The calculation is shown in Table 13.1.

The chances are 1 in 24 that these parents will produce an albino child.

1. Three pedigrees are shown in the Evaluation section of this laboratory activity. Study them carefully and complete the necessary information for each one.

Figure 13.3 Punnett squares showing calculations that intended father carries recessive allele

Figure 13.4 Punnett square showing calculations that intended mother carries recessive allele

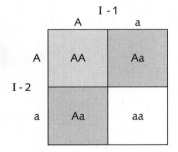

(a) p that intended father's father (II - 2) is heterozygous = $\frac{2}{3}$

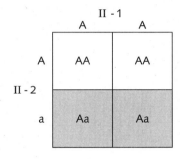

(b) p that intended father (III - 1) is heterozygous = $\frac{1}{2}$

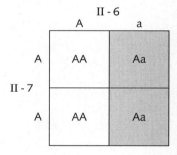

p that intended mother is heterozygous = $\frac{1}{2}$

Table 13.1

Probability that:								
Father's father is carrying allele		Intended father is carrying allele		Intended mother is carrying allele		Intended parents will produce albino child		Couple will produce albino child
II-2		III-1		III-5		III-1 × III-5		
2/3	×	1/2	×	1/2	×	1/4	=	1/24

EVALUATION 13

Human Inheritance: Pedigree Analysis

1. The following pedigree shows the inheritance of a trait in four generations of a family. Answer the questions that relate to this pedigree.

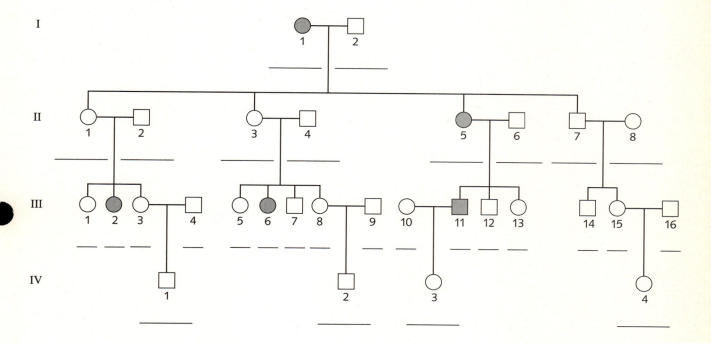

(a) How is this trait inherited?

(b) Use the letters B (dominant allele) and b (recessive allele) for this trait and list the genotypes for each individual on the blank lines under the symbols for male and female.

(c) What is the probability that the children from the following marriages will show the trait?

III-1 × III-12 _____   IV-1 × IV-4 _____

IV-2 × IV-3 _____   III-6 × III-13 _____

(d) If we assume that the frequency of heterozygous persons for this trait in the normal population is $\frac{1}{71}$, then the chances of two heterozygotes marrying and producing a child with the trait are:

heterozygous parent		heterozygous parent		probability of having child with trait		
_____	×	_____	×	_____	=	_____

Use the information in parts (c) and (d) above to make a statement about the problem associated with consanguineous (closely related) marriages.

2. The following pedigree shows the inheritance of a sex-linked trait in three generations of a family. Study the pedigree carefully and, if necessary, review the section in your textbook concerning sex-linked inheritance before doing this problem.

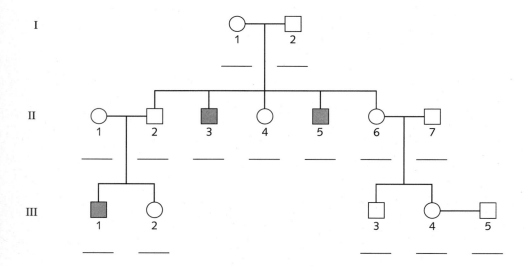

(a) Use the symbols X^T and X^t for this trait and list the genotypes for each individual on the blank lines under the symbols for male and female.

(b) Which of the parents in generation I is a heterozygote? _____

(c) From whom must individual III-1 have received the allele? _____

(d) What is the probability that:

II-4 is a carrier? _____

II-6 is a carrier? _____

the male children of a marriage between III-4 and III-5 will have the

trait? _____

3. The following family history shows the inheritance of a type of anemia
(pyruvatekinase deficient hemolytic anemia) in an Amish family in
Pennsylvania (McKusick, 1964).[1] Study the pedigree carefully and answer
the questions that follow.

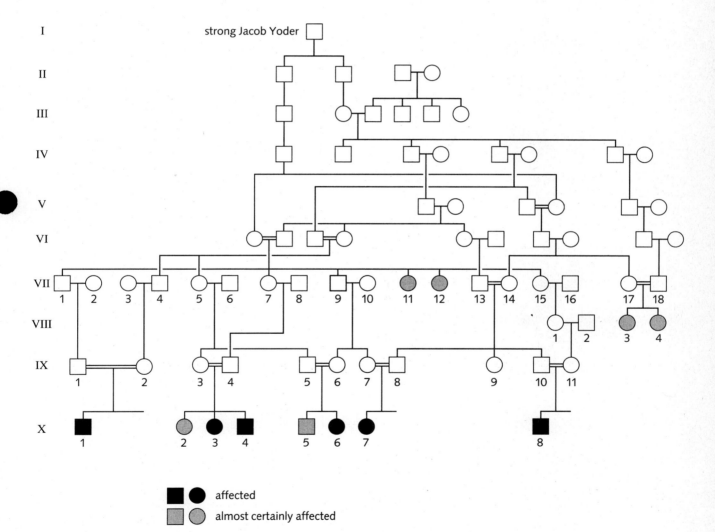

affected

almost certainly affected

1. McKusick, V. A., et al., "The Distribution of Certain Genes in the Old Order Amish," Cold
Spring Harbor Symposium of Quantitative Biology, 29:104, 1964.

(a) What type of inheritance is shown in this family history?

(b) As you follow this family history through ten generations, what appears to be the major factor leading to the increased frequency of affected individuals?

(c) Based on the fact that the individuals in generation X have the trait, what conclusion can you draw about the parents of these children?

(d) To which individual can you trace all of the affected persons in this pedigree?

(e) Assuming that complete penetrance occurs, what is the probability that the fourth child of parents IX-3 and IX-4 will have the trait?

(f) Considering parents VII–1 and VII–2, which one is most likely to be a carrier for this type of anemia?

Population Genetics

OBJECTIVES

At the end of this laboratory activity you should be able to:

- explain the Hardy-Weinberg principle.
- use the Hardy-Weinberg formula to predict the genotypes and phenotypes resulting from random mating within a given population.

INTRODUCTION

The genetics of individual organisms is of obvious importance to biologists. But in addition to knowledge of the genetic makeup of individuals, biologists are also interested in the **gene pool,** or the total genetic material present in all the individuals of an interbreeding population. Even more specifically, biologists often wish to know the frequency of occurrence of certain genes within a population; these are the concerns of **population genetics.**

In 1908 G. H. Hardy, an English mathematician, and Wilhelm Weinberg, a German physician, independently developed a formula for predicting the frequency of alleles in large interbreeding populations. Using the label p for the dominant allele in the gene pool and q for the recessive allele, the Hardy-Weinberg formula is expressed as

$$p^2 + 2pq + q^2 = 1.$$

In the formula, p^2 represents the homozygous dominant population, $2pq$ represents the heterozygous population, and q^2 represents the homozygous recessive population. Mathematically, this formula may also be expressed as

$$(p + q)^2 = 1 \quad \text{or} \quad (p + q)(p + q) = 1.$$

To illustrate the Hardy-Weinberg principle, consider the following example. If we know that in a population the frequency of the dominant allele p is 80% and the frequency of the recessive allele q is 20%, then we can determine the frequency of the homozygous and heterozygous individuals. In this interbreeding animal population, 80% of the sperm will contain a p allele and 20% of the sperm will contain a q allele. The same ratio holds for the eggs that are produced. The Punnett square in Figure 14.1 illustrates the frequency of genotypes of offspring when the dominant and recessive alleles occur in such an 80-to-20 ratio.

The Punnett square indicates that 64% (0.64) of the population will be homozygous recessive, 32% (0.16 + 0.16 = 0.32) will be heterozygous, and 4%

Figure 14.1 Calculation of genotype frequency in a population

	sperm	
	0.8 p	0.2 q
0.8 p	0.64 pp	0.16 pq
0.2 q	0.16 pq	0.04 qq

eggs

(0.04) will be homozygous dominant. To check, we can substitute the values from the Punnett square in the formula.

$$p^2 + 2pq + q^2 = 1$$
$$0.64 + 2(0.16) + 0.04 = 1$$

The Hardy-Weinberg formula may be used to predict the frequency of heterozygotes in a population. This is particularly useful because heterozygotes may not be easily observed in a population. The following activities will illustrate the application of the Hardy-Weinberg principle.

MATERIALS

Calculator
Pencil and paper

PROCEDURE

An Illustration

Sickle-cell anemia is a serious health problem within certain populations, including Americans of African descent (see Figure 14.2). This type of anemia is a genetic disorder caused by a mutant gene that produces an abnormal hemoglobin molecule. The normal gene (S) will produce normal hemoglobin, and the mutant gene (s) will produce an abnormal hemoglobin. Heterozygotes for this condition have a mild form of this anemia while homozygous recessives die from the disorder. In a given population, 3% (0.03) are born homozygous recessives (ss). Using the Hardy-Weinberg principle, we can figure out what percentage of the population is heterozygous for this condition.

We want to find the value of $2pq$ in the Hardy-Weinberg formula. We already know that q^2 represents the homozygous recessive population, or

Figure 14.2 Red blood cells
(a) normal

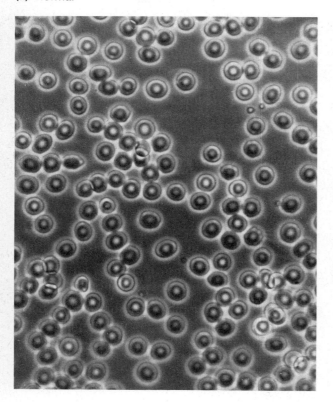

(b) sickle

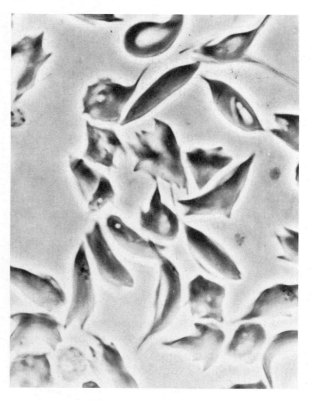

0.03 in our case. So, if

$$q^2 = 0.03,$$

then

$$q = \sqrt{0.03} = 0.173.$$

Next, since the Hardy-Weinberg formula can be written as

$$(p + q)(p + q) = 1,$$

we can take the square root of each side to get

$$p + q = 1$$

or

$$p = 1 - q.$$

Substituting 0.173 for q gives

$$p = 1 - 0.173 = 0.827.$$

Finally, we can substitute 0.827 for p and 0.173 for q in $2pq$ to yield

$$2(0.827)(0.173) = 0.286142.$$

Thus, approximately 28.6% of the individuals in the population are heterozygotes.

Analyzing a Gene Pool

The Old Order Amish of Lancaster County, Pennsylvania were the basis of a study by Dr. Victor McKusick of the Johns Hopkins University. The Amish are well suited for genetic research because:

- they are a defined group
- they are a closed group
- their origin is known
- they have extensive genealogies
- they have a high standard of living
- they have no illegitimacy

The Amish of Pennsylvania are a hardworking, conservative religious group that accepts very few outsiders into their ranks. They record their own genealogies completely and accurately, and illegitimacy is all but unknown.[1] Emigrating from Switzerland, most of their forefathers arrived in Lancaster County around 1720.

Their isolation has resulted in much inbreeding. For instance, 23% of the group share the last name Stolzfus and are descendants of one Nicholas Stolzfus, an original settler, while 50% of the others

1. McKusick, V. A., et al., "The Distribution of Certain Genes in the Old Order Amish," Cold Spring Harbor Symposium of Quantitative Biology, 29:104, 1964.

share the family names of King, Fisher, Beiler, Lapp, and Zook. In such an inbred group a mutant gene has a better chance of expressing itself phenotypically because it is more likely that heterozygotes will marry than in the greater population. McKusick found quite a few genetic deviations in this group, one of the most serious being the Ellis–van Creveld syndrome.

The Ellis–van Creveld syndrome is caused by a recessive autosomal gene that is **pleiotrophic,** that is, a single gene producing more than one phenotypic effect. In this case a single mutant gene in the homozygous condition causes dwarfism, six fingers on each hand, and—in 50% of the cases—a heart defect. Until Dr. McKusick's study, only 50 cases of this syndrome had been documented worldwide. In 1972, however, McKusick discovered 75 cases among the Amish in Lancaster County. He found the frequency of the Ellis–van Creveld syndrome gene to be 0.07 or 7% ($q = 0.07$). The genealogical record showed all of the afflicted persons to be descendants of a single couple, Samuel King and his wife. One of them was a carrier of this disorder.

1. Turn to Question 2 in the Evaluation.

Field Activity

Population genetics can be quite exciting. The following exercise permits you to go into the field to collect some original data on one of two human genetic characteristics (see Figure 14.3).

Earlobe attachment is determined by a single autosomal gene. The dominant allele is for unattached earlobes, while the recessive allele is for attached lobes. Thus, individuals with attached lobes are homozygous recessives.

Tongue rolling is also determined by a single autosomal gene. The dominant allele is for tongue rolling, while the recessive allele is for the inability to roll the tongue. Therefore, individuals who cannot roll their tongues are homozygous recessives.

1. Collect data concerning earlobe attachment or tongue rolling from at least 10 students on campus.

2. Pool your data with classmates when you return to the laboratory; the more data you collect the better.

3. Calculate the percentage of individuals in your population that are heterozygotes for the condition that you chose to study.

4. Record your data and show your calculations in Question 3 in the Evaluation.

Figure 14.3   Genetic characteristics
(a) Earlobe attachment

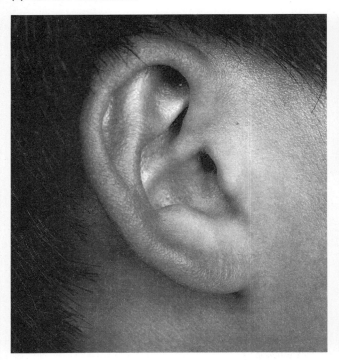 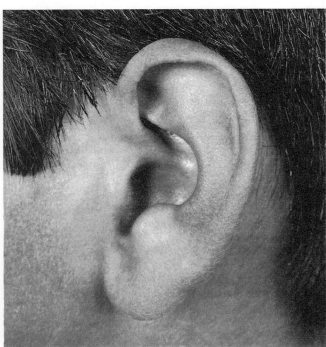

(b) Tongue rolling

EVALUATION 14

Population Genetics

1. Sickle-cell anemia presents a serious health problem to Americans of African descent. In Africa, however, the situation is quite different. Heterozygotes have a greater immunity to malaria, a disease that claims many lives in this region of the world. Heterozygotes can be identified by microscopic examination of the erythrocytes (red blood cells). How might this phenomenon affect the frequency of this allele in American and in African risk populations? Explain your answer.

2. (a) Use the Hardy-Weinberg formula to determine what percentage of the Amish population is heterozygous for the Ellis–van Creveld syndrome. Show your calculations.

   (b) Given the strict controls it administers on itself, how might the Amish group reduce the incidence of Ellis–van Creveld syndrome?

3. (a) Record below the number of homozygous recessive individuals you counted within the population you studied.

Trait	Homozygous recessives	
	Number	Percent
Tongue rolling		
Attached earlobes		

(b) Using the Hardy-Weinberg law, calculate the heterozygotes that are carriers for this recessive allele.

Animal Diversity

Experts in the area of phylogenetics disagree on exactly how the different phyla are related, but they generally agree on the overall relationships indicated in the phylogenetic tree in Figure 1 on the following page. This phylogenetic tree reinforces the general idea that once living organisms develop a useful structure, they pass it on to other organisms descending from their group. In other words, each new group does not start from scratch.

After carefully studying the phylogenetic tree, you should read the sections in your textbook that are devoted to the phyla shown in this figure.

The initial organization of the tree is based on the presence or absence of the **coelom** (body cavity) between the outer wall and the intestinal tract (see Figure 2 on the following page). **Acoelomates**—as their name indicates—have no body cavity. In all of these animals, **ectodermal** cells produce the outer body covering and **endodermal** cells produce the wall of the gut. Some of the acoelomates have a third layer of cells, the **mesoderm,** which lies between the ectoderm and endoderm. Nematodes (roundworms) have developed a **pseudocoelom,** or cavity, between the gut and the body wall. The mesoderm lines the outer wall but not the wall of the gut, which is why biologists call it a *pseudo*coelom. In animals with a true coelom, the entire cavity is lined with mesodermal cells. The three germ layers (endoderm, mesoderm, and ectoderm) allow for the development of all body organs; the coelom provides a space for their placement.

Further branching of the tree is based on certain early embryological characteristics. As an animal embryo develops, it produces a hollow ball of cells. This early stage of development is the **blastula,** and the inner cavity is the **blastocoel.** During later development, certain cells begin to move inward through an opening, the **blastopore.** In protostomes the blastopore of the developing embryo becomes the mouth, whereas the blastopore of deuterostomes becomes the anus.

The representative animals you will be dissecting and studying in the following three activities are indicated in the boxed areas of the phylogenetic tree and represent important developments in the evolution of animal phyla.

Figure 1 A phylogenetic tree

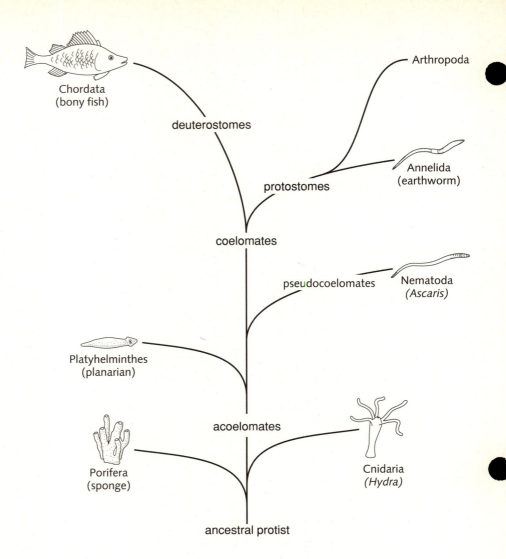

Figure 2 Location of ectoderm, mesoderm, and endoderm in acoelomates, pseudocoelomates, and coelomates

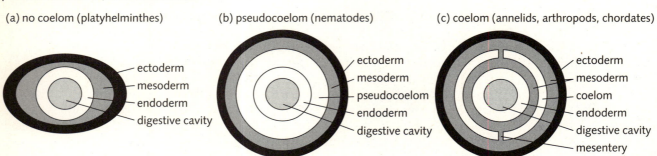

(a) no coelom (platyhelminthes)

- ectoderm
- mesoderm
- endoderm
- digestive cavity

(b) pseudocoelom (nematodes)

- ectoderm
- mesoderm
- pseudocoelom
- endoderm
- digestive cavity

(c) coelom (annelids, arthropods, chordates)

- ectoderm
- mesoderm
- coelom
- endoderm
- digestive cavity
- mesentery

The Acoelomate Animals

OBJECTIVES

At the end of this laboratory activity, you should be able to:

- describe the distinguishing characteristics of phyla Porifera and Cnidaria.
- state the function of the major structures of each animal studied.
- describe the level of organization of each phylum studied.

INTRODUCTION

Of the major phyla whose members zoologists recognize as acoelomates, you will study two of the more common ones as an introduction to animal diversity. These phyla, Porifera and Cnidaria, comprise the simplest multicellular animals known (see Table 15.1).

Over 5000 species of sponges make up phylum Porifera. The group is unique in that only one animal, the sponge, is found in this phylum. Early investigators mistook sponges for plants because they are **sessile** (permanently attached to some object) and because many are green in color. The source of this coloring was later found to be algae growing on the sponge body. Sponges usually inhabit the shallow areas of the oceans but sometimes may be found in deep water farther away from shore. Freshwater sponges are also known.

The body plan of the sponge is **radially symmetric** (having a circular shape when viewed from above) although many are so irregular that they appear asymmetric. The outward appearance of the sponge is due mainly to the internal supporting material secreted by ameboid cells wandering throughout the sponge body. In some sponges the supporting material is **collagen,** a protein. After this type of sponge dies, the supporting material remains and forms the common bath sponge. In others the supporting material is composed of **spicules** (needles) of calcium or silicon.

Water enters the **spongocoel** (body cavity) of the sponge through many **incurrent pores,** or openings, and exits through an **osculum,** or excurrent pore (see Figure 15.1). This arrangement of openings makes the sponge an excellent filter feeder.

The cells are arranged loosely into tissue or cell layers, showing some division of labor. No organs are present. The outer layer of cells is the epidermis, beneath it is a fluid layer, the **mesenchyme.** Three types of cells are embedded in the mesenchyme: choanocytes, porocytes, and amoebocytes. **Choanocytes** (collar cells) line the incurrent canals and move water through the sponge by means of beating flagella. The collar cells use ameboid movement to engulf food particles. After digestion, nutrients are distributed to other cells by diffusion. **Porocytes** open and close to regulate flow in the incurrent canals. **Amoebocytes** are found in the mesenchyme and have many functions, including storage of materials, aiding digestion, secreting skeletal materials, and forming gametes.

Members of phylum **Cnidaria** include the *hydra*, jellyfish, coral, and anemone. You will study two representatives of this group in order to see a more advanced level of multicellular organization than was evident in the sponges.

Table 15.1 Phyla Porifera and Cnidaria

Phylum	Major features	Example
Porifera	Skeleton composed of fibers or crystals System of pores for filtering water Loose aggregation of cells Cell-tissue level of organization	Sponges: *Basket sponge* *Leucosolenia*
Cnidaria	Radially symmetric Usually two body forms: a sessile polyp and a free-floating medusa Tissue level of organization Stinging cell (cnidocyte) containing a stinging capsule (nematocyst) Nerve net Gastrovascular cavity or coelenteron Found mostly in salt water	*Hydra* Jellyfish Coral Anemone

Figure 15.1 Sponge anatomy

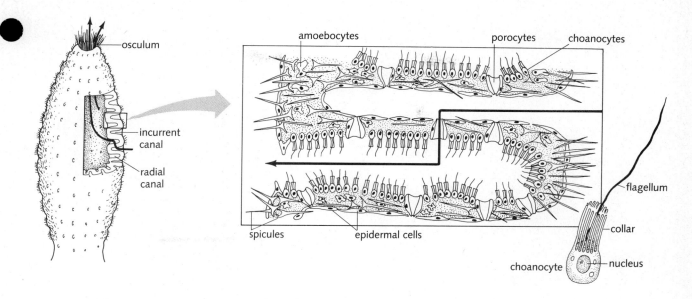

Most cnidarians have tentacles armed with cells called **cnidocytes** (see Figure 15.2). Each cnidocyte has an organelle, the **nematocyst** (stinging capsule), which is important in capturing food, in locomotion, and in attachment. Frequently the nematocysts are armed with barbs or spines and may even secrete toxins that paralyze a prey animal on contact.

In addition to the cells that are specialized for catching prey and reproducing, the cnidarians have other specialized cells. Unlike sponges, these animals have an **epidermis** (an outer layer of cells), which is mainly protective, and a **gastrodermis** (an inner layer of cells), which forms the gastrovascular cavity, or gut. Between these two layers is a jelly-like layer called the **mesoglea.**

Although there is no centralization of nerve tissue into a brain, a nerve net extends throughout the body. Other cells, the **epitheliomuscular cells,** possess contractile fibers. One set of fibers is arranged like tiny circular muscles and the other is arranged in a longitudinal direction. The coordinated action of the nerve net and the epitheliomuscular cells enables the cnidarians to move, catch prey, and respond to a variety of stimuli in their environment.

Figure 15.2 *Hydra* anatomy

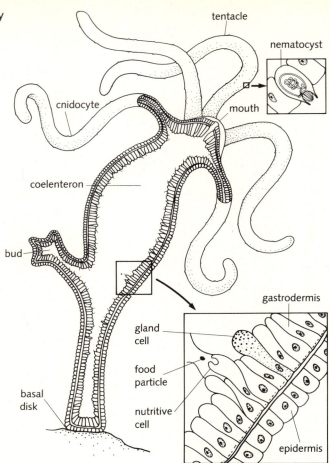

Figure 15.3 Body plans in cnidarians

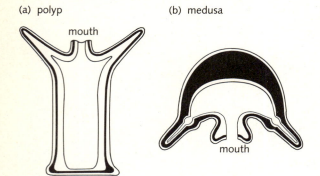

(a) polyp

(b) medusa

Animals in this phylum usually have two body forms (see Figure 15.3) during their life cycle. One form, the **polyp**, is a sessile stage and is characterized by an elongated hollow body that attaches to the substrate by a **basal disk.** This is the form commonly demonstrated by the *Hydra.* The second body form, the **medusa,** is the free-floating form familiar in jellyfish. The life cycle of most animals in this phylum is unique in the animal kingdom. It is referred to as **alternation of generations.**

The life cycle of the *Obelia* (diagramed in Figure 15.4) is a typical example of this alternation of gen-

Figure 15.4 Life cycle of the *Obelia*

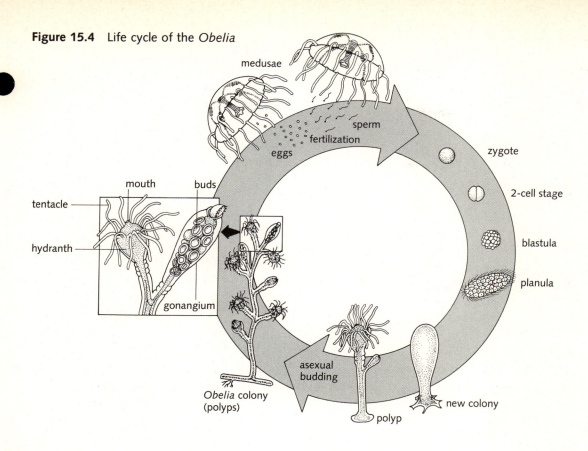

medusae

sperm

fertilization

eggs

zygote

2-cell stage

blastula

planula

mouth

buds

tentacle

hydranth

gonangium

asexual budding

Obelia colony (polyps)

polyp

new colony

Figure 15.5 The *Obelia*

erations. The *Obelia* is a colonial hydrozoan with two types of polyps growing from one central stalk (see Figure 15.5). The **hydranth,** a polyp with tentacles, is responsible for feeding the colony. The other polyp, the **gonangium,** reproduces the medusa stage asexually by budding. These medusa buds eventually leave the gonangium and, while drifting in the water, produce either eggs or sperm. This is the sexual phase in the cycle. After fertilization, a small larva develops, the **planula,** attaches to a substrate, and eventually matures into a new polyp.

MATERIALS

Prepared slides of:
  Leucosolenia
  Grantia
  Hydra
  Obelia
Compound microscope
Living Hydra
Watch glass
Dissecting microscope
Living Daphnia
Brine shrimp (Artemia)
Pasteur pipette

PROCEDURE

Phylum Porifera

During this activity you will examine two sponges of the class Calcarea: Leucosolenia and Grantia.

1. Examine a prepared slide of a whole mount of Leucosolenia. Diagram this whole mount in Question 1 of the Evaluation and label the following structures: osculum, spicules, and base.

2. Examine a prepared slide of a longitudinal section of Grantia. Locate the radial canals, which will appear circular in your specimen because the sponge has been cut longitudinally. The many collar cells that line the canal project toward the center. Their flagella may not be visible. The other circular areas you will see are the incurrent pores, which are not lined with collar cells.

3. Make a diagram of this longitudinal section in Question 2 of the Evaluation and label collar cells, radial canals, and incurrent pores.

Phylum Cnidaria

During this activity you will examine two members of the class Hydrozoa: the freshwater Hydra and the saltwater Obelia, a more complex member of this phylum.

Hydra

1. Examine a prepared slide of a Hydra polyp under the lowest power of a compound microscope. Make a diagram of the Hydra in Question 5 in the Evaluation and label the following structures (refer to Figure 15.2 as necessary).

tentacles	cnidocytes
coelenteron	basal disk
mouth	epidermis
gastrodermis	mesoglea

2. Obtain a living Hydra from your instructor and place it in fresh water in a watch glass. Place the watch glass on a dark surface for better visibility and examine with a dissecting microscope. When disturbed, the Hydra contracts into a ball; allow your Hydra to acclimate for several minutes and it will extend itself.

3. With a Pasteur pipette, add a drop of water containing Daphnia or washed brine shrimp (Artemia) to the watch glass.

4. Observe the feeding behavior of the Hydra and record your findings in Question 6 in the Evaluation.

5. Touch one tentacle of the Hydra with a pin. Describe its response in Question 7 in the Evaluation.

Obelia

1. Obtain a prepared slide of an Obelia and examine under low power of the compound microscope.

2. In Question 8 in the Evaluation, diagram a representative section of your Obelia colony, labeling the following structures (refer to Figure 15.4 as necessary).

hydranth	tentacles
gonangia	medusa buds

EVALUATION 15

The Acoelomate Animals

1. Make a diagram of the whole mount of the sponge *Leucosolenia*. Label the osculum, spicules, and base.

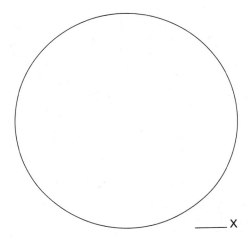

_____ X

2. Diagram the longitudinal section of *Grantia* and label the radial canals, collar cells, and incurrent pores.

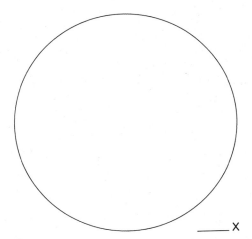

_____ X

3. Study Figure 15.1 and describe the path of water through the sponge body. What is the function of the flagella in the collar cells? Why is it necessary for large quantities of water to pass through the sponge every day?

4. What is meant when it is said that the sponges are at the tissue level of organization?

5. Diagram and label the prepared slide of the *Hydra*. Label the tentacles, gastrovascular cavity, mouth, gastrodermis, cnidocytes, basal disk, epidermis, and mesoglea.

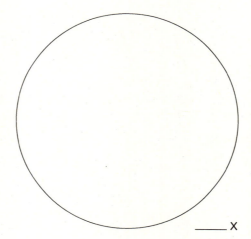

_____ X

6. Describe the manner in which the *Hydra* catches and ingests its prey. What cell-tissues are involved in this process?

7. Describe the *Hydra*'s response when it is touched with a pin.

8. Diagram and label a representative section of your *Obelia* colony. Include the following structures in your diagram: hydranth, tentacles, gonangia, and medusa buds.

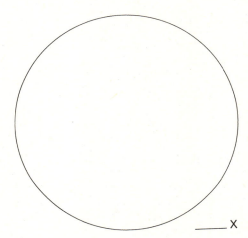

_____ X

9. The diagram of the *Obelia* in Figure 15.4 illustrates a reproductive cycle known as alternation of generation. Explain in your own words what this means.

10. The animals you have observed during this activity are acoelomates. What observations have you made of these animals that relate to this classification? Be specific.

The Emergence of the Coelom

OBJECTIVES

At the end of this laboratory activity, you should be able to:

- describe the distinguishing characteristics of the members of phyla Platyhelminthes, Nematoda, and Annelida.
- state the function of the major structures of each animal studied.
- describe the level of organization of each phylum studied.

INTRODUCTION

The three phyla studied in this laboratory activity demonstrate the evolution of the basic animal body plan to at least the organ level of development. The development of organs requires the presence of three embryonic germ layers: **endoderm, mesoderm,** and **ectoderm.** In the developing embryo, ectoderm produces the external covering, endoderm produces the gut, and mesoderm produces most other tissues.

A major development that can be seen when studying animals in these phyla is the emergence of the coelom. Members of the first phylum to be studied, Platyhelminthes, do not have a coelom. In Nematoda a pseudocoelom has developed, and in Annelida a true coelom is present. The major characteristics of the phyla studied in this laboratory activity are summarized in Table 16.1.

The animals in phylum Platyhelminthes are the simplest organisms to demonstrate **cephalization,** or the formation of a head. Along with this development they show a new type of body plan: **bilateral symmetry,** which means that the animal can

be cut longitudinally into left and right mirror images. This streamlined linear body with its anterior head fitted with sense organs is well adapted for directional movement. The level of complexity of these organisms appears to have progressed as far as possible without a coelom (see Figure 1 in the introduction to this section).

The presence of three embryonic **germ** layers gives the members of this phylum the ability to form complex organs. The ectoderm produces the epithelium that covers the surface of the animal. The mesoderm develops into excretory organs, muscle fibers, and reproductive organs, while the endoderm lines the digestive gut.

Members of this phylum lack respiratory and circulatory systems, which probably contributes to the flatness of these worms. The flatness allows materials to enter and leave the body by simple diffusion. Even though these animals are considerably more advanced than those of the phyla studied previously, they still lack a coelom. The only internal cavity present is the digestive cavity.

A representative of this phylum is the planarian (*Dugesia tigrina*) (see Figure 16.1). This common flatworm lives under rocks, wood, and leaves in freshwater streams and lakes. A distinctive head is present with two eyespots and **auricles** (lateral projections). The nervous system consists of two long ventral nerve cords connected anteriorly by a bilobed **ganglion** (primitive brain).

These animals prey upon small invertebrates if available, otherwise on dead organic material. Feeding occurs in an interesting manner, employing a unique structure, the **pharynx.** The pharynx is located along the midline of the animal's ventral surface. During feeding the pharynx extends through the mouth and sucks food up into the digestive cavity. Since the planarian has only one

Table 16.1 Phyla Platyhelminthes, Nematoda, and Annelida

Phylum	Major features	Example	
Platyhelminthes	Cephalization Bilateral symmetry Organized at the organ level of development Ectoderm, mesoderm, endoderm present Coelom absent	Planarian (*Dugesia*) *Ascaris* Sheep liver fluke	
Nematoda	Nonliving, flexible cuticle Longitudinal muscles only Cylindrical body Organized at organ level of development Psuedocoelom present	Tapeworm *Trichinella*	
Annelida	Segmented body Thin, flexible cuticle containing setae Protostome pattern of development Closed circulatory system True coelom	Earthworm (*Lumbricus*) Leech (*Hirudinea*)	

Figure 16.1 A planarian

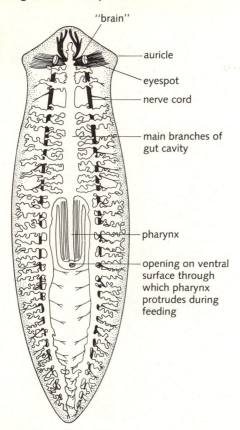

"brain"
auricle
eyespot
nerve cord
main branches of gut cavity
pharynx
opening on ventral surface through which pharynx protrudes during feeding

Figure 16.2 Asexual reproduction and regeneration in the planarian

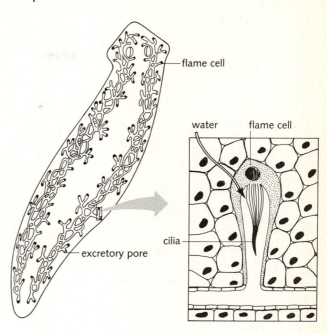

flame cell
water
flame cell
cilia
excretory pore

opening to its digestive tract, any undigested food is discharged through the pharynx.

Because the planarian is flat and thin, most of the metabolic wastes simply diffuse out of the body. However, flatworms have evolved a system to maintain water balance. This process, **osmoregulation,** is accomplished by specialized **flame cells,** which have long cilia that move water out of the body (see Figure 16.2). This regulation of water balance is a step toward the evolution of an excretory system.

The planarian can reproduce sexually by the production of both sperm and eggs in the same animal. However, cross-fertilization normally occurs, providing more genetic variability in the population. The animals can also reproduce asexually by transverse fission, wherein the animal attaches itself firmly to the substrate and simply fragments into two parts. Each part regenerates any lost parts.

The animals in phylum Nematoda have developed a pseudocoelom, which differs from a true coelom primarily in its development rather than in

its final appearance. The pseudocoelom is bounded on one side by tissue of mesodermal origin and on the other side by the gut wall, which is derived from endoderm.

Organ development within the nematodes includes a well-differentiated digestive tract with a mouth, an anus, and specialized regions for digestion, absorption, and excretion (see Figure 16.3). This gut represents a true advance over the flatworms, which have only a single opening. Because the gut wall lacks mesodermal tissue, it does not develop a muscle layer (muscle develops from mesoderm). Nematodes thus depend on movements of the entire body to push food through the digestive tract. Furthermore, because nematodes have no circulatory system, nutrients must move by way of diffusion, a slow process that limits the diameter of the body. Locomotion is not precise or elaborate in these animals; their characteristic whiplike movements are caused by the longitudinal arrangement of their muscles.

Figure 16.3 Internal anatomy of the nematode *Ascaris*

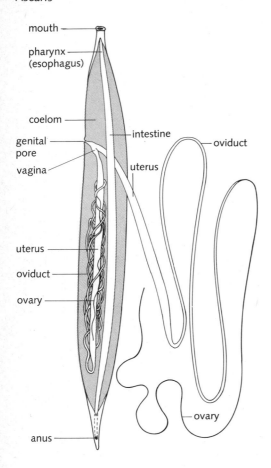

- mouth
- pharynx (esophagus)
- coelom
- genital pore
- vagina
- intestine
- oviduct
- uterus
- uterus
- oviduct
- ovary
- ovary
- anus

Nematodes are found in nearly every known habitat, as long as sufficient moisture is present. They are also among the most ubiquitous and persistent pests in our world. They can cause humans untold misery by attacking our food crops, livestock, and pets. Two of the most dangerous types are *Ascaris,* an intestinal parasite of mammals, and *Trichinella,* which cause trichinosis by migrating from the intestine to the muscles.

The famous French biologist Jean Baptiste Lamarck (1744–1829) chose the name for phylum Annelida, deriving it from the French word *annelide,* meaning "ringed." As this name implies, the prominent characteristic of the members of this phylum is their many ringed body segments. The phylum is relatively small, containing only about 8,000 species that inhabit fresh water, salt water, and the soil. Three classes of segmented worms make up the phylum: Oligochaeta, Polychaeta, and Hirudinea, which are represented by the earthworm, the sandworm, and the leech, respectively.

Annelida is the lowest phylum whose members have a true coelom, lined completely with mesoderm, and a segmented body. Both developments are of considerable evolutionary importance. In more evolved phyla, the coelom is used for many purposes in development. Segmentation allows the structures for locomotion, circulation, and excretion to be duplicated in each segment. In some cases segments can be highly specialized, bearing antennae, mouth parts, reproductive structures, or sense organs.

Another important development in annelids is a **closed circulatory system.** In this type of system, the blood is pumped into a closed circuit of vessels and returned to the heart, providing more blood to the body tissues. This significant development in animals allows for larger body size and increased rates of respiration.

As segmentation and a closed circulatory system provide for larger body size, the problem of waste disposal presents itself. Annelids have solved this problem with the **nephridia** that collect waste and excess water in the coelom and excrete them.

Annelids reproduce by the same means as earlier worms employing both asexual and sexual methods. Many species are **hermaphroditic,** possessing both male and female reproductive organs.

A representative of phylum Annelida is the earthworm (*Lumbricus terrestris*) (see Figure 16.4). The most obvious external feature of the earthworm is its segmentation. An earthworm emerges from its cocoon with the number of segments it will have throughout its lifetime; growth occurs by the enlargement of each segment. The number of segments varies among the species. *Lumbricus terrestris* has approximately 150 segments.

The surface epithelium of the earthworm contains 8 **setae** (stiff bristles) per segment. These setae are provided with both retractor and protractor muscles, which enable the earthworm to move through the soil. If you have ever attempted to remove a night crawler from its tunnel in the soil you know the effectiveness of these setae. The surface epithelium contains several specialized cells, including those that secrete mucus and others that are sensitive to light.

Directly under the epidermis is a layer of circular muscle and, deeper still, a layer of longitudinal muscle. When the circular muscle contracts the worm becomes thinner and longer. As the longitudinal muscles contract the worm becomes shorter and stout.

Figure 16.4 External anatomy of the earthworm (*Lumbricus*)

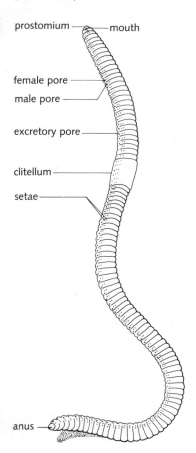

prostomium — mouth
female pore
male pore
excretory pore
clitellum
setae
anus

A sexually mature earthworm has a beltlike band encircling its body in the anterior. This band, the **clitellum,** secretes a protein material that moves over and off the worm to form a cocoon for fertilized eggs. The earthworm is hermaphroditic.

The digestive tract of the earthworm has evolved into a tube with many specialized areas along its length for storage, mechanical break up of food, digestion, and absorption. The digestive tract is enclosed within the tubelike outer wall, and the space between is a true coelom lined with specialized mesodermal cells, the **peritoneum.**

The earthworm has a closed circulatory system with arteries, veins, and capillaries. A portion of the main dorsal artery and the aortic arches can contract, forcing blood through the closed circuit.

An excretory system is present in the earthworm to rid the body of nitrogenous wastes and to maintain water balance. The functioning unit of

this system is the **nephridium,** a long meandering tube surrounded by numerous blood vessels. The materials in this tube are filtered; some are reabsorbed and others are excreted.

A well-advanced nervous system, including a small brain located in the anterior portion of the body, is present to coordinate the movements of the earthworm.

MATERIALS

Goggles
Living *Dugesia sp.*
Watch glass
Pipette
Daphnia sp.
Washed brine shrimp (*Artemia sp.*)
Penlight
Microscope slides
Cover slips
Petri dish
Dissecting microscope
Prepared slide of planarian showing digestive tract
Prepared slide of planarian cross section
Compound microscope
Latex gloves
Preserved *Ascaris sp.* female cross section
Dissecting pins
Dissecting tray
Scissors
Living vinegar eels (*Anquillula aceti*)
Living earthworm (*Lumbricus terrestris*)
Ethanol (10%)
Scalpel
Prepared slide of earthworm cross section

PROCEDURE

Phylum Platyhelminthes

Living Planarian (*Dugesia sp.*)

1. Obtain a living planarian and place it in a watch glass with several drops of water. Observe its movement and record your observations in Question 1 in the Evaluation.

2. With a pipette, add several drops of water containing *Daphnia* or washed brine shrimp (*Artemia*). Carefully observe the feeding process. Record your observations in Question 1 in the Evaluation.

3. Shine a small penlight on one side of the planarian. Describe whether the worm moves toward or away from the light (phototaxis). Record your observations in Question 1 in the Evaluation.

4. Move your planarian to the center of a microscope slide and add a drop of water. Place a cover slip at each end of the slide and then add a second microscope slide on top.

5. Mount this "sandwiched" planarian under the dissecting microscope, ventral side up, on a petri dish (with lid in place) filled with ice water. This will serve as an insulator and prevent heat damage to the planarian. Carefully observe the ventral surface as the worm moves.

6. Locate the mouth and pharynx and label the diagram in Question 2 in the Evaluation.

Preserved Specimen

1. Obtain a prepared slide of a whole mount of a planarian. It should be stained to show the digestive system.

2. Study the digestive system and diagram the branched gut. Label the pharynx and branches of the gut in Question 2 in the Evaluation.

3. Obtain a prepared slide of a planarian cross section and observe under low power of the compound microscope. Identify the following structures:
   - **epidermis:** outer covering
   - **dorsal/ventral surface:** the flat side is the ventral surface
   - **adhesive glands:** appear as a small section of straight lines at the extreme sides of the ventral surface
   - **branched gut:** appears as several cavities lined with endodermal cells
   - **parenchyma cells:** cells between the gut and epidermis. Note the absence of a body cavity.

4. After you have identified these structures, diagram and label them in Question 3 in the Evaluation.

Phylum Nematoda

▶ **CAUTION: Be certain to wear protective gloves and goggles when you complete this dissection. Some Ascaris eggs can remain alive even after being preserved. When finished, wash your gloved hands and instruments thoroughly. After removing the gloves wash hands again.**

Ascaris sp.

Obtain a preserved specimen of a female *Ascaris* and place it in a dissecting tray.

External Anatomy

1. Examine the external structure of the *Ascaris* under the dissecting microscope. Locate the following structures:
   - **mouth:** located at the anterior end and surrounded by three lips
   - **anus:** a small opening at the posterior end
   - **genital pore:** located about one-third of the way from the anterior end
   - **cuticle:** the smooth, nonliving covering of the body

2. Turn to the Evaluation and answer Question 4.

Internal Anatomy

1. Pin the anterior and posterior ends of the worm to the wax in a dissecting tray.

2. With scissors carefully make an incision along the entire length of the body.

▶ **CAUTION: Be careful not to cut too deeply into the body cavity.**

3. Carefully examine the internal anatomy and identify the following structures:
   - **uterus:** two thick round tubes located toward the posterior half of the body
   - **oviduct and ovary:** thin round tube coiled around the uterus. (It is difficult to distinguish between the two organs by external examination alone.)
   - **intestine:** a long ribbonlike structure
   - **coelom:** the cavity between the gut and the body wall normally filled with fluid in the living worm

Vinegar Eel (*Anguillula aceti*)
The vinegar eel is a free-living nematode commonly found in raw or unprocessed vinegar. It feeds on the bacteria and yeast that settle to the bottom of the container.

1. Add a drop of vinegar eel culture to a microscope slide and examine it with the compound microscope.

2. Observe the movements of the worms on your slide and describe this locomotion in Question 5 in the Evaluation.

Phylum Annelida
Earthworm (*Lumbricus terrestris*)
Obtain a living earthworm from your instructor and place it in a dissecting tray.

External Anatomy

1. Observe the earthworm and note the following structures:
   - **cuticle:** thin and moist
   - distinctly **segmented** body
   - **dorsal** side (dark, round) and **ventral** side (light, flat)
   - **clitellum:** about one-third back from the anterior end
   - **prostomium:** tip of the anterior end overhanging the mouth

2. Run your finger over the ventral side and feel the setae.

3. Allow the worm to move and observe its method of locomotion. Describe its movement in Question 6 in the Evaluation.

Internal Anatomy
Prior to dissection your instructor will anesthetize the worm by placing it in 10% ethanol for several minutes. After the worm has stopped moving, place it in the tray. Certain internal organs are difficult to find, and you will study only selected systems. Refer to Figure 16.5 as you do the dissection. Complete the diagram in Question 7 in the Evaluation.

1. Pin the anterior and posterior ends of the earthworm to the tray, dorsal side up.

2. With a scalpel or scissors, carefully make a shallow incision just to the left or right of center. As you cut, gently separate the body wall and pin it to the tray. Place the pins every ten segments to make it easier to find the internal organs as you examine each system. The segment just posterior to the prostomium is number one.

3. Note the large coelom, which is divided by many crosswalls, or **septa.**

Figure 16.5 Internal anatomy of *Lumbricus*, dorsal view

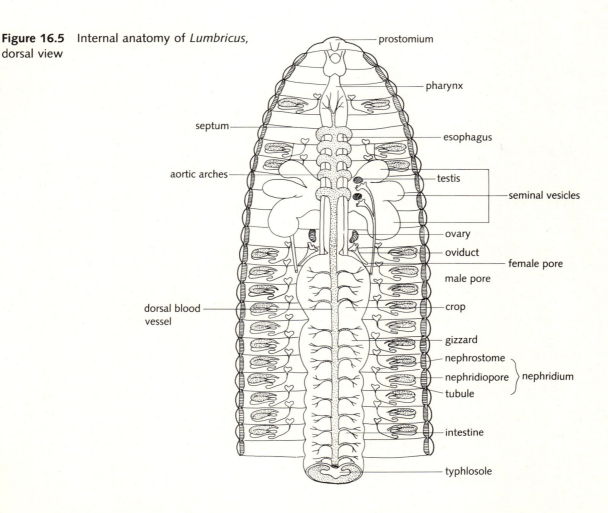

> CAUTION: The earthworm's organs are soft, delicate, and easily destroyed. Be careful! Keep the worm moist throughout the dissection.

Digestive System

1. The large digestive tract extends from the mouth to the anus. Trace it from anterior to posterior and locate the following structures. (The numbers in parentheses refer to the segment(s) where the structure is located.)
   - **pharynx** (4–6): muscular tube that pushes food through the esophagus
   - **esophagus** (7–14): carries food to crop
   - **crop** (15–17): stores food temporarily
   - **gizzard** (18–20): grinds food with aid of ingested soil particles
   - **intestine** (21–anus): digestion and absorption of food

2. Turn to the Evaluation and answer Question 8 concerning the digestive system.

   As an alternative activity, remove a piece of the intestine about 1 cm long and place in a test tube with 1 mL of water. Shake well and make a wet mount of this suspension. Examine under the compound microscope. Patient observation should reveal ciliated protozoans (one-celled animals) and nematodes. How do they survive in the gut?

Circulatory System

On top of the earthworm's digestive tract is the dorsal blood vessel, which carries blood anteriorly to the "hearts" or aortic arches. The five pairs of aortic arches are located around the esophagus (7–12) and connect to the ventral blood vessel.

1. Observe the aortic arches under the dissecting microscope and note their rhythmic contractions.

Reproductive System

1. Cut across the intestine about 1 cm posterior to the gizzard. Carefully remove the intestine with a pair of forceps up to segment 4. Do not remove any round, whitish structures.

2. Identify the following structures:
   - **seminal vesicles**: large three-lobed structures (9–13) that store sperm
   - **testes**: for sperm production, located inside the seminal vesicles
   - **ovaries** (13): a pair of very small sacs for egg production

Excretory System

A pair of nephridia are located in every segment except segments 3 or 4 and the last segment.

1. Use a dissecting microscope or the low power of a compound microscope and examine a segment for the coiled **nephridium**.

Prepared Slide of Cross Section

1. Obtain a prepared slide of an earthworm cross section and examine it under the compound microscope. As you identify each structure, label the cross section diagram in Question 9 in the Evaluation. The following structures are described from exterior to interior.
   - **cuticle**: nonliving outer covering
   - **setae**: bristles for locomotion
   - **epidermis**: thin layer of cells under cuticle
   - **circular muscles**: thin layer of muscle running around the body wall
   - **longitudinal muscle**: thick layer of muscle under circular muscles
   - **coelom**: large body cavity
   - **nephridium**: coiled tube in coelom
   - **dorsal blood vessel**: dorsal to the intestine
   - **intestinal wall and lumen** (cavity)
   - **typhlosole**: invagination of intestinal wall projecting from dorsal side

EVALUATION 16

The Emergence of the Coelom

1. Record your observations concerning the planarian's behavior. Include movement, feeding, and response to light.

2. Label the pharynx in the following diagram of the planarian and sketch in the branched gut.

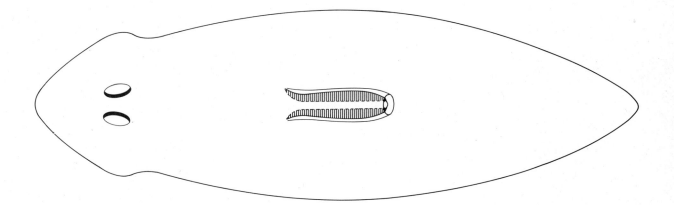

3. Label the following cross section of the planarian.

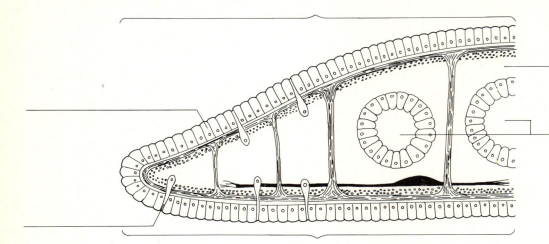

4. (a) Why does *Ascaris* need a protective cuticle?

(b) In *Ascaris* the mouth and anus exemplify a "tube-within-a-tube" body plan. What does this mean?

5. Describe the movement of the vinegar eel and explain this movement in terms of the animal's musculature.

6. Describe the earthworm's locomotion. What is the function of setae in this process? What sets of muscles are responsible for this movement and how do they contract with regard to each other?

7. Diagram the internal structures of the earthworm within the outline below. Label the following: septa, pharynx, esophagus, crop, gizzard, intestine, aortic arches, seminal vesicles, testes, nephridium.

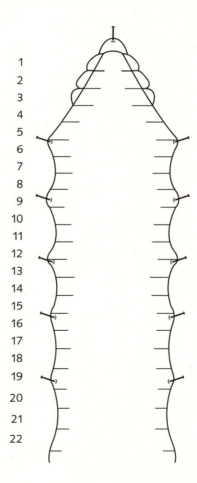

1
2
3
4
5
6
7
8
9
10
11
12
13
14
15
16
17
18
19
20
21
22

8. Compare the digestive systems of the earthworm and the *Ascaris*. Which is the more advanced? Explain.

9. Label the following cross section of the earthworm.

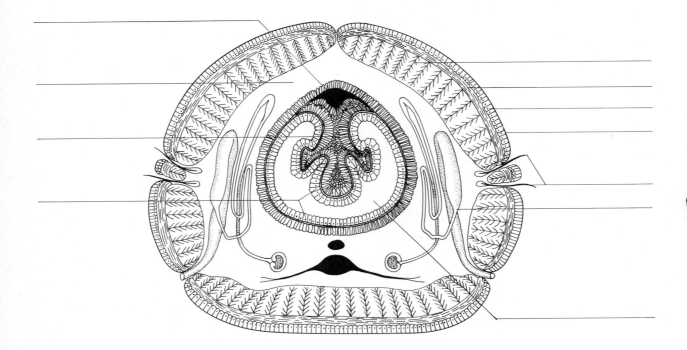

10. Compare the internal anatomy of the planarian with that of the *Ascaris* and earthworm. What developments have occurred due to the presence of a coelom?

A Coelomic Deuterostome: The Fish

OBJECTIVES

At the end of this laboratory activity, you should be able to:

- describe the distinguishing characteristics of phylum Chordata.
- describe the major organs/systems found in a fish and explain how they have allowed fish to adapt to an aquatic environment.

INTRODUCTION

Your investigations of the representative animals studied in the previous two laboratory activities have helped you trace the evolution of the three germ layers, bilateral symmetry, the coelomic cavity, and segmentation. All these characteristics were present in the earthworm, the animal you studied representing phylum Annelida. Many of the characteristics found in annelids appear again in arthropods and mollusks with only minor modifications. The animals in this evolutionary branch are called **protostomes** because their mouth, if present, was derived from the embryonic **blastopore** (opening into the blastula).

A second pattern of embryological development occurred in the coelomates and gave rise to the chordates. In these organisms, the **deuterostomes,** the blastopore develops into the anus. The animal you will study in this activity is a deuterostome and a representative of phylum Chordata (Table 17.1).

In the most advanced members of this phylum the nerve cord has developed into a sophisticated spinal cord and brain. The **notochord,** a flexible rod located on the dorsal side of the gut, is present throughout the life of the invertebrate chordates (lacking a vertebral column) but in the vertebrates is replaced during embryological development by the vertebral column. **Gill slits** are present in the embryo but lost during later development of land forms. These gill slits give evidence to the aquatic ancestry of the chordates. The genes causing the formation of gill slits still express themselves although gill slits are present for only a short period of the animal's life.

We will give our general attention to the class Osteichthyes (bony fish) and more specifically to the **teleosts** (modern bony fish). Bone tissue appeared for the first time in an ancient group of fish, the **ostracoderms,** or jawless fish (see Figure 17.1). This new type of tissue was composed of **osteocytes** (bone cells), which were capable of secreting hard "bony" material rich in phosphate. One hypothesis states that this method may have evolved as a means of storing phosphate, which is limited in aquatic environments. If this is true, bone tissue used for support evolved from tissue that originally had a totally different function.

Another important animal characteristic that evolved in the ancient fish is the jaw. The jaw present in modern fish and in other higher vertebrates evolved from this basic design. The importance of the evolution of the jaw can hardly be overstated!

Figure 17.1 A jawless fish

Table 17.1 Phylum Chordata

Phylum	Major features	Examples	
Chordata	Dorsal notochord present at some time in life cycle	Fish	
	Dorsal tubular nerve cord	Snake	
	Paired pharyngeal clefts (gill slits) at some stage in life cycle	Bird	
		Man	

Jaws changed the food chain greatly by enabling animals to capture, hold, and eat other organisms more efficiently.

The bony fish with jaws that you are about to dissect is a member of a highly adaptive and successful group of fish. One reason for this success is the presence of scales that cover the body surface. These scales are formed by the mesoderm and are covered with a thin layer of living cells (see Figure 17.2).

The **swim bladder** (or air bladder) is a major adaptation that gives a fish the ability to regulate its **specific gravity** or density compared to water. Whenever air is added to the bladder it increases in size and the fish becomes more buoyant. If air is removed from the bladder, the fish will sink. The swim bladder also enables the fish to hover in the water or remain almost motionless.

Another significant adaptation is the gill. The fish gill is an extremely efficient organ for absorbing dissolved oxygen from water. The dissolved oxygen concentration of water is usually measured in parts per million, a very small quantity. The gill is

Figure 17.2 Fish scales

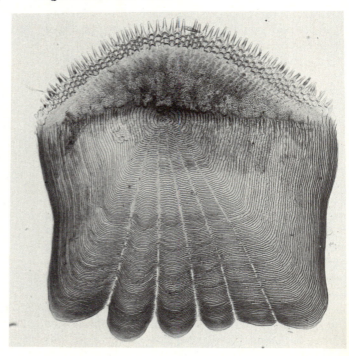

Figure 17.3 Water flow over gills

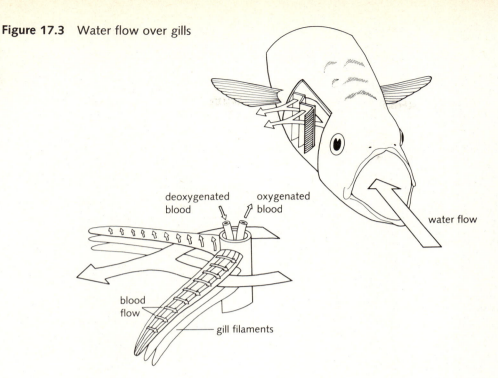

efficient because gill filaments provide an enormous surface area and are highly vascularized (well supplied with blood). In addition, water passes over the gill filaments in a one-way direction by entering the mouth, passing over the gill and exiting through the operculum (see Figure 17.3).

The yellow perch (*Perca flavescens*) is typical of modern bony fish and is a good example to study if one is to appreciate the many ways in which the members of this class have adapted to life in the water.

MATERIALS

Latex gloves
Goggles
Preserved yellow perch (*Perca flavescens*)
Live fish (optional)
Glass slide
Cover slip
Compound microscope
Forceps
Scissors
Dissecting tray
Dissecting needles

PROCEDURE

During this laboratory activity you will examine selected systems and organs of the perch that show the adaptions that allow for life in fresh water.

▶ **CAUTION: Wear goggles and latex gloves during this dissection.**

External Anatomy

1. Obtain a preserved specimen of a perch and study its overall body plan. Note the shape of the body from anterior to posterior and from dorsal to ventral surfaces. How is the body shaped? How does this shape facilitate movement through the water?

2. Examine the head region and locate the **nostrils** just above the mouth and below the eyes. Find the eyes and determine if eyelids are present. Examine the mouth and locate the teeth. Note their position in the mouth.

3. On the side of the head find the **operculum,** a hard structure that covers the gills. Which edge of the operculum is open? How does this facilitate the movement of water over the gills?

4. If your instructor has a live fish available, observe its breathing movements. How are the opening and closing of the mouth and operculum coordinated?

5. Fish move and stabilize themselves in water by the use of fins and a swim bladder. Examine the fish and locate the following fins:
   - unpaired: two **dorsal,** one **caudal** (tail), one **anal** (on ventral side anterior to caudal fin)
   - paired: **pectoral** (just posterior to the operculum), **pelvic** (below the pectoral fin)

6. Locate the **fin rays** that support the fins.

7. Locate the **anal opening** just anterior to the anal fin and the **urogenital opening** between the anus and the anal fin.

8. Locate the **lateral lines** on each side of the body midway between the dorsal and ventral surface and extending from the eyes to the base of the tail. These lines sense vibrations in the water and appear to aid the fish in orientation and avoiding objects. In this respect they seem to function in a manner similar to which sense organ in man?

9. Examine the scales. Describe their arrangement.

10. Fish scales grow continuously throughout the life of the fish and are not replaced if lost. During growth, concentric rings are formed. With your forceps, remove a single scale, prepare a wet mount slide and examine it under the compound microscope. Examine the concentric rings and look for the teeth at the base of the scale.

11. Complete Question 3 in the Evaluation.

Internal Anatomy

1. Hold the fish ventral side up and use scissors to make a shallow incision from the anal opening to between the opercula. Place the fish in the dissecting tray and make a vertical cut behind the operculum up to the top of the body cavity. Cut along the dorsal side back to the anal opening. Carefully remove the body wall and expose the internal organs. Refer to Figure 17.4.

2. The **liver** is located in the anterior body cavity just behind the operculum. The **stomach** is located under the liver and can be found after removing the liver. Trace the alimentary tract from the stomach to the **intestine,** ending at the anus.

3. The **swim bladder** is a large sac at the top of the body cavity.

4. Carefully remove the stomach and the intestine. Under the swim bladder find the **ovary** (single sac with eggs) or the **testes** (a pair of white thin bodies).

5. Above the swim bladder are two long dark colored bodies, the **kidneys.**

6. Locate the two-chambered **heart** behind the gills and anterior to the liver. Locate the **ventral aorta,** which extends anteriorly from the heart and carries blood to the gills.

7. Use scissors to remove the operculum and expose the gills. Note the way the gills are positioned. Dissect out one gill and examine it carefully. Locate the following structures:
   - **gill arch:** long bone to which the soft gill filaments are attached
   - **gill filaments:** fingerlike extensions from the gill arch where gas exchange occurs
   - **gill rakers:** bony projections extending from the gill arch that prevent the passage of coarse material through the gills

8. Complete Question 4 in the Evaluation.

Figure 17.4 Internal anatomy of the yellow perch

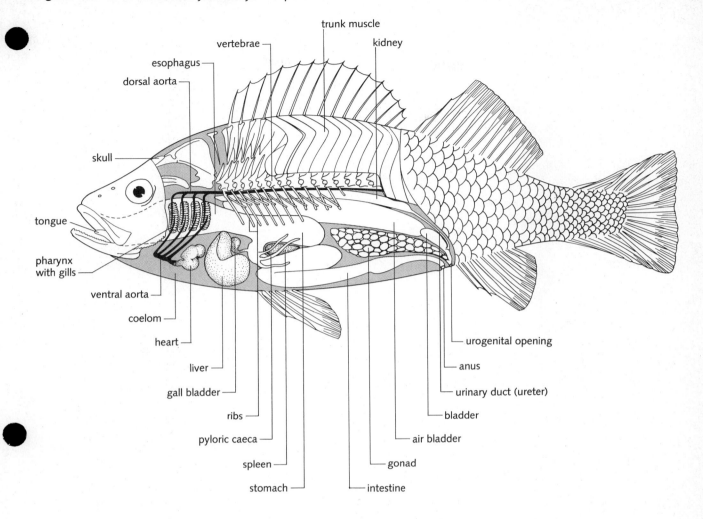

EVALUATION 17

A Coelomic Deuterostome: The Fish

1. Write a paragraph describing the external anatomy of your fish, indicating how each structure has adapted fish to life in water.

2. Label the following diagram showing the external anatomy of the fish.

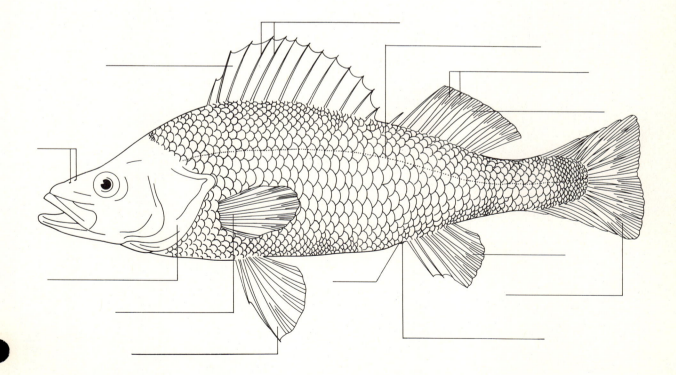

3. Diagram a fish scale. Label the teeth and concentric rings.

4. List the internal organs that have been most instrumental in adapting fish to an aquatic life and explain how they were instrumental.

Organ   Adaptation

Diversity of Soil Organisms

I. SOIL MICROORGANISMS

OBJECTIVES

At the end of this laboratory activity, you should be able to:

- grow and isolate microorganisms found in the soil.
- identify these soil organisms as filamentous fungi, yeasts, or bacteria.
- list the general characteristics of the bacteria isolated.

INTRODUCTION

Soil is a complex mixture of nonliving materials and living organisms. Of the many organisms present in the soil, none rank higher in importance than the microorganisms. Many of these microorganisms are classified as **decomposers** because they are responsible for the breakdown of dead plants and animals that ultimately become part of the soil. Other microorganisms consume the organic and inorganic products of this decomposition to obtain energy. In this manner nitrogen, sulfur, phosphorous, and other elements are cycled through the soil to become nutrients for plants. In this activity you will grow, isolate, and study a few of these important organisms.

MATERIALS

Soil sample
Balance
Sterile filter paper
Sterile water bottles (99 mL)
Glass marking pencil or pen
Sterile pipettes (1 mL)
Pipette pump
Petri dishes
Tubes of melted nutrient agar
Glass spreading rod
Ethanol (95%)
Bunsen burner
Inoculating loop
Wash bottle of distilled water
Staining tray
Hand lens or dissecting microscope (optional)
Microscope slides
Crystal violet stain
Paper towels
Beaker
Gram's iodine stain
Safranin stain

PROCEDURE

Session 1

Your instructor has either assigned you to bring in a sample of soil or will provide you with a sample. Note where the soil sample was obtained. Was it from a fertile field, a well-maintained lawn, or maybe from a rich garden? Any such information about the site may enlighten your ultimate findings. It might be interesting to compare the kinds of organisms isolated from different types of soil.

1. Record any information you have about your soil source in Question 1 in the Evaluation.

2. Carefully weigh out 1 g of your soil sample on a piece of sterile filter paper.

3. Obtain two bottles of sterile water that contain 99 mL of water each. Label the bottles #1 and #2.

4. Add your soil sample to bottle #1. Shake vigorously and allow the particles to settle. Any microorganisms in the soil should now be distributed in the water.

5. Transfer 1 mL of the liquid from bottle #1 to bottle #2 with a sterile pipette. Do not mouth pipette; use the pipette pump provided. This gives you two dilutions; a 1:100 dilution in bottle #1 and a 1:10,000 dilution in bottle #2.

6. Obtain two sterile petri dishes. Taking care to keep the lids in place, label the bottom of each with the following information: your name, date, and dilution (1:100 or 1:10,000).

7. Obtain two tubes of melted nutrient agar. Pour a tube of the melted agar into each of the petri dishes and set aside until the agar has hardened. You may need to refer to the procedures for preparing a pour plate given in Appendix F.

8. Using a sterile pipette, transfer 1 mL of the dilution from bottle #1 onto the surface of the hardened agar in the appropriately labeled petri dish.

9. Obtain a glass spreading rod that has been standing in a beaker of 95% ethanol and flame off any excess alcohol. Allow the rod to cool for a minute or so. Using the flamed portion of the rod, spread out the 1 mL of diluted inoculum across the entire surface of the agar. Replace the lid on the petri dish as quickly as possible. See Figure 18.1.

10. Place 1 mL of the dilution from water bottle #2 onto the surface of the agar in the second pour plate and repeat the spreading procedure described above.

Many soil microorganisms grow well at room temperature; you can incubate them in any part of the laboratory where they will be undisturbed. A

Figure 18.1 Spread plate technique

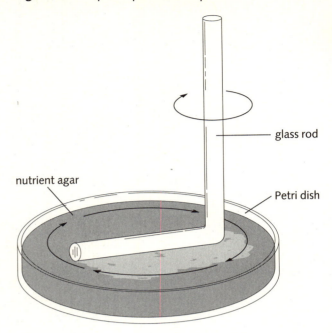

drawer would be excellent if one is available. The petri dishes should be stored upside-down to prevent condensation from developing on the lid and falling into the culture.

11. Examine your spread plate cultures in 48 hours. Aren't you impressed with the amount of growth of the organisms on the surface of the agar?

In Session 2 you will determine what forms are growing in these colonies. The following procedures should help you with this determination.

Session 2 Colony Characteristics

The "patches of growth" on the surface of the agar are more correctly known as colonies. Notice on the flowchart in Figure 18.2 that the next step in identification of soil microorganisms is the macroscopic examination of these colonies. They are species specific, meaning that each colony contains organisms of a single kind. That means that each colony could have grown from a single organism. Notice that some colonies may have grown over and crowded out others. You may see the apparent retardation of a particular colony's growth as it approached another. This may be due to the production of an antibiotic substance (see Figure 18.3). Antibiotics used to cure disease in humans are ex-

Figure 18.2 Flowchart for identifying soil microorganisms

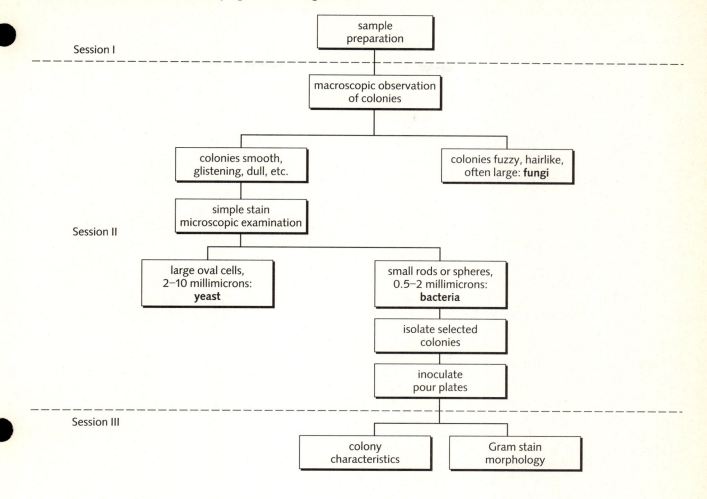

Session I

sample preparation

macroscopic observation of colonies

colonies smooth, glistening, dull, etc.

colonies fuzzy, hairlike, often large: **fungi**

simple stain microscopic examination

Session II

large oval cells, 2–10 millimicrons: **yeast**

small rods or spheres, 0.5–2 millimicrons: **bacteria**

isolate selected colonies

inoculate pour plates

Session III

colony characteristics

Gram stain morphology

Figure 18.3 Soil microorganism colony exhibiting antibiotic production

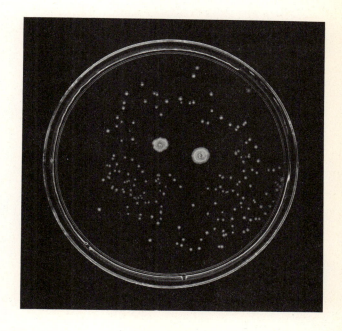

tracted from such organisms. These colonies could be yeast, filamentous fungi, or bacteria. Before you start examining your colonies read the sections in your textbook describing fungi and bacteria.

1. Examine each colony carefully with the lid in place. A hands lens or a dissecting microscope may be used. Using a glass marking pencil, draw a circle on the outside bottom of the petri dish around each fuzzy, hairlike colony; these are fungi colonies. These fungi are important decomposers in the soil.

2. Circle and number all the remaining colonies growing on the surface of the agar plates. As indicated in the flowchart in Figure 18.2, these colonies are either yeast or bacteria.

3. Record the characteristics for each colony in Question 2 of the Evaluation. See Table F.1 in Appendix F to aid you in recording these colony characteristics.

4. Select for further investigation four of your circled and described colonies that are not fungi. Make a simple stain of the microorganisms in each of your selected colonies. See Appendix F for the instructions for preparing a simple stain.

5. In Question 3 in the Evaluation, make a sketch of the microorganisms found in each colony. Based on individual size, decide if each microorganism is a bacterium or a yeast (refer to Figure 18.2). Record your observations in Question 3 of the Evaluation.

At this point you have completed all the work necessary to identify the fungi and yeast colonies. The remainder of your investigation will be concerned with your selected bacterial colonies. Isolate these selected bacterial colonies according to the following procedure.

1. Melt a tube of nutrient agar for each pour plate needed (one for each bacterial colony selected). Allow the agar in the pour plate to harden sufficiently before continuing.

2. Follow the instructions in Appendix F to prepare a streak plate of each bacterial colony selected. Label the petri dish with your name, colony number, and date. Incubate upside-down at room temperature for 48 hours.

Session 3

1. Examine each colony on the original plate and compare it with the isolated bacterial colonies prepared from it on the streak plate. The colonies should appear the same on both plates if you have prepared the streak plate correctly.

2. Prepare a Gram stain of each of the bacteria colonies that you have isolated according to the directions given in Appendix F. Record your results in Question 3 in the Evaluation.

You now have isolated pure colonies of soil bacteria. You know their colony characteristics, their morphology, and Gram-staining characteristics. In a given situation it may be important to identify the particular bacterium that has been isolated. In such cases the bacteriologist would complete additional biochemical tests on the unknown bacterium and use the appropriate keys to complete the identification.

II. SOIL INVERTEBRATES

OBJECTIVES

At the end of this laboratory activity, you should be able to:

- describe the operation of the Berlese funnel.
- name the major groups of cryptozoan organisms that inhabit soil communities.
- calculate a diversity index for each site studied.
- compare the cryptozoan populations of several different sites and offer an explanation for any differences in population size, species composition and diversity.

INTRODUCTION

Of all of the animal phyla known today, only a few are known to live all or part of their lives in the soil environment. Collectively they are called **cryptozoa** and include macroscopic organisms as well as the microscopic protozoa. Compared to life above ground, the soil habitat offers a rather uniform and stable environment for the organisms that live there. Because of this stability, these organisms have not evolved elaborate structures, colors, or complicated ways of life in order to survive. For example, many cryptozoans have either reduced eyes or no eyes whatsoever. Most have a simple body structure and dark coloration with no distinctive markings.

Another example of adaptation to this environment is found in the flow of energy among the cryptozoans. Above the soil, organisms carry out their many functions: ingesting or making food, reproducing, excreting, and eventually dying. The result of all of this activity is the deposition of organic matter on and in the soil. Since little light penetrates the soil environment, photosynthesis cannot occur and the cryptozoa must depend on this supply of organic material from above to provide the energy they need to survive. This material is known as **detritus** and it forms an important part of terrestrial food webs. The cryptozoans live in and on this detritus and many have adapted to literally eat their way through it to extract the energy they need to run their metabolism.

As the cryptozoans feed on this organic material, they in turn may be eaten by underground predators. Predators' wastes (and eventually their dead remains) are also consumed by other soil inhabitants, and this process continues until only an indigestable residue is left. This remaining organic material is termed **humus** and is an important component of soil.

Soil ecosystems are important to humans. Without the many biological and chemical activities that occur there, the rich soils that support our major ecosystems and supply us with our basic source of food would not exist.

During this exercise you will collect soil organisms, arrange them into look-alike groups, and calculate a diversity index as you did in the stream study in Laboratory Activity 7. You should select more than one site for your collection and compare the diversity indexes and the types of organisms. For example, you might select a grassy area of the campus and compare it to a wooded area. A recently tilled field could be compared either with one of these or with a field undergoing ecological succession (becoming overgrown with the natural vegetation of the area). You could also compare an area that has been treated with pesticide to a similar area that was left untreated. Use your own knowledge of the area around you and develop your own ideas concerning what would be a meaningful comparison.

MATERIALS

Coffee can (10 cm diameter)
Trowel
Plastic bag
Marking pencil or pen
Gummed labels
White enamel tray
Forceps
Isopropyl alcohol with glycerin
Cheesecloth
Berlese funnel apparatus
Light bulb
Collecting vial
Dissecting microscope
Watch glass
Dissecting needles or fine brush

PROCEDURE

Field Collection

Sampling for soil organisms will require that you choose 5–10 locations within your site and collect samples of soil along with the litter (loose organic material on the surface).

1. After you have decided on the sites you wish to study, describe them in Question 5 in the Evaluation.

2. In soft soils, collect samples in a coffee can (10 cm diameter). Push it into the soil to a depth of 5–6 cm. Use a trowel to dig out the soil, and then place the soil sample with litter in a plastic bag for transport to the laboratory.

3. In harder soils measure an area 10 cm in diameter (or square) and dig out the sample with a trowel.

4. Label the bags with the appropriate information: date, collection site, and your name. Samples may be stored in the refrigerator for a week if necessary.

In the Laboratory

1. Spread the contents of the bag on a white enamel tray. Using a pair of forceps, remove the larger organisms and place in a jar of isopropyl alcohol with several drops of glycerin.

If the soil is particularly wet you should allow it to dry until it is crumbly but not totally dry and hard. If the soil is too dry, add enough water to make it moist but not muddy.

2. To separate the smaller organisms from the soil, place the sample in a piece of cheesecloth (one layer is sufficient) and place it on top of the hardware cloth inside the Berlese funnel apparatus (see Figure 18.4).

3. Suspend a light bulb several centimeters above the sample to drive the organisms into the collecting vial containing alcohol. Allow 1–2 days for collecting.

Observation and Classification

After collecting the specimens, classify them into look-alike groups. Since most soil organisms are nearly microscopic, use a dissecting microscope.

1. Place your sample of organisms in a watch glass and, using dissecting needles or a fine brush,

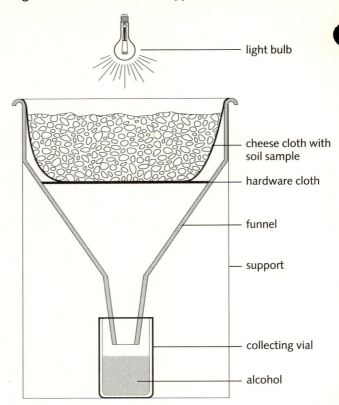

Figure 18.4 Berlese funnel apparatus

- light bulb
- cheese cloth with soil sample
- hardware cloth
- funnel
- support
- collecting vial
- alcohol

sort them into groups according to their structure. This may seem difficult at first because they are quite small, but be patient and study each organism carefully. Once you have sorted the organisms, make a sketch of a representative of each group, including the larger organisms that have been preserved in alcohol. Identify each group with a letter or name. Make your diagrams in Question 6 in the Evaluation.

2. Count the number in each group, and record the data in the frequency column in Question 7 in the Evaluation.

Although look-alike groups are sufficient to calculate a diversity index, you should be able to classify many of the organisms into the major taxonomic groups found in the soil. The following dichotomous key (similar to the one used to classify algae in Laboratory Activity 3) will help in doing this. Refer back to Activity 3 if you need to refresh your memory on how a dichotomous key works.

Dichotomous Key to the Major Phyla of Soil Organisms

1. (a) body spindle shaped, smooth, slender, usually light colored, approximately 0.5–1.5 mm long: Phylum Nematoda (roundworms)

   (b) body soft or covered with a hard skeleton, may be divided into several regions or segments: 2

2. (a) body soft and smooth, may have calcareous spiral shell, two eyes on stalks, gliding movement: Phylum Mollusca (snails, slugs)

   (b) body with two or more segments, which may be numerous, small, and ringlike or wider and fewer in number: 3

3. (a) segments numerous, small, and ringlike, body soft: Phylum Annelida (earthworms)

   (b) body covered with hard skeleton (exoskeleton), always three or more pairs of jointed legs: Phylum Arthropoda

The following key can be used to classify the members of phylum Arthropoda into the next lower taxonomic group or class. Refer also to Figure 18.5.

Dichotomous Key to the Classes of Arthropoda

1. (a) body with three pairs of legs, three body divisions, one pair of antennae: Class Insecta (springtails, most common insect found in soil)

   (b) body with more than three pairs of legs: 2

2. (a) body with four pairs of legs, no antennae, two body regions: Class Arachnida (mites, pseudoscorpions)

   (b) body with more than four pairs of legs, one or two pair of antennae: 3

3. (a) body with two pairs of antennae, more than five pairs of legs: Class Crustacea (sow bugs)

   (b) body long with one pair of antennae, numerous body segments with one or two pairs of legs on each segment: 4

4. (a) each body segment with one pair of legs: Class Chilopoda (centipedes)

   (b) each body segment with two pairs of legs: Class Diplopoda (millipede)

Figure 18.5 Common arthropods found in the soil

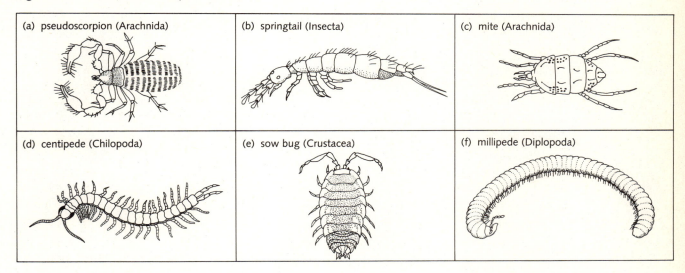

(a) pseudoscorpion (Arachnida)

(b) springtail (Insecta)

(c) mite (Arachnida)

(d) centipede (Chilopoda)

(e) sow bug (Crustacea)

(f) millipede (Diplopoda)

EVALUATION 18

Diversity of Soil Organisms

I. SOIL MICROORGANISMS

1. Describe the source of your soil sample.

2. Diagram the colonies that have grown on your petri dishes in the "plates" below. Since there will probably be many duplicate colonies, you may use the same number more than once. Complete the data table below for each of the numbered colonies.

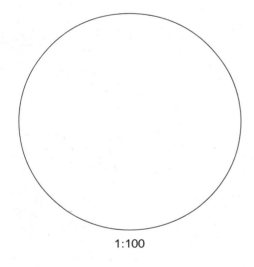

1:100

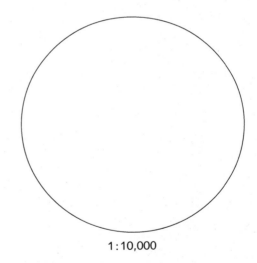

1:10,000

Colony #	Surface	Form	Elevation	Margin

3. After doing a simple stain and a Gram stain, make a drawing of the organisms found in each of the colonies selected for study. Complete the table under each diagram.

Dilution	_____
Colony #	_____
Morphology	_____
Size (μm)	_____
Bacteria/yeast	_____
Gram stain	_____

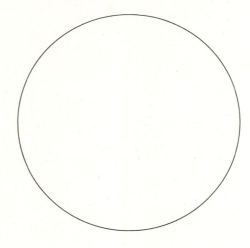

Dilution	_____
Colony #	_____
Morphology	_____
Size (μm)	_____
Bacteria/yeast	_____
Gram stain	_____

Dilution	_____
Colony #	_____
Morphology	_____
Size (μm)	_____
Bacteria/yeast	_____
Gram stain	_____

Dilution	_____
Colony #	_____
Morphology	_____
Size (μm)	_____
Bacteria/yeast	_____
Gram stain	_____

NAME ______________________ SECTION __________ DATE ______________

4. Why were you able to grow your bacteria at room temperature rather than in an incubator at 37°C?

II. SOIL INVERTEBRATES

5. In the chart below describe the sites you chose for your investigation and the reason for selecting these areas.

	Description	Reason for choosing
Site #1		
Site #2		
Site #3		

6. In the spaces below, make an outline diagram of a representative organism from each major taxonomic group in each site studied.

Site #1

Group: Group: Group:

Group: Group: Group:

Group: Group: Group:

Site #2

Group: Group: Group:

Group: Group: Group:

Group: Group: Group:

Site #3

Group: Group: Group:

Group: Group: Group:

Group: Group: Group:

Site #4

Group: _____ Group: _____ Group: _____

Group: _____ Group: _____ Group: _____

Group: _____ Group: _____ Group: _____

7. In the following tables, enter your data for each site and calculate the Shannon-Weaver diversity index (see Appendix B).

Site #1:

Organism	Frequency (P_1)	Natural log (P_1)	$(P_1)(-\ln P_1)$

Site #2:

Organism	Frequency (P_1)	Natural log (P_1)	$(P_1)(-\ln P_1)$

Site #3:

Organism	Frequency (P_1)	Natural log (P_1)	$(P_1)(-\ln P_1)$

$$N = \underline{\hspace{2cm}}$$

8. On a separate piece of paper, write a summary of your findings for each site studied. Compare each site with regard to the population size, type of organisms found, and their diversity indexes. What does this information tell you about the relative stability and/or environmental quality of each site?

Form and Pattern in Plants

OBJECTIVES

At the end of this laboratory activity, you should be able to:

- give several examples of plants demonstrating a spiral arrangement of certain structures.
- describe the Fibonacci number series.
- relate this mathematical principle to the arrangement of certain plant structures.
- explain the possible influence these patterns in nature may have had on human art and architecture.

INTRODUCTION

Since Charles Darwin and Alfred Wallace introduced their concept of evolution in 1858, biologists have discovered that animals and plants display a vast number of characteristics that demonstrate evolutionary relationships. Many examples indicate that if a molecule, a particular structure, or an arrangement of parts functions well, it may be maintained in living forms as they evolve.

In the following activity you will observe a particular form and pattern in plants that may be another example of the evolutionary process.

MATERIALS

Pinecone
Pineapple
Sunflower seed head
Pincushion cactus
Masking tape

PROCEDURE

1. Obtain a pinecone, a pineapple, a sunflower seed head, and a cactus plant from your instructor. Examine them carefully.

2. Describe the one common characteristic that you note with regard to the arrangement of parts in the pinecone, pineapple, sunflower seed head, and cactus plant. Record your answer in Question 1 of the Evaluation.

Pinecone

1. Study the pattern displayed by the **bracts** (modified leaflike structures) covering the seeds of the pinecone. Hold the pinecone in such a way that you can see the clockwise and counterclockwise spirals of bracts.

2. Count the rows of bracts arranged in a clockwise manner around the cone until you have turned the cone completely. Use masking tape to mark your starting point. See Figure 19.1.

3. Repeat this procedure, this time counting the rows arranged in a counterclockwise manner.

4. Record your data in Question 2 in the Evaluation. Calculate the simplest ratio of clockwise to counterclockwise rows.

Pineapple

Repeat the process used above in studying the spiral patterns produced by the ovules of a pineapple (see Figure 19.2).

1. Use a small piece of masking tape to mark your starting point for counting the clockwise and counterclockwise spirals.

Figure 19.1 Pinecone showing spirals

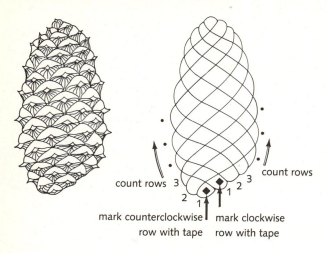

count rows 3
2
1
count rows 1 2 3

mark counterclockwise row with tape   mark clockwise row with tape

2. Record your data in the chart in Question 2 in the Evaluation.

3. Calculate the simplest ratio of clockwise to counterclockwise rows.

Cactus Plant

If you examine the pincushion cactus carefully you will be able to count the rows of needles that spiral in each direction as you did previously (see Figure 19.3).

1. Record your data in the chart in Question 2 in the Evaluation.

2. Calculate the simplest ratio of clockwise to counterclockwise rows.

Figure 19.2 Pineapple showing spirals

Figure 19.3 Pincushion cactus showing spirals

Sunflower Seed Head

1. Obtain a dwarf sunflower head from your instructor (see Figure 19.4). Count the rows of seeds that spiral in the clockwise and counterclockwise directions.

2. Record your data on the chart in Question 2 in the Evaluation.

3. Calculate the simplest ratio of clockwise to counterclockwise rows.

The simplest ratios calculated for these spirals are impressively similar. Let's examine this similarity further.

In A.D. 1202, in a book entitled *Liber Abaci*, Leonardo de Pisa (Fibonacci) introduced the Hindu-Arabic numeration system to Europe. The *Liber Abaci* contained a brief chapter on a series of numbers that came to be known as the Fibonacci series. A Fibonacci series is listed below.

1, 1, 2, 3, 5, 8, 13, 21, 34, 55, 89, 144, 233, . . .

What is the pattern of this series?

Look at Question 2 in the Evaluation at the ratios you have calculated. Notice that all of the spiral ratios you have examined are sets of adjacent numbers (2:3, 3:5, 5:8, . . .) of this Fibonacci series. Also note that the adjacent numbers in this series constitute a simple ratio of 1:1.6.

1. Meet with several of your laboratory partners and discuss the following questions. Why are these plant parts arranged in this strange manner? Could it be that this pattern, so often repeated in plants, is of some importance or advantage to plants? Could this be the result of evolution?

At this point you need to collect some additional data. Your instructor has placed three rectangles on the wall of the laboratory.

1. Look at the rectangles for a moment. Which one of them is most pleasing to you: A, B, or C?

2. Collect data for your class and record it in Question 3 in the Evaluation.

3. Review all of the data collected so far and then answer Question 4 in the Evaluation.

Figure 19.4 Sunflower seed head showing spirals

EVALUATION 19

Form and Pattern in Plants

1. What common structural arrangement did you observe in a pinecone, a pineapple, a sunflower seed head, and a cactus plant?

2. Fill in the following data summary chart.

| | Number of rows | | Simplest ratio |
	Clockwise	Counterclockwise	
Pinecone (bracts)			
Pineapple (ovules)			
Cactus (needles)			
Sunflower (seeds)			

3. Record your data concerning the rectangles below.

Rectangle	Number of students selecting	Students selecting (%)	Simplest ratio of width to length
A			
B			
C			

4. In this laboratory activity you have collected data that (we hope) has produced many questions. This is the very "stuff" of science. From such questions, scientists develop hypotheses for investigation. Before the next laboratory period, state at least one hypothesis concerning today's activity and attempt to collect information that would support or reject this hypothesis. Here is an example:

> Some artists believe that this ever-recurring pattern in nature, of spirals to the left and spirals to the right, has had an effect on humans. They hypothesize that we have derived our principles of aesthetics from our environment, that our artistic sense has been modified by these patterns. You could research this hypothesis in the library and perhaps by talking to an art teacher.

Think big and give your imagination free reign!

Water and Life

OBJECTIVES

At the end of this laboratory activity, you should be able to:

- describe the bonding in a water molecule and how this bonding affects its behavior.
- define adhesion and cohesion.
- explain how capillary movement occurs.
- describe the cause of surface tension.
- explain how temperature affects water density.
- relate all of the above physical characteristics of water to various biological phenomena.

INTRODUCTION

Of the many important molecules studied in biology, probably the one most taken for granted is water. This ubiquitous substance, which makes up three-quarters of the earth's surface and about the same percentage of an organism's mass, is absolutely essential for life. Indeed, it is somewhat tautological but nevertheless correct to say that where there is life there is water, and where there is water there is usually life.

Biologists believe that life began in an aqueous environment and evolved for millions of years bathed in water. With the move to land, organisms left the comfort of a water environment behind, but all terrestrial organisms still cling tenaciously to the water molecule. The reason for this is simply that water has many unique properties that allow it to play a paramount role in the numerous physical and chemical events that must occur if life is to continue.

Water's unique properties can be fully appreciated only if one understands the structure of the water molecule. The hydrogen and oxygen atoms in a water molecule are covalently bonded to each other, meaning that two hydrogen atoms share electron pairs with the oxygen atom. This sharing, however, is not equal. The attraction of the oxygen nucleus for electrons is stronger than that of the hydrogen nucleus and, as a result, the electrons spend more time near the oxygen atom. This results in a lopsided, or polar, molecule with a distinctly negative charge at the oxygen end and a positive charge at each hydrogen end. Figure 20.1 indicates the electron configuration of a water molecule.

Figure 20.1 A water molecule

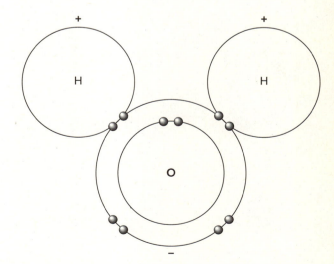

Several properties of water are described below, along with their biological importance. During this laboratory activity, you will conduct experiments that demonstrate these properties.

Water molecules form hydrogen bonds with many kinds of molecules, including other water molecules. **Cohesion** refers to the bonding of water molecules to each other, whereas **adhesion** occurs when water molecules are attracted to other kinds of molecules. The cohesiveness of water results in another property called **surface tension.** In a body of water, the molecules are attracted to other water molecules equally in all directions. However, at the surface there is a net force on the molecules pulling them to the bulk of the fluid, as illustrated in Figure 20.2(a). Because surface molecules have attraction in only three directions instead of four, the overall force holding them together is stronger. This inward pulling force is surface tension. For examples of surface tension at work in nature, water striders walk on water (see Figure 20.2(b)), and the dry fly used by a fly fisherman sits high on the water. Surface tension in the alveoli of the human lung is so strong that the lungs would collapse if a surfactant (soaplike material) were not present to reduce this tension. Soap or soaplike materials reduce the attraction of water molecules for each other, thereby increasing their attraction for other substances.

Capillarity, which comes from the Latin *capillus* (hair) is the tendency of liquids to be forced into small pores or tubes. Capillarity can explain the swelling of wood placed in water, the movement of water across a paper towel, and water rising through soil to the surface of a plowed field. This interesting characteristic of water is due to an interaction between the forces of adhesion and cohesion. You have probably noticed the meniscus, or curved upper surface of a column of liquid in a glass tube. Water rises up a tube with a small inside diameter because water is attracted to the glass by adhesion. Once the "climbing molecules" have moved up the sides of the tube, these water molecules pull other water molecules up by cohesion. This climbing and pulling of water molecules produces capillarity. This movement will stop when the weight of the water column overcomes the forces of capillarity (see Figure 20.3).

One of the most interesting characteristics of water is its ability to maintain a relatively stable temperature. Raising the temperature of a liquid is a matter of increasing the average kinetic energy, the rate of motion of its molecules. In substances, such as water, that are subject to hydrogen bonding, this motion is retarded; when heat is added to the system, much of the energy is needed to break the countless hydrogen bonds. As a result, less energy is available to increase the kinetic energy of the molecules and the temperature will not show as much of a rise as it would if hydrogen bonds were not being broken. Chemists refer to the amount of heat needed to raise the temperature of a given substance as **specific heat.** Thermal energy (heat) is measured in **calories.** A calorie is defined as the amount of heat needed to raise the temperature of 1 g of water 1°C. Biologists use the term large Calorie, which is the amount of heat needed to raise the temperature of 1000 g (1 kg) of water 1°C. Thus, one large Calorie is equal to 1000 small calories. Compared with other substances,

(a)

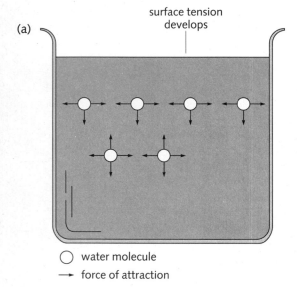

surface tension develops

○ water molecule
→ force of attraction

Figure 20.2 Surface tension
(a) Illustration of surface tension
(b) A water strider

(b)

Figure 20.3 Capillary movement
(a) Adhesive forces are stronger than cohesive forces in water.
(b) Cohesive forces are stronger than adhesive forces in mercury.

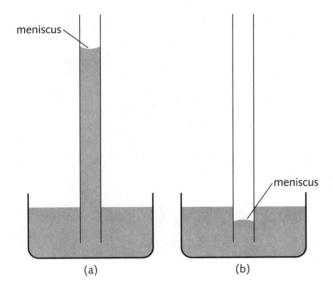

meniscus

meniscus

(a)   (b)

water has a very high specific heat. Table 20.1 lists the specific heat for several common substances.

The high specific heat of water is important to organisms. For example, metabolic activity generates considerable heat within an organism, and this heat can be readily absorbed by water with little corresponding change in body temperature. Water's high specific heat also helps to regulate the temperature of water in the environment. Large bodies of water have relatively stable temperatures, making them rather hospitable environments for such organisms as fish, which are unable to regulate their body temperatures.

When water goes from the liquid to the gaseous state, the molecules at the surface must have sufficient kinetic energy to escape as a vapor. Just as in specific heat, the hydrogen bonding of the water molecules requires that considerable energy be added to the system in order for them to break away from each other. The **heat of vaporization** for 1 g of water is 540 calories, considerably higher than any other substance. Table 20.2 lists the heat of vaporization for several substances.

Because of water's high heat of vaporization, evaporating water molecules remove considerable heat from the surface when they escape into the atmosphere. Animals make use of this in order to keep their body temperatures relatively constant. When human body temperature rises, the hypothalamus detects this increase and signals the sweat glands to produce more perspiration. Evaporation then cools the body. On a larger scale, we find that when water evaporates from a large body, such as an ocean or lake, it takes an enormous amount of heat with it. This evaporation moderates the temperature of the surrounding area and is part of the heat distribution process that is constantly occurring on Earth.

One of the many remarkable characteristics of water is the way its density varies with temperature fluctuations. In most substances other than water, **density** (mass per unit of volume) increases as temperature decreases; thus, these substances are more dense in the solid state than they are in the liquid state. Water, however, reaches its maximum density at 4°C (39°F). Below this temperature, the water's hydrogen bonds cause its molecules to move apart and form a crystal lattice that is completed at 0°C. Because the molecules in ice are less closely packed than in liquid water, ice is less dense and floats.

This ability of water to maintain a relatively stable temperature enables many aquatic organisms

Table 20.1 Specific heat for several common substances

Substance	calories/g/°C
Water	1.000
Acetic acid	0.468
Ether	0.517
Ethyl alcohol	0.456
Benzene	0.400
Olive oil	0.471
Turpentine, oil	0.411

Table 20.2 Heat of vaporization for several common substances

Substance	calories/g/°C
Water	540.0
Acetic acid	96.8
Ether	83.9
Ethyl alcohol	204.0
Benzene	94.3
Turpentine, oil	68.3

to survive throughout the year in deep bodies of water. During even the coldest of winters, aquatic forms can find refuge at the bottom of a lake, where water remains at or above 4°C. Although surface water in a lake may warm during the summer, temperatures beneath the surface rise less in direct relationship to depth.

MATERIALS

Plastic or glass rod
Fur or cloth
Buret
2 beakers (400 mL)
Microscope slide
Medicine dropper
Ethyl alcohol (95%)
Petri dish
Lens tissue
Insect mounting pins
Toothpicks
Diluted detergent
Capillary tubes of varying inside diameters
Colored water
Metric ruler
2 Erlenmeyer flasks (125 mL)
Dry sand
One-hole stopper, vented and fitted with a
  thermometer (°C)
Boiling chips
Hot plate
2 alcohol thermometers
Cheesecloth
Thread
Electric fan
Beaker (1000 mL)
White paper
Beaker (250 mL)
Red food coloring
Ice cubes
Pipette
Ice cubes with blue food coloring

PROCEDURE

You can observe the polar nature of water (see Figure 20.4) with the following setup.

1. Take a plastic or glass rod and rub it with a piece of fur or cloth; this will develop a static charge on the rod.

Figure 20.4 The polar nature of water

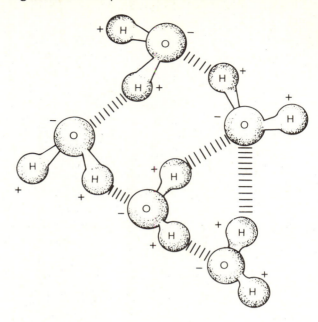

2. Fill a buret with water and hold the tip of the rod about 1 cm away from the end of the buret. Allow the water to run out of the buret into a beaker and observe what happens to the stream.

3. Repeat the process, this time placing the rod on the opposite side of the stream.

4. Complete the diagram in Figure 20.5, showing how the stream of water behaved in each case. Assuming that the rod developed a negative charge, draw a water molecule as you think it would orient itself in the stream in relation to the rod.

Figure 20.5 Demonstrating the polar nature of water

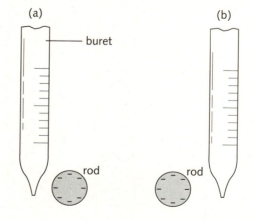

To demonstrate the properties of adhesion and cohesion, complete the following activity.

1. Wash a microscope slide until it is "squeaky" clean and dry it completely. Carefully place one drop of water on one end of the slide and one drop of ethyl alcohol onto the other end. Immediately observe the two drops and describe them in terms of relative diameter, elevation, shape of margin (edge), and the relative rate of evaporation (which substance evaporates more slowly).

2. Fill out the following chart and then complete Question 2 in the Evaluation.

	Water	Ethanol
Diameter		
Elevation		
Margin		
Relative rate of evaporation		

Surface tension[1] allows insects to walk on water. You can demonstrate this effect in the following activity.

1. Fill a petri dish with water. Very carefully place a piece of lens tissue about 1 cm × 3 cm on the surface of the water.

2. Gently place a straight pin on the lens paper. With two toothpicks, carefully push the paper to the bottom of the dish, allowing the pin to rest on the surface. Describe the surface of the water where the pin is resting. _____

3. Add a drop of diluted detergent to the surface of the water as far away from the pin as possible. Describe and explain what happens.

_____

_____

4. Rinse the dish and pin until completely free of detergent.

1. Surface tension is measured in newtons per meter. The newton is a unit of force in the meter-kilogram-second system.

The following exercise illustrates capillarity.

1. Place several short pieces of glass tubing of varying inside diameters in a culture dish containing about 0.5 cm of colored water.

2. Compare the height that the water rises in each tube and record in the chart below.

Diameter of tube	Height of water column (mm)

3. What relationship exists between the inside diameter of the tubing and the height to which the water rose?

4. Explain this phenomenon.

The temperature stability of water is due to its high specific heat. You can demonstrate this by comparing the rate of cooling of water with that of another substance, such as sand.

1. Take two 125-mL Erlenmeyer flasks and fill one with water and the other with *dry* sand. Fill both to just below the bend in the neck of the flasks.

2. Your instructor has provided a rubber stopper fitted with a thermometer. Insert this stopper into the flask so that the tip of the thermometer is about two-thirds of the way in the water or sand. The rubber stopper should also be vented with a small hole to prevent the buildup of pressure in the flask.

3. Add about 200 mL of water to a 400-mL beaker, drop in a few boiling chips, and then place one of the flasks in the beaker. The water level in the beaker should come approximately to the bend in the neck of the flask (see Figure 20.6). Repeat this procedure for the second flask.

4. Place each beaker on a hot plate and heat until the temperature reaches approximately 75°C. Do *not* boil. Remove each flask from the beaker of hot water and allow to cool. If one flask reaches 75°C before the other, you should remove it from the beaker and begin recording your data. It is not necessary to collect data from both flasks at the same time.

Figure 20.6 Apparatus for measuring specific heat

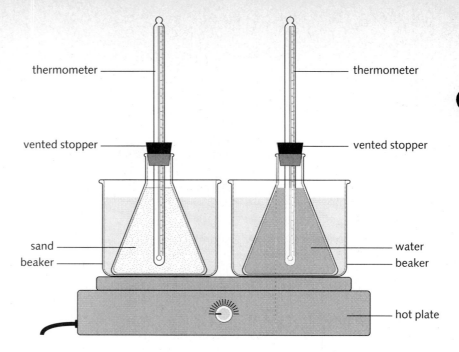

5. In the table below, record the starting temperature for each flask, and then record the temperature at two-minute intervals until a temperature of approximately 50°C is reached.

6. Graph this data on the paper provided in the Evaluation. You should consult Appendix A for the correct way to graph data. Below your graph, write an explanation of your results.

Time (min.)	Sand temperature (°C)	Water temperature (°C)

Heat of vaporization is easily demonstrated by the following activity.

1. Take two alcohol (*not* mercury) thermometers and wrap the bulb of each one with several layers of cheesecloth. Secure with a piece of thread.

2. Wet one bulb with water and squeeze gently to remove excess water. Keep the other bulb dry.

3. Record the temperature of each thermometer and then wave the thermometers in the air or place in front of a fan for 5 minutes, recording the temperature of each thermometer at one-minute intervals in the following table.

	0	1	2	3	4	5
Dry bulb						
Wet bulb						

Account for any difference in temperature between the two thermometers.

To demonstrate how temperature affects water density, complete the following exercise.

1. Heat approximately 800 mL of water in a 1000-mL beaker to 30°C and place the beaker on a sheet of plain white paper.

2. In a 250-mL beaker add approximately 150 mL of water add enough red food coloring to produce a deep red color. The food coloring will act as a tracer.

3. Add sufficient ice to the red water in the 250-mL beaker to bring the temperature of the water to 4°C. (Your instructor may supply you with water at 4°C from the refrigerator.)

4. With a pipette, gently add approximately 10 mL of the red 4°C water to the bottom of the 1000-mL beaker of 30°C water (see Figure 20.7).

5. Observe from the side of the beaker. Explain what happens to the 4°C water.

6. Obtain an ice cube made from blue-colored water.

7. Gently place the ice cube on the surface of the water in the 1000-mL beaker and observe the melting process and any water movement that occurs.

8. Why is the ice floating?

9. Explain the movement of the blue ice water.

10. Describe the final condition of the water in the beaker.

11. Record your observation in Question 6 in the Evaluation.

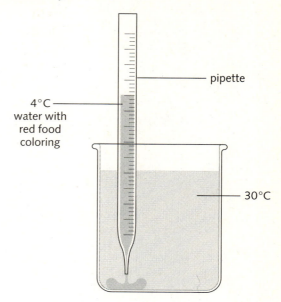

Figure 20.7 Demonstrating how temperature affects water density

pipette

4°C water with red food coloring

30°C

EVALUATION 20

Water and Life

1. When a negatively charged rod is placed near a falling stream of water, the water is attracted to the rod. Using your knowledge of the structure of the water molecule, explain why this occurs.

2. In one of your activities you observed and compared a drop of water and a drop of alcohol. Explain your observations in terms of adhesion and cohesion.

3. The surface tensions of some liquids in contact with air are given in the following table.

Liquid	Temperature (°C)	Surface tension (N/m)
Benzene	20	29.0×10^{-3}
Ethanol	20	22.3×10^{-3}
Glycerin	20	63.0×10^{-3}
Mercury	20	465.0×10^{-3}
Water	0	75.6×10^{-3}
Water	20	72.8×10^{-3}
Water	100	58.9×10^{-3}

(a) Why does the surface tension of water vary with the temperature of water?

(b) Which liquid would best be able to support the weight of a penny?

4. In plants, xylem cells form vessels that conduct water and minerals upward. Many forces are involved in this transport, including capillarity. Ignoring the other forces, attempt to calculate the height to which water would rise in these vessels based on the following information. If a section of capillary tubing with an inside diameter of 0.56 mm was placed in a dish of water, the column of water would rise 30 mm. If a xylem cell had an inside diameter of 5 μm, how far would the water be lifted by capillary action? Show your calculation below.

5. Interpret your graph concerning the temperature stability of water (high specific heat).

6. If a substance has a high heat of vaporization, what does this mean in terms of the quantity of heat needed for the evaporation of this liquid?

7. Write a detailed observation of the water density experiment and explain all the water movements. How would these water movements help to cycle minerals and oxygen in a deep lake?

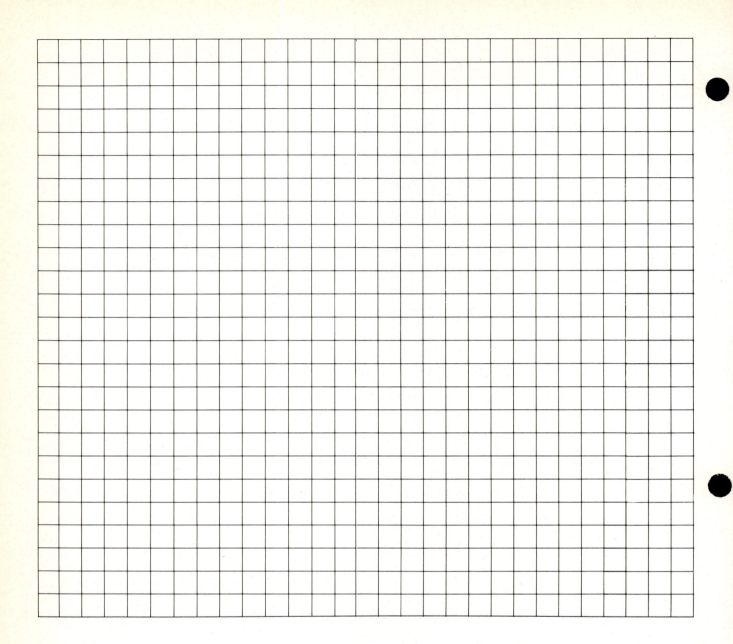

Enzymes: Biological Catalysts

OBJECTIVES

At the end of this laboratory activity, you should be able to:

- state why enzymes are essential to the normal functioning of cells.
- explain why different types of tissue have varying concentrations of catalase.
- explain the operation of the manometer used to measure catalase activity.
- state the relationship between substrate concentration and enzyme activity.
- state the relationship between enzyme concentration and enzyme activity.

INTRODUCTION

Of all the elements and compounds that help make life possible, none ranks higher in importance than the group of molecules we call enzymes. Enzymes are catalysts, and the following example will help you understand the nature of this class of compounds. If you mix hydrogen and oxygen gas in a container you will probably not be too excited by what you see. However, if you apply a spark to this mixture you will be impressed by the violent reaction that occurs as these two gases combine to form water. This reaction occurs in part because the reactants need a push to get started. This push is called **activation energy** and is needed to overcome an energy barrier that frequently exists in chemical reactions. Once the reaction is initiated,

it becomes self-sustaining, using the heat generated in the reaction to continue the reaction.

An alternative to the spark is to add powdered platinum. By interacting in a specific manner with the reactants, the platinum greatly lowers the heat of activation and allows the reaction to occur at room temperature. Once completed, the platinum can be recovered and used again. Platinum acts as a **catalyst,** that is, a substance that increases the rate of a chemical reaction without being permanently consumed itself.

Living organisms survive for many reasons. One of the most important is their remarkable assembly of diverse chemical reactions, each of which contributes to the success of the total organism. When carrying out these many reactions, most organisms must operate under rather severe constraints, most importantly the need to initiate and sustain chemical reactions in a very narrow temperature range (approximately 10°C–50°C). The chemical reactions in cells that involve organic compounds have significant energy barriers to surmount if they are to be initiated. In cells, it is not possible to overcome the activation energy barrier with the addition of heat (the spark in $H_2 + O_2$), because raising the temperature of an organism may have severe consequences. This problem was solved with the evolution of organic catalysts or **enzymes** (see Figures 21.1 and 21.2).

Most enzymes are protein molecules, and each enzyme has at least one cleft or groove known as the **active site,** where the **substrate** (reactant) attaches and forms an enzyme-substrate complex. Once this complex is formed, chemical bonds are

Figure 21.1 Activation energy with and without catalyst

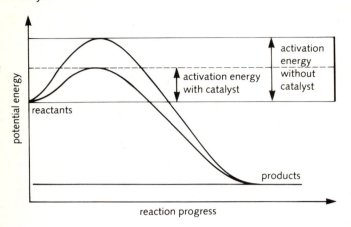

Figure 21.2 Enzymes and activation energy

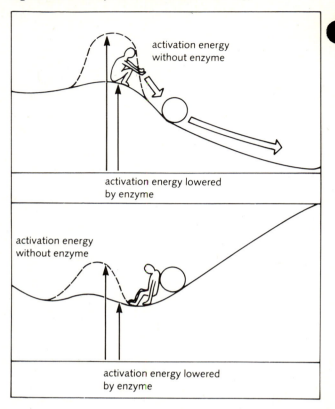

made or broken between the atoms of the substrate, resulting in the formation of the product. The process can be illustrated by the following equation.

$$E + S \longrightarrow ES$$
enzyme   substrate   enzyme-substrate complex

$$\longrightarrow EP$$
enzyme-product complex

$$\longrightarrow E + P$$
enzyme   product

The formation of the enzyme-substrate complex allows chemical bonds to be formed or broken at a much lower energy level than if the enzyme were absent (see Figure 21.3).

The enzyme you will work with in this laboratory exercise is called catalase. Catalase is one of the most effective catalysts known and is able to decompose high concentrations of hydrogen peroxide into water and oxygen. Hydrogen peroxide is toxic to cells and must be removed. In spite of its toxicity, hydrogen peroxide may be used in low concentrations as a germicide. You may have used it to clean a cut or abrasion, and probably noticed oxygen gas bubbling from the wound.

In this laboratory activity you will do several exercises that illustrate the activity of catalase in living tissues. You will use an enzyme that can be extracted from living tissue to reinforce the idea that enzymes have their origin in living cells and not in a chemist's laboratory.

MATERIALS

Deep well slides
Marking pencil or pen
Potatoes (fresh)
Adipose tissue (fresh)
Liver (fresh)
Glass stirring rod
Dissecting microscope
Dropping bottles of hydrogen peroxide (3.0%, 1.5%, 0.75%, 0.375%)
Dropping bottle of HCl (1 N)
Dropping bottle of NaOH (1 N)
Blender
Distilled water
Filter paper
Buchner funnel
Beaker (150 mL)
Beaker (1000 mL)
Ice
Test tubes (10 mm × 100 mm)
Pipette (1 mL)
Manometer apparatus with 500-mL Erlenmeyer flask
Graduated cylinder (100 mL)
Pipette (10 mL)

Figure 21.3 The enzyme-substrate complex

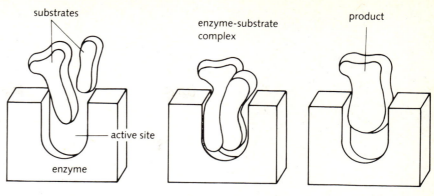

1. Substrate molecules approach active site.

2. Substrate molecules bind to active site; reaction takes place.

3. Product molecules leave active site; enzyme is unaltered.

PROCEDURE

Decomposition of Hydrogen Peroxide by Different Tissues

In this activity you will observe qualitatively the activity of catalase found in different tissues.

1. Obtain four deep well slides and with a marking pencil label them "normal," "pH acid," "pH base," and "boiled." Notice that each slide has three wells.

2. To the slide labeled "normal," add a small portion of potato tissue to one well, a small piece of adipose (fat) tissue to the second well, and a small piece of liver to the third well. Make sure that you use similar quantities of each of these tissues. Use a glass stirring rod to crush the tissue in each well to release the enzymes contained in the cells. Wash the rod thoroughly after crushing each tissue or use a different rod.

3. Repeat this process with the slides labeled "pH acid," "pH base," and "boiled." Make certain to obtain tissue that has been boiled for the last slide.

4. Place the slide labeled "normal" on the stage of a dissecting microscope and add a drop of 3.0% hydrogen peroxide to the tissue in each well. Observe carefully and record your results in the table provided in Question 1 in the Evaluation.

➤ **CAUTION: Hydrochloric acid and sodium hydroxide are skin irritants. If you spill either of these solutions on your clothing or skin, immediately flush with water and notify your instructor.**

5. Add two drops of 1 N hydrochloric acid (HCl) to the tissue in each well of the slide marked "pH acid." Allow it stand for one mintue and then add one drop of 3.0% hydrogen peroxide to each well. Observe under the dissecting microscope and record your results in Question 1 in the Evaluation.

6. Repeat this process with the slide labeled "pH base," adding two drops of 1 N sodium hydroxide (NaOH) in place of the hydrochloric acid. Record your results in Question 1 in the Evaluation.

7. Add a drop of 3.0% hydrogen peroxide to each well of the slide labeled "boiled" and again record your results in Question 1 of the Evaluation.

Semiquantitative Analysis of Catalase Activity

Catalase is an enzyme in living tissues that decomposes hydrogen peroxide into water and oxygen gas:

$$2\ H_2O_2 \longrightarrow 2\ H_2O + O_2$$

The quantity of oxygen bubbles generated when peroxide is added to tissue containing catalase indicates the level of catalase activity. This exercise will show how the concentration of catalase determines the rate of its activity. Before you can continue on to the next two activities you need to obtain an extract containing this enzyme. The potato tissue used in the previous activity decomposed hydrogen peroxide, indicating the presence of the enzyme catalase. Use the following procedure to obtain a crude catalase extract from potato tissue.

1. Peel a medium-sized potato and weigh it carefully. You will need at least 50 g of potato tissue. Dice the potato into small pieces and place in a blender. Add 2 mL of distilled water per gram of tissue. Blend at high speed for 30 seconds.

2. Add a piece of filter paper to a Buchner funnel and filter the homogenate. The resulting filtrate will be pink to white in color and will get darker as time passes. Pour this crude enzyme extract into a small beaker and place it into a larger beaker of water containing ice to prevent the decomposition of the enzyme.

3. Obtain 5 small test tubes (10 mm × 100 mm) and use the following table to prepare test tubes of different concentrations of enzyme. Use a clean 1-mL pipette for each tube.

Tube	Enzyme (mL)	Water (mL)	Dilution ratio
1	1.00	0	1:0
2	0.75	0.25	3:1
3	0.50	0.50	1:1
4	0.25	0.75	1:3
5	0	1.00	0:1

4. Add 5 drops of 3.0% hydrogen peroxide to each tube and observe the relative production of bubbles in each tube. Make this observation over the next 30 minutes. Record your results in Question 2 in the Evaluation.

Substrate Concentration and Catalase Activity

In this activity you will measure the amount of oxygen produced during catalase activity by direct measurement with a manometer (see Figure 21.4). Your manometer must be in perfect working condition before you start this experiment. Examine it carefully in the following manner.

1. Observe that the flask is fitted with a two-hole stopper. A vent tube should be installed in one hole and a manometer tube filled with fluid in the other. When the C-clamp is closed on the vent tube and the stopper is fitted in the neck of the flask properly, the system should be airtight. Notice that the needle of the syringe pierces through the stopper completely. Air can be added or removed from the flask by carefully moving the plunger of the syringe up or down.

Test your manometer before attempting the activity.

2. Open the vent, secure the rubber stopper in the mouth of the flask, push the plunger into the syringe as far as possible, and then close the vent. Your system should be airtight. To test for this condition, carefully pull the plunger out so that the fluid in the manometer moves toward the flask (careful; not too far). The fluid should "hang" in the tube when the plunger is not moving. Return the plunger to the starting position.

If your apparatus is not working, check for leaks. Be sure you are able to read the calibrations on the syringe before you start.

3. With a 10-mL pipette, add 10 mL of cold potato enzyme extract to the flask and then add 10 drops of 3.0% hydrogen peroxide. Quickly put the stopper into the mouth of the flask. Make certain that the vent is open and that the plunger is completely inserted into the syringe. Close the vent tube and record the starting time in Question 3 in the Evaluation.

4. In order not to heat the flask, hold it at the very end of the neck with the tips of your fingers. Gently rotate the flask as bubbles of gas are given off. At this point, oxygen that is being released is pushing the manometer fluid away from the flask.

5. After 30 seconds carefully pull the plunger out to return the manometer fluid to the starting position. Read the number on the syringe at the bottom of the plunger to obtain a reading of the volume of air you removed from the flask. This quantity is the same as the volume of oxygen produced during that 30-second period. Have your partner record this datum in Question 3(a) in the Evaluation. Repeat the process for another 30 seconds.

6. Continue collecting readings every 30 seconds for 10 minutes or until no more oxygen is being produced. You may have to vent your system during this period of time if the plunger is pulled out to its limit. Do this by loosening the clamp on the vent tube and returning the plunger to the starting position.

7. When you are finished, calculate the milliliters of oxygen produced per minute in Question 3b in the Evaluation.

Figure 21.4 Manometer apparatus

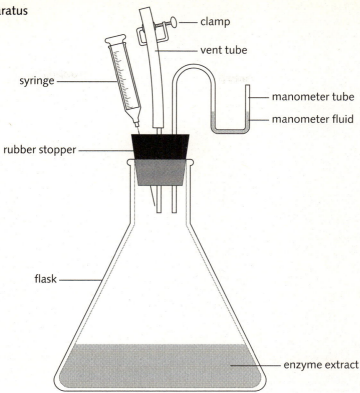

- clamp
- vent tube
- syringe
- manometer tube
- manometer fluid
- rubber stopper
- flask
- enzyme extract

8. Repeat this entire process using 1.5%, 0.75%, and 0.375% hydrogen peroxide solutions provided by your instructor. Record all data and calculations in Questions 3(a) and 3(b) in the Evaluation.

OPTIONAL ACTIVITY

By now you have learned about the enzyme catalase specifically and therefore about enzymes in general. You may also have some questions about the factors that can influence activity in cells. For example, altering the pH of the enzyme solution may have a profound effect on its activity. The enzyme pepsin, which is found in the stomach, works well in an acidic condition but its activity stops when it reaches the alkaline environment of the small intestine. You also know that increasing or decreasing temperature will alter the rate of a chemical reaction. Use one of these conditions or any other that seems reasonable for stating a hypothesis and then design an experiment to test your hypothesis.

Record your hypothesis, experimental results, and conclusion in the Optional Activity section in the Evaluation.

EVALUATION 21

Enzymes: Biological Catalysts

1. (a) Use the following chart to record the results from the first experiment. On a scale of 0 (no activity) to 3 (highest level of activity), rank the relative ability of each tissue to decompose hydrogen peroxide.

Condition	Potato tissue	Adipose tissue	Liver tissue
Normal			
pH: Acid			
pH: Base			
Boiled			

(b) On the slide labeled "normal," which tissue showed the most catalase activity?

Read in your textbook about this tissue and then explain why it showed so much activity.

(c) What is the effect of each of the following on catalase activity?

pH _____

temperature _____

(d) Hydrogen peroxide will not decompose rapidly into water and oxygen unless your skin is broken open and tissue fluids are exposed. Can you explain this based on your observations in this laboratory activity?

2. (a) After approximately 30 minutes record the relative quantity of bubbles in each of the 5 test tubes. Use a scale of 5 (most bubbles) to 1 (least bubbles) to rank each tube with regard to amount of enzyme activity.

	Tube (Dilution ratio—enzyme:water)				
	1 (1:0)	2 (3:1)	3 (1:1)	4 (1:3)	5 (0:1)
Relative rate of activity					

(b) What is the effect of catalase concentration on the rate of the decomposition of hydrogen peroxide?

3. (a) In the table below, record the milliliters of oxygen produced at 30-second intervals for each concentration of substrate.

| | Substrate concentration | | | |
Time	3.0%	1.5%	0.75%	0.375%

(b) Calculate the rate of oxygen production per minute for each of the substrate concentrations.

Concentration: 3.0%

_____ ÷ _____ = _____ mL/min
  total mL   total
   O_2 minutes

Concentration: 1.5%

_____ ÷ _____ = _____ mL/min
  total mL   total
   O_2 minutes

Concentration: 0.75%

_____ ÷ _____ = _____ mL/min
  total mL   total
   O_2 minutes

Concentration: 0.375%

_____ ÷ _____ = _____ mL/min
  total mL   total
   O_2 minutes

(c) Use the graph paper provided at the end of this Evaluation to graph the data in Question 3(a), using time as the independent variable and milliliters of oxygen produced as the dependent variable. Your graph should have four lines, one for each substrate concentration. (Your instructor may want you to use class data for this graph.)

(d) Describe the slope of each line produced by your data. Did the rate of oxygen production level off? If so, explain why this occurred.

OPTIONAL ACTIVITY

1. State why you chose to do this particular experiment.

2. State your hypothesis for this experiment.

3. Prepare a table in the space below and record your data for this experiment.

4. Graph your data using the graph paper at the end of the Evaluation.

5. Describe the slope of the line in your graph.

6. Was your hypothesis substantiated? Explain your answer.

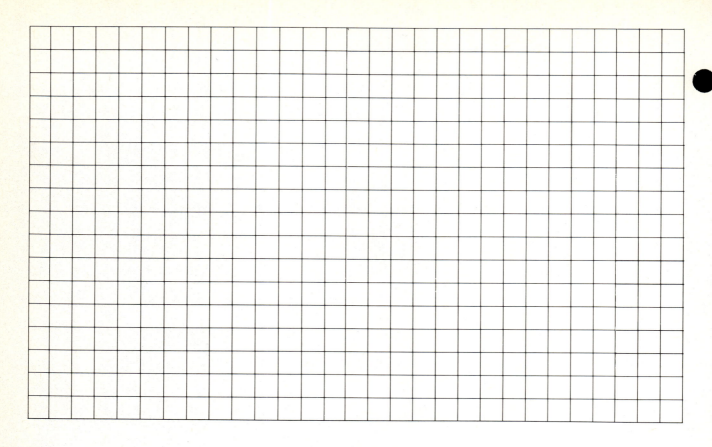

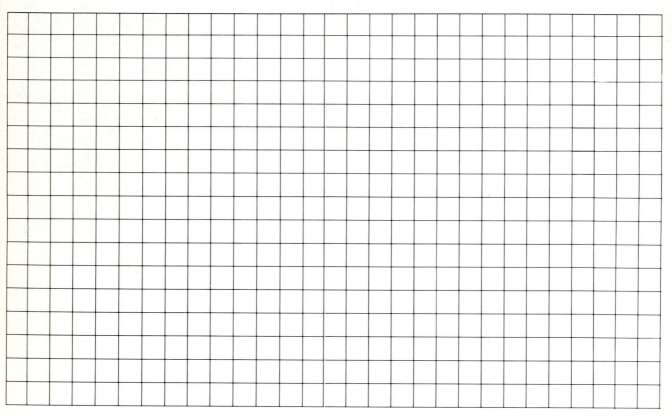

Model Building and the Isolation of DNA

OBJECTIVES

At the end of this laboratory activity, you should be able to:

- construct a paper model of the DNA molecule.
- list the major components of the DNA molecule.
- describe the three-dimensional structure of the DNA molecule.
- isolate DNA from a tissue sample.

INTRODUCTION

Biologists have known since 1944 that the genes of organisms are composed of deoxyribonucleic acid (DNA). But until James Watson and Francis Crick uncovered its structure in 1953, biologists did not know how the architecture of this molecule allowed it to encode information and duplicate itself.

Watson, a young American zoologist, and Crick, an English biochemist, met at Cambridge University in 1951. Despite their unusual backgrounds and the skepticism of their colleagues, this unlikely pair unraveled the structure of DNA (see Figure 22.1) in just two years.

The story of how these men synthesized a theory of DNA structure is one of the most exciting in the history of science. Watson and Crick constructed a model based on previous studies completed by many researchers and synthesized these studies in a single 900-word paper! This paper was published in *Nature,* a scientific journal, and subsequently changed the world of biology. You can best learn the basic structure of this molecule in much the same way as Watson and Crick: by building a model. They based their model on studies conducted by others on the chemistry of DNA and

followed certain rules of bonding concerning the atoms contained in the molecule. In this activity you will see the structure of the molecule unfold as you build your model from the cutouts in this manual.

Before beginning the construction of your molecule, read the original paper by Watson and Crick reprinted at the end of this laboratory activity. After you have constructed the DNA model, you will attempt to extract DNA from living tissue.

Figure 22.1 Model of DNA

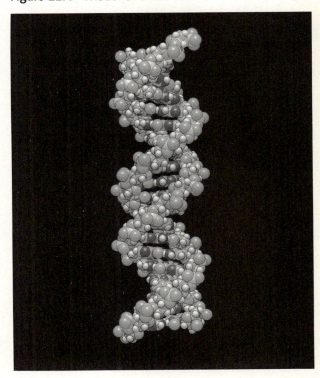

MATERIALS

Model Building:

Base pair cutouts
Cardboard
Rubber cement
Scissors
Crayons or felt-tipped markers (green, red, blue, orange)
Hole punch
Side chain cutout
Razor blade
Tape (1 in. wide)
Wood stand with dowel

DNA Isolation:

Frozen mouse tails
Polypropylene centrifuge tubes with caps
Sterile pipettes
Sterile buffer solution
Proteinase K (10 mg/mL)
Vortex stirrer
Temperature-controlled water bath
Phenol
Centrifuge
Chloroform isoamyl alcohol (24:1)
Sterile polystyrene tubes (clear) with caps
Ethanol (95%)
Sealed blood capillary tube
Sterile test tubes (10 mm × 40 mm)

PROCEDURE

Building Your Model of DNA

The following four pages contain diagrams representing purine and pyrimidine base pairs.

1. Remove the pages of base pairs and paste them on a piece of poster board or other stiff cardboard with rubber cement.

2. Cut out each base pair along the dark border. You should have a total of 20 base pairs when finished.

3. Use a crayon or felt-tipped marker and color each of the bases according to the following color scheme.
   - Pyrimidines (single ring)
     cytosine: red   thymine: blue
   - Purines (double ring)
     adenine: green   guanine: orange
   (You may use other colors; just remember that a particular base must always be the same color.)

4. Using a hole punch or cork borer, remove the circle in the center of the base pairs.

5. Stack the base pairs so that all of the base names are facing in the same direction and set them aside.

6. Cut out the nine strips of paper containing the diagrams of the phosphate (PO_4) molecules.

7. Using a razor blade, cut along each dotted line to make a slit in the paper band.

8. Tape each strip to the end of another strip with just enough overlap to keep the slits evenly spaced.

9. Select at random one of the cutout base pairs.

Notice that the base pairs are either adenine/thymine, thymine/adenine or cytosine/guanine, guanine/cytosine. Each base is held to its complementary base by hydrogen bonds (see Figure 22.2). Bonded to each base of the base pairs is a five-carbon sugar, deoxyribose (see Figure 22.3). This bond is formed between the nitrogen of the base and the number-one carbon of the sugar. The sugar molecule is in the form of a ring, and the sugar on the left side shows carbon number five bonded to a phosphate molecule (see Figure 22.4) of the side chain. The bond from the number-five carbon is drawn with a solid line to indicate that it will extend up to the phosphate molecule in the assembled model. The number-three carbon of the same sugar is bonded to another phosphate molecule. The bond from this carbon is represented by a dashed line to indicate that it extends down to the phosphate molecule in the assembled model. On the right side of the cutout, note that the sugar molecule is inverted and that the bonding is reversed.

When Watson and Crick determined the structure for the DNA molecule they relied heavily on x-ray diffraction studies by Wilkins and Franklin. X-ray diffraction photographs showed the molecule to be a double helix (see Figure 22.5). You can build a similar model in this configuration with your cutout materials. In the DNA molecule that you will build, the deoxyribose sugar and base molecules are already joined. The phosphate portion of the model is symbolized by the repeating units, PO_4^{-1} on the 11-inch strips of paper.

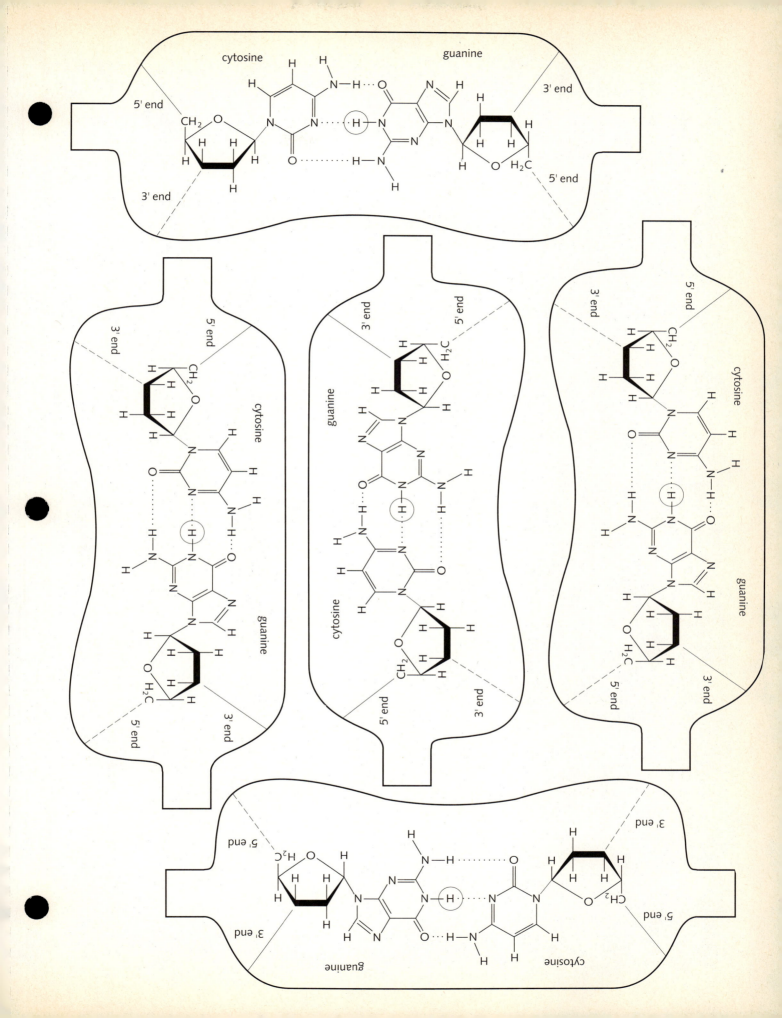

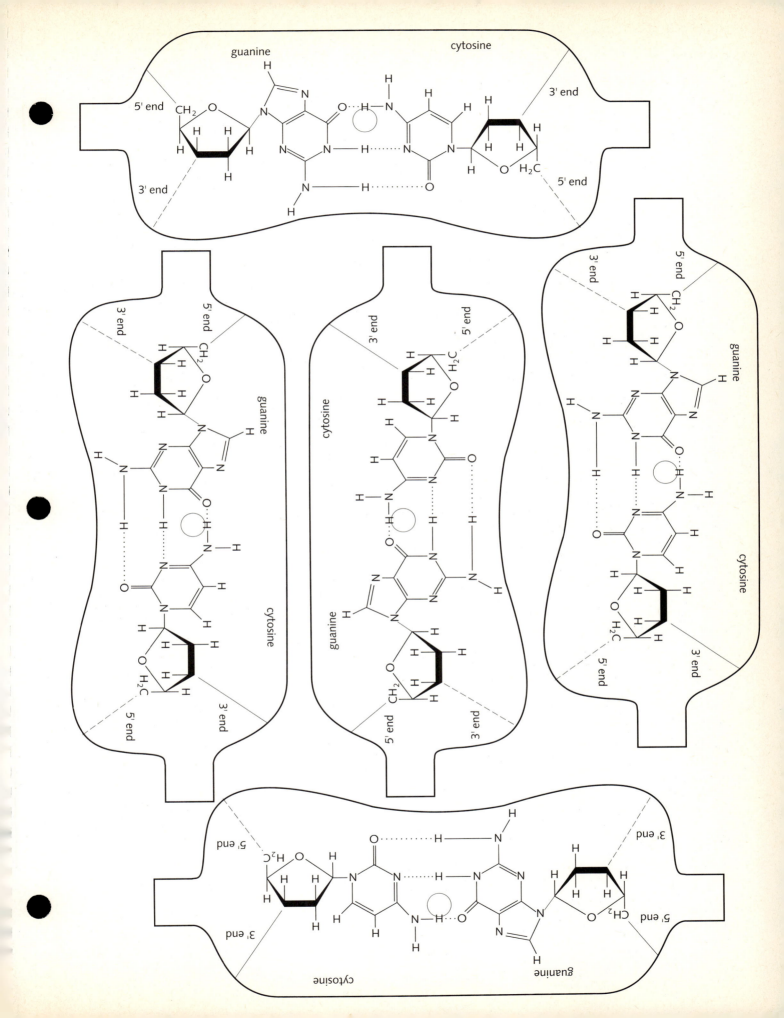

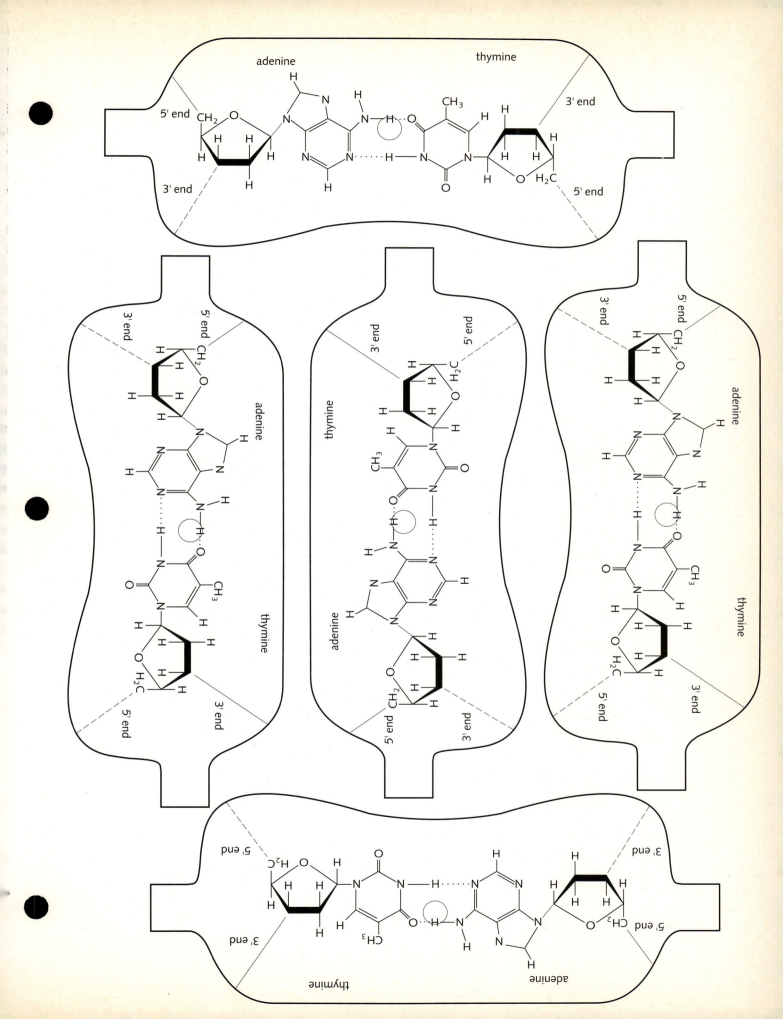

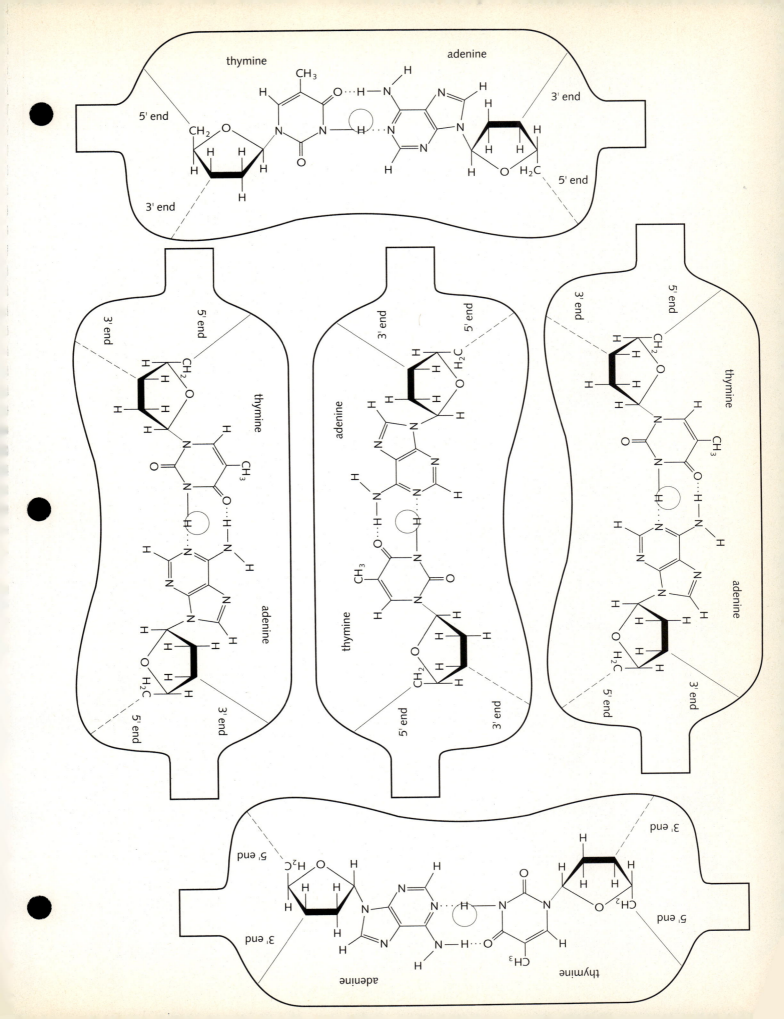

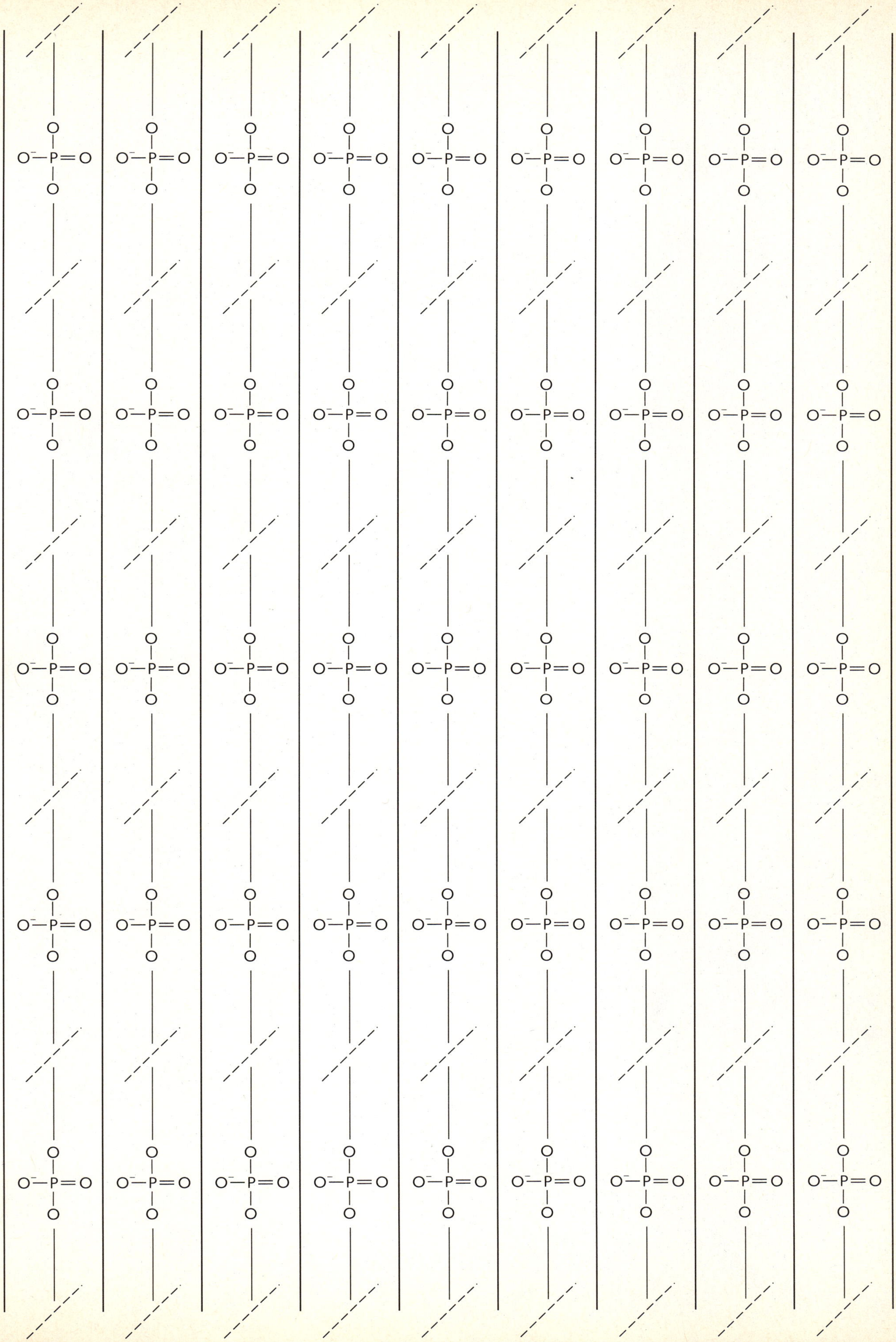

Figure 22.2 Hydrogen bonds between base pairs in DNA

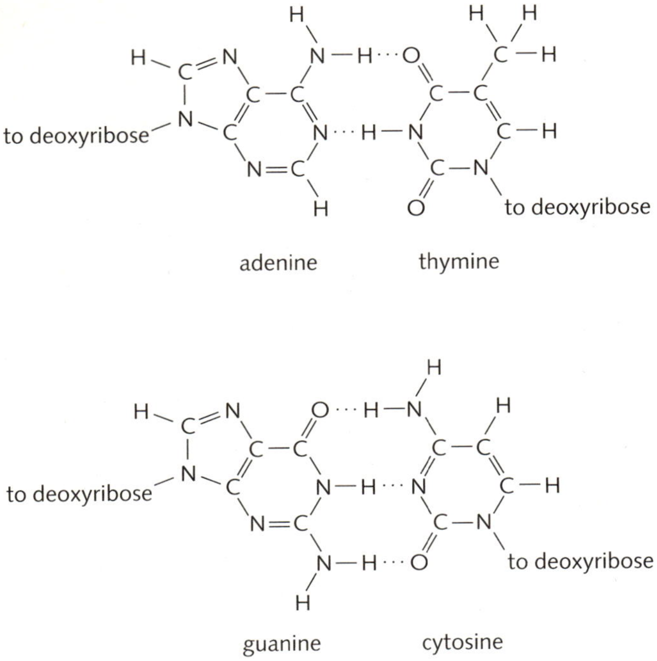

adenine   thymine

guanine   cytosine

Figure 22.3 Simplified structural formula of deoxyribose

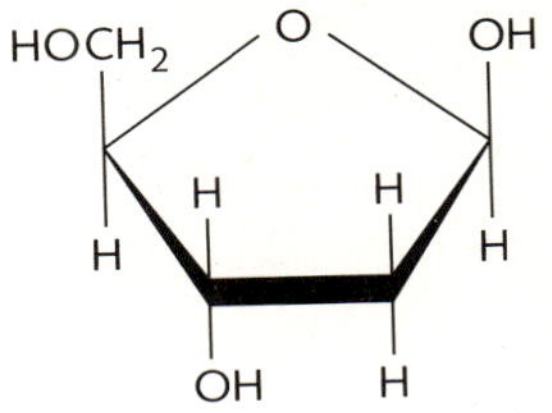

Figure 22.4 Structural formula of phosphate group in DNA

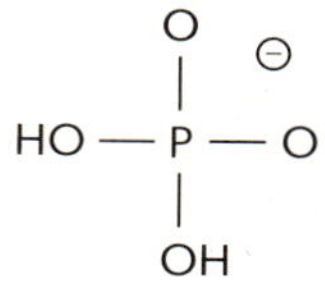

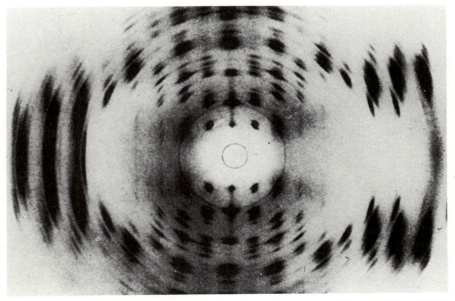

Figure 22.5 X-ray diffraction of DNA

1. Obtain a stand consisting of a base and a 24-inch-high dowel.

2. Wrap a piece of one-inch-wide masking tape around the dowel 2–3 times just above the point where it joins the base.

3. Select a DNA base pair at random and slide it on the dowel through the hole in the base pair until it rests on top of the masking tape.

4. Wrap a second piece of tape above this base pair and add a second base pair on top of this tape (see Figure 22.6). Continue this process until all 20 base pairs are assembled on the dowel. The one-inch masking tape will keep the base pairs evenly separated.

5. Starting with the bottom base pair, insert one of the tabs of the base pair into the first slot of the side chain (see Figure 22.7).

6. Rotate the next base pair above to allow its tab to fit into the next slot of the side chain. Continue until the phosphate strip has been attached to one side of the model. Repeat this process on the other side of the model.

If you have assembled the model properly, it should assume the shape of a double helix and look like the one in Figure 22.8.

7. Turn to the Evaluation and complete Questions 1–4.

Figure 22.6 Building a model of DNA

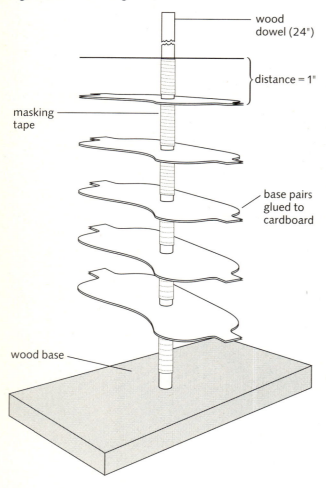

wood dowel (24")

distance = 1"

masking tape

base pairs glued to cardboard

wood base

Figure 22.7 DNA model close-up

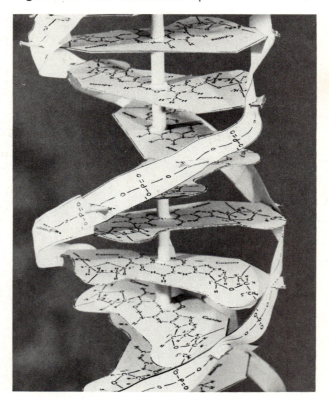

Figure 22.8 Finished DNA model

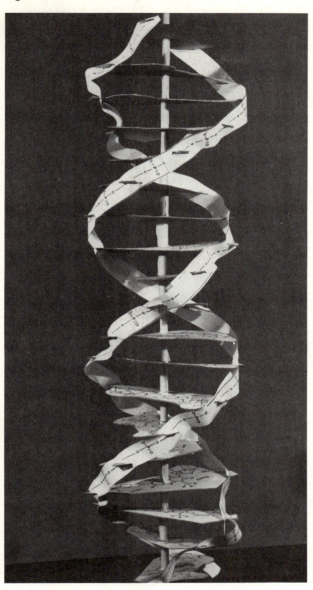

The next activity will be to determine the amino acid sequence represented by your DNA model. Before doing this activity you should read the section in your textbook about the genetic code and protein synthesis.

1. Begin with the base at the top right side of your molecule and record the first letter of its name (A, T, G, C) in the first blank space in Question 4 in the Evaluation. Continue down this side of the molecule, in the 5' carbon to 3' carbon direction, recording each base until you have entered 18 bases. When you are finished, complete the remaining parts of Question 4 in the Evaluation.

The Isolation of DNA

The following procedure enables you to isolate a relatively pure extract of DNA, as done in a research laboratory. You will work with mouse tails since they are relatively easy to obtain in the fresh state and eliminate the need to sacrifice the animal.

Your instructor has clipped approximately a half inch from the tail of a mouse and frozen it immediately to inactivate the enzymes (endogenous nucleases) that would digest the DNA.

1. Obtain a piece of frozen tail tissue and place it in a sterile polypropylene tube with a cap.

2. With a sterile 1-mL pipette add 0.7 mL (700 μL) of sterile buffer solution to the polypropylene tube. This buffer will maintain an alkaline condition in which DNA remains stable. In addition, the buffer contains a **chelating agent** that ties up metal ions needed to activate the nucleases, preventing the breakdown of DNA. A second compound in the buffer assists **lysis** (disruption of cell membranes).

3. Add 0.035 mL (35 μL) of 10 mg/mL proteinase K solution to this tube with a sterile pipette. Mix the contents with a vortex stirrer or by rolling the tube between your hands.

4. Incubate this tube in a water bath at 55°C for 8 hours or overnight. During this incubation the proteinase will digest the protein in the tail.

(The above steps may have been done by your instructor before class. In this case, you will be provided with a test tube of digested tail tissue.)

5. To the tube containing the digested tail tissue, add an equal volume of phenol and mix completely with a vortex stirrer or by rolling between your hands.

6. Centrifuge for 5 minutes at 3000 rpm (minimum) to separate into distinct layers. The DNA will remain in the top layer. The proteins are rendered insoluble by the phenol and are found in the bottom layer.

7. Decant the top aqueous layer into a sterile polypropylene tube. Discard the lower phenol layer.

8. Add an equal volume of chloroform-isoamyl alcohol (24:1) to the aqueous layer to separate out any remaining proteins. Mix completely with a vortex stirrer or by rolling between the hands.

9. Separate by centrifuging at 1500 g (gravity) for 2 minutes.

10. Decant the aqueous layer that contains the DNA into a clear polystyrene tube.

11. Add a quantity of 95% ethanol that is twice the volume of the liquid in the test tube, cap the tube, and gently invert to mix the layers. The DNA may be seen as a white, stringy precipitate visible throughout the solution.

12. Remove the DNA by inserting a thin sealed blood capillary tube into the polystyrene tube and twirling gently (see Figure 22.9). The DNA should attach to the glass tube. Examine the DNA carefully as it wraps around the tube.

13. Discuss the observed characteristics of DNA with your laboratory partner. Answer Question 5 in the Evaluation.

Figure 22.9 White, stringy DNA on glass tube

MOLECULAR STRUCTURE OF NUCLEIC ACIDS; A STRUCTURE FOR DEOXYRIBOSE NUCLEIC ACID

J. D. Watson
F. H. C. Crick
Medical Research Council Unit for the Study of the Molecular Structure of Biological Systems
Cavendish Physical Laboratory
Cambridge University
Cambridge, England

We wish to suggest a structure for the salt of deoxyribose nucleic acid (D.N.A.). This structure has novel features which are of considerable biological interest.

A structure for nucleic acid has already been proposed by Pauling and Corey.[1] They kindly made their manuscript available to us in advance of publication. Their model consists of three intertwined chains, with the phosphates near the fibre axis, and the bases on the outside. In our opinion, this structure is unsatisfactory for two reasons: (1) We believe that the material which gives the X-ray diagrams is the salt, not the free acid. Without the acidic hydrogen atoms it is not clear what forces would hold the structure together, especially as the negatively charged phosphates near the axis will repel each other. (2) Some of the van der Waals distances appear to be too small.

Another three-chain structure has also been suggested by Fraser (in the press). In his model the phosphates are on the outside and the bases on the inside, linked together by hydrogen bonds. This structure as described is rather ill-defined, and for this reason we shall not comment on it.

We wish to put forward a radically different structure for the salt of deoxyribose nucleic acid. This structure has two helical chains each coiled round the same axis (see diagram). We have made the usual chemical assumptions, namely, that each chain consists of phosphate diester groups joining β-D-deoxyribofuranose residues with 3',5' linkages. The two chains (but not their bases) are related by a dyad perpendicular to the fibre axis. Both chains follow right-handed helices, but owing to the dyad the sequences of the atoms in the two chains run in opposite directions. Each chain loosely resembles Furberg's[2] model No. 1; that is, the bases are on the inside of the helix and the phosphates on the outside. The configuration of the sugar and the atoms near it is

Reprinted by permission from *Nature*, Vol. 171, pp. 737–738. Copyright © 1953 by Macmillan Magazines Ltd.

We are much indebted to Dr. Jerry Donohue for constant advice and criticism, especially on interatomic distances. We have also been stimulated by a knowledge of the general nature of the unpublished experimental results and ideas of Dr. M. H. F. Wilkins, Dr. R. E. Franklin and their co-workers at King's College, London. One of us (J. D. W.) has been aided by a fellowship from the National Foundation for Infantile Paralysis.

1. Pauling, L., and Corey, R. B., *Nature* **171**:346 (1953); *Proc. U.S. Nat. Acad. Sci.* **39**:84 (1953).
2. Furberg, S., *Acta Chem. Scand.* **6**:634 (1952).

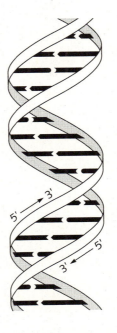

This figure is purely diagrammatic. The two ribbons symbolize the two phosphate-sugar chains, and the horizontal rods the pairs of bases holding the chains together. The vertical line marks the fibre axis.

close to Furberg's 'standard configuration,' the sugar being roughly perpendicular to the attached base. There is a residue on each chain every 3.4 A. in the z-direction. We have assumed an angle of 36° between adjacent residues in the same chain, so that the structure repeats after 10 residues on each chain, that is, after 34 A. The distance of a phosphorus atom from the fibre axis is 10 A. As the phosphates are on the outside, cations have easy access to them.

The structure is an open one, and its water content is rather high. At lower water contents we would expect the bases to tilt so that the structure could become more compact.

The novel feature of the structure is the manner in which the two chains are held together by the purine and pyrimidine bases. The planes of the bases are perpendicular to the fibre axis. They are joined together in pairs, a single base from one chain being hydrogen-bonded to a single base from the other chain, so that the two lie side by side with identical z-co-ordinates. One of the pair must be a purine and the other a pyrimidine for bonding to occur. The hydrogen bonds are made as follows: purine position 1 to pyrimidine position 1; purine position 6 to pyrimidine position 6.

If it is assumed that the bases only occur in the structure in the most plausible tautomeric forms (that is, with the keto rather than the enol configurations) it is found that only specific pairs of bases can bond together. These pairs are: adenine (purine) with thymine (pyrimidine), and guanine (purine) with cytosine (pyrimidine).

In other words, if an adenine forms one member of a pair, on either chain, then on these assumptions the other member must be thymine; similarly for guanine and cytosine. The sequence of bases on a single chain does not appear to be restricted in any way. However, if only specific pairs of bases can be formed, it follows that if the sequence of bases on one chain is given, then the sequence on the other chain is automatically determined.

It has been found experimentally[3,4] that the ratio of the amounts of adenine to thymine, and the ratio of guanine to cytosine, are always very close to unity for deoxyribose nucleic acid.

It is probably impossible to build this structure with a ribose sugar in place of the deoxyribose, as the extra oxygen atom would make too close a van der Waals contact.

The previously published X-ray data[5,6] on deoxyribose nucleic acid are insufficient for a rigorous test of our structure. So far as we can tell, it is roughly compatible with the experimental data, but it must be regarded as unproved until it has been checked against more exact results. Some of these are given in the following communications. We were not aware of the details of the results presented there when we devised our structure, which rests mainly though not entirely on published experimental data and stereochemical arguments.

It has not escaped our notice that the specific pairing we have postulated immediately suggests a possible copying mechanism for the genetic material.

Full details of the structure, including the conditions assumed in building it, together with a set of co-ordinates for the atoms, will be published elsewhere.

3. Chargaff, E., for references see Zamenhof, S., Brawerman, G., and Chargaff, E., *Biochim. et Biophys. Acta* **9**:402 (1952).
4. Wyatt, G. R., *J. Gen. Physiol.* **36**:201 (1952).

5. Astbury, W. T., Symp. Soc. Exp. Biol. 1, *Nucleic Acid*, **66** (Camb. Univ. Press, 1947).
6. Wilkins, M. H. F., and Randall, J. T., *Biochim. et Biophys. Acta* **10**:192 (1953).

EVALUATION 22

Model Building and the Isolation of DNA

1. Study your model carefully and list the appropriate bases in the blanks below. Be sure to pair them correctly.

	Pyrimidines	Number of hydrogen bonds joining bases	Purines
Base names			
Number of rings in base			

2. Knowing the role DNA has in the life of an organism, speculate on the reason that the double helical shape is advantageous. Consult your textbook as you answer this question.

3. Follow one full turn of the molecule. Even though this is a paper rendition of the DNA molecule, you should be able to count the number of bases in one full turn.

   (a) How many are there?

   (b) Knowing the number of bases in one full turn, and that a full turn contains 360 degrees, calculate the angle that each base is offset from each adjacent base.

4. (a) After you have recorded the DNA bases (read from 5′ to 3′ end) in the space below, record on the second line the mRNA bases that are complementary to these DNA bases. Each group of three bases in the RNA molecule is known as a **codon** and will code for a specific amino acid.

DNA base sequence	__ __ __	__ __ __	__ __ __	__ __ __	__ __ __	__ __ __
RNA base sequence (codons)	__ __ __	__ __ __	__ __ __	__ __ __	__ __ __	__ __ __
Amino acid sequence	_____	_____	_____	_____	_____	_____

(b) To determine the specific amino acid sequence specified by your model, use the genetic code in Table 22.1. To read the table, begin at the left column and find the first base in the codon. Next, find the second base in the codon across the top. Finally, locate the third base in the codon in the right column. Where these three base letters intersect in the table, you will find the name for the amino acid that is coded for by this sequence of bases. Repeat this process for each codon and record the abbreviation of the amino acid in the appropriate space.

Table 22.1 The genetic code (mRNA codons)

First position (5′ end)	Second position				Third position (3′ end)
	U	C	A	G	
U	phenylalanine	serine	tyrosine	cysteine	U
	phenylalanine	serine	tyrosine	cysteine	C
	leucine	serine	stop*	stop*	A
	leucine	serine	stop*	tryptophan	G
C	leucine	proline	histidine	arginine	U
	leucine	proline	histidine	arginine	C
	leucine	proline	glutamine	arginine	A
	leucine	proline	glutamine	arginine	G
A	isoleucine	threonine	asparagine	serine	U
	isoleucine	threonine	asparagine	serine	C
	isoleucine	threonine	lysine	arginine	A
	(start) methionine	threonine	lysine	arginine	G
G	valine	alanine	aspartate	glycine	U
	valine	alanine	aspartate	glycine	C
	valine	alanine	glutamate	glycine	A
	valine	alanine	glutamate	glycine	G

*The codons UAA, UAG, and UGA signal the end of the message. They do not code for any amino acid.

(c) A segment of DNA that contains 45 bases may code for as many as
_____ amino acids.

5. Explain why each of the following processes is important in the isolation
of DNA.

(a) Digestion

(b) Precipitation

(c) Centrifugation

(d) Deproteinizing

6. Is the shape of the DNA model that you constructed compatible with the
characteristics of the DNA you isolated? Explain.

The Cell

OBJECTIVES

At the end of this laboratory activity, you should be able to:

- prepare wet mount slides of selected tissues.
- describe cyclosis.
- identify cell walls, nuclei, chloroplasts, and mitochondria.
- discuss the structure/function relationships found in selected cells.

INTRODUCTION

The **cell theory** states that the cell is the basic unit of life and that all organisms are composed of cells and materials produced by cells. In addition, the theory states that all cells come from previously existing cells.

The cell theory was not developed by any one person or demonstrated by any one single experiment. Rather, dozens of scientists contributed to its development over several centuries. One of the first scientists to contribute to the cell theory was the English scientist Robert Hooke. In 1662, Hooke was appointed curator of the Royal Society of London, a leading forum for the exchange of scientific ideas at the time. As curator Hooke demonstrated experiments to the members, including experiments using his improved compound microscope. As you may recall from Laboratory Activity 2, Hooke used the microscope to study cork tissue and observed that the tissue was composed of tiny units he named cells. In addition, Hooke observed the presence of pith cells, the soft tissue present in the center of certain plant stems. You will duplicate Hooke's observation in one of the activities below.

Ten years after Hooke's microscopic examinations of cork tissues, the Royal Society began to receive reports and drawings from the Dutch microscopist Anton van Leeuwenhoek. One such report was on his attempt to discover whether the spiciness of pepper could be viewed microscopically. van Leeuwenhoek soaked ordinary peppercorns in fresh water and observed them under the microscope. While the pepper itself contained no surprises when viewed microscopically, van Leeuwenhoek was amazed by the plethora of tiny organisms that swarmed in a single drop of the peppercorn water. Like many scientists responsible for important discoveries, van Leeuwenhoek had stumbled on the existence of bacteria and protozoa while searching for something entirely different. This discovery was key to the development of the cell theory because it showed that even the smallest of living forms are also cellular.

While the development of the cell theory may be attributed to many scientists, certain researchers deserve credit for major discoveries that provided important insights. The work of Jean Baptiste de Lamarck merits such credit as he was responsible for directing the study of cells to their interior. In 1809, while writing his *Philosophie Zoologique*, Lamarck stated, "Every living body is essentially a mass of cellular tissue in which more or less complex fluids move more or less rapidly." Robert Brown (1773–1858), a physician-botanist, was one of many scientists who continued this investigation of the cell's interior. Brown examined many types of what he called "vegetable cells," noticing that every cell had a clear spot within the cytoplasm. Brown called this spot the *nucleus*, and is credited with its discovery.

While the quality of the light microscope was improved and refined, scientists continued to be confined by the limits of light resolution. However, by the early 1950s, the electron microscope was being used for the study of cells. This allowed the viewer to see structures smaller than 1 nm, giving biologists new insight into cell structure.

The following activities will enable you to duplicate some of the hallmark investigations in the development of the cell theory and familiarize you with several different kinds of cells. In addition, you will study an electron micrograph to observe structures that are beyond the resolution of the light microscope. Make sure you have read carefully the section of your text pertaining to the cell theory and that you are familiar with the name, shape, and function of the major cell organelles labeled in Figure 23.1.

MATERIALS

Razor blade
Plant stem
Dissecting microscope
Microscope slides
Dissecting needles
Cover slips
Compound microscope
Peppercorn water
Celery stalk
Tweezers
Sucrose solution (5%)
Living moss plant
Raw beef
Methylene blue stain

PROCEDURE

Hooke was amazed when he discovered the cellular makeup of cork tissue. He continued his investigations with other plant tissues. For example, he examined the pith cells (parenchyma) of the elder, carrot, burdock, and reed. These pith cells occupy the center of plant stems and are used for storage. Interestingly, pith is soft like cork and may have been the reason Hooke investigated this type of tissue at that time.

Figure 23.1 An animal cell
(a) Electron micrograph

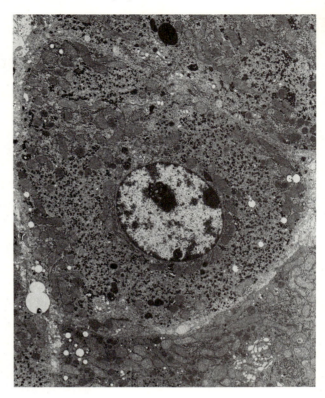

Pith Cells

1. With a razor blade, cut off about 1–2 cm of stem tissue from the stems provided. Examine the cross section of the stem to identify the pith located in the central portion of the stem. It will usually appear white and soft.

2. View the pith cells in the stem cross section under a dissecting microscope.

3. Place a drop of water on a microscope slide. Remove a small portion of this pith tissue with a dissecting needle and place it in the drop of water. Seal with a cover slip.

4. Using the compound microscope, examine under low and high power and prepare a drawing of these cells in Question 1 in the Evaluation.

(b) Illustration

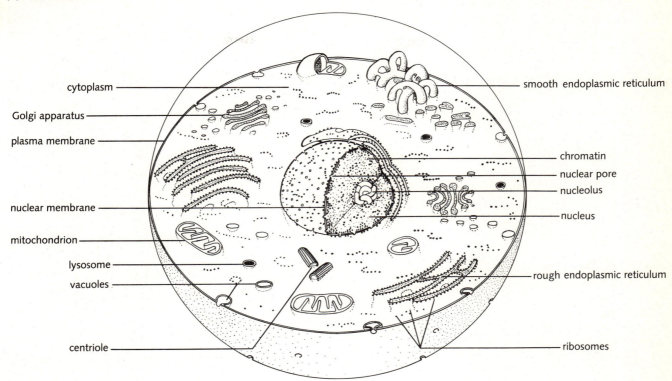

cytoplasm

Golgi apparatus

plasma membrane

nuclear membrane

mitochondrion

lysosome

vacuoles

centriole

smooth endoplasmic reticulum

chromatin

nuclear pore

nucleolus

nucleus

rough endoplasmic reticulum

ribosomes

Peppercorn Water

The following activity will allow you to see what Leeuwenhoek saw as he studied peppercorns soaked in water. (What he saw was considered miraculous in the 1600s!) Your instructor will provide you with peppercorns that have soaked in water for several weeks.

1. Prepare a wet mount of a drop of this "peppercorn water."

2. Examine under the light microscope, using high power.

3. Make a drawing of your observations in Question 2 in the Evaluation. The bacteria will appear as extremely tiny rods and spheres. Protozoans, even though they are composed of single cells, will be many times larger.

Celery Epidermis Cells

Cytoplasmic streaming and several cell organelles can be seen by careful examination of living celery epidermal cells.

1. Using a razor blade, make a transverse cut across two strands on the surface of a 3-cm-long portion of a celery stalk. Make a second transverse cut across the same two strands about 1 cm from the first.

2. Using tweezers, secure both strings at once and carefully pull them away from the stalk, allowing the thin layer of epidermis between the strands to remain in place.

3. Place the tissue in several drops of 5% sucrose on a microscope slide and cover with a cover slip.

4. Examine the cells of the epidermis under low and high power. You should be able to see the **chloroplasts** (small green bodies) moving around the perimeter of the cell. The streaming cytoplasm (**cyclosis**) will be evident if the cells are alive.

5. Locate a clear **nucleus**. It is difficult to see because light can pass right through it and because it might be covered with other organelles, but if you examine enough cells you will have success.

6. Locate the small vibrating rods or spheres in a clear area of cytoplasm. These are the **mitochondria**. These extremely small bodies contain many enzymes on their membranes, which produce ATP as an important step in cellular respiration.

7. Diagram a cell showing cyclosis in Question 4 of the Evaluation.

Moss Leaf Cells

As you observe these cells, note what organelles are present and relate them to the function of the tissue being studied. Your textbook will contain valuable information concerning the structure of plant cells, which will help you with this task.

1. With tweezers obtain a single leaf from a moss plant. Normally these leaves are only one cell thick, making this an ideal plant tissue to study.

2. Place the moss leaf in a drop of water on a microscope slide and cover with a cover slip.

3. Observe and diagram a typical cell of this leaf in Question 5 in the Evaluation.

Striated Muscle in Beef

The tissues you have studied in the previous activities have illustrated the structure of plant tissue. The next activity will introduce you to an animal tissue.

1. In a drop of water on a clean slide, tease apart a tiny portion of raw beef muscle tissue using two dissecting needles.

2. Place a few of the smallest visible strands of muscle from this slide in a drop of water on another slide. Add a drop of methylene blue stain and cover with a cover slip.

3. Observe under high power and make a sketch of your observation in Question 6 in the Evaluation. Along the length of the muscle fiber (cell) you should see alternating light and dark bands. These bands are composed of two proteins (actin and myosin) involved in muscle contraction. Read the section on these two proteins in your text.

Electron Microscope

The electron micrograph shown in Figure 23.1(a) reveals several cellular structures not seen under the light microscope.

1. Study the micrograph carefully, using the outline sketch in Figure 23.1(b) to help you find the organelles.

2. Turn to Question 7 in the Evaluation and label the electron micrograph provided.

EVALUATION 23

The Cell

1. Make a diagram of pith cells. Label the cell wall and cytoplasm.

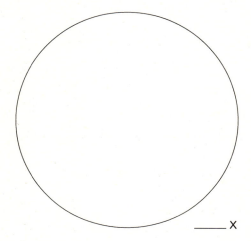

_____ X

2. Draw a representative of each organism found in the "peppercorn water." Identify each form as specifically as possible.

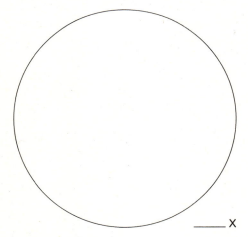

_____ X

3. Your "peppercorn water" was prepared by using sterile water and dry peppercorns. Where did the living organisms come from?

4. Prepare a drawing of the celery epidermal cells. Label cell wall, cytoplasm, chloroplasts, and mitochondria. Show the direction of cyclosis with an arrow.

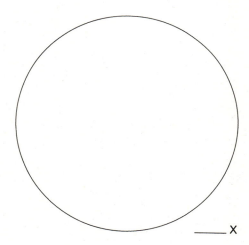

_____ X

5. Diagram a typical cell found in the moss leaf. Label any organelles that you are able to observe.

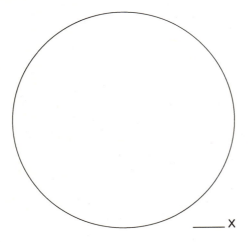

_____ X

6. Diagram and label a portion of a muscle fiber.

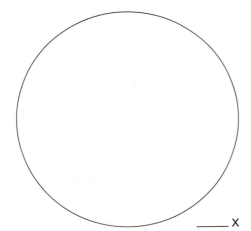

_____ X

7. Label the following electron micrograph.

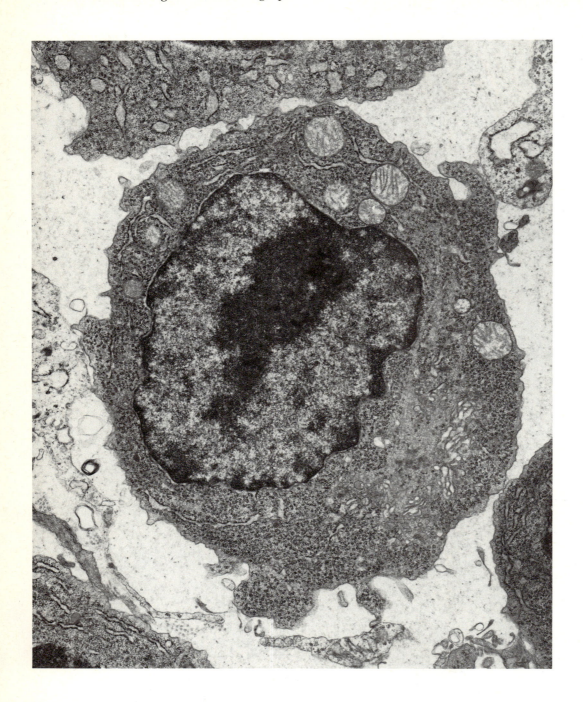

Cell Organelles:
Their Separation and Observation

OBJECTIVES

At the end of this laboratory activity, you should be able to:

- define the terms *filtrate*, *supernatant*, and *homogenate*.
- list the cell organelles studied in order of decreasing mass.
- state which cell organelles are most prevalent in carrot root and pea seeds and why.
- explain how the centrifuge and blender are used to separate cell components.

INTRODUCTION

As you have already observed in previous laboratory activities and from reading your textbook, cells are not perfectly homogeneous in their content. Within the cell boundary and suspended in the cytoplasm are numerous organelles such as the nucleus, plastids, mitochondria, starch grains, and vacuoles, to name a few. In order to study cells and their organelles, cell biologists have had to rely on certain techniques involving both biochemistry and the physical characteristics of cell parts, in addition to the microscope. Much of our information concerning cell ultrastructure and cell function has evolved from such studies. In this laboratory activity you will isolate and identify several of the cell organelles found in the carrot root and pea seed.

MATERIALS

Balance scale
Diced carrot root
Blender
Sucrose-buffer solution
Cheesecloth
Beakers (400 mL)
Centrifuge tubes
Centrifuge
Microscope slides
Cover slips
Compound microscope
Dropping bottle of iodine-potassium iodide stain (I$_2$KI)
Absorbent paper
Dropping bottle of methylene blue stain
Spatula
Pea seeds (soaked)
Test tubes (10 mm × 100 mm)
Tetrazolium chloride solution (TTZ)
Test tube rack
Pasteur pipette
Water bath (37°C)

PROCEDURE

The organelles that appear in the cells of the carrot root and pea seed will differ slightly in number and type. This is a reflection of the function of the plant organ in which these cells are found. Work in teams of four or more students to do the initial separations, but carry out all microscope observations individually or in pairs.

Carrot Root

1. Weigh out 10–15 grams of diced carrot. Place in a blender with 100 mL of sucrose-buffer solution and blend at high speed for one minute to prepare a **homogenate.** Clean blender thoroughly after use.

2. Filter this homogenate through four layers of cheesecloth and collect the **filtrate** in a beaker. The material in the cloth is the **residue** (see Figure 24.1).

3. Distribute filtrate equally between 2 15-mL centrifuge tubes and centrifuge at about 1300 g for 10 minutes. The second tube will balance the centrifuge and will also serve as a backup in case the first tube breaks.

4. While you are separating the filtrate, take a *very small* sample of the residue in the cheesecloth and make a wet mount by adding a drop of water and cover slip to a slide.

5. Mount the slide on the stage of your microscope and add a drop of iodine-potassium iodide stain (I$_2$KI) to one edge of the cover slip and touch the opposite side with a piece of absorbent paper. The I$_2$KI will move under the cover slip and stain the starch grains purple-black.

6. Add a drop of methylene blue stain to one edge of the cover slip and again touch the opposite side with a piece of absorbent paper. This dye will stain the cell walls.

Figure 24.1 Preparing carrot homogenate and filtrate

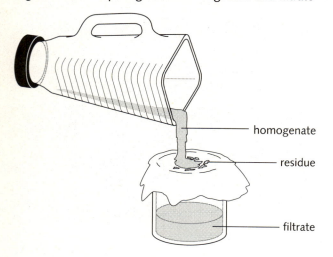

homogenate

residue

filtrate

7. Observe under low and high power and identify the following structures:
   - whole cells: have a continuous cell wall, rectangular in shape
   - cell fragments: pieces of cell walls
   - starch grains: stained purple-black and oval-shaped
   - vessel elements: these support the cell wall and appear like a stretched-out spring or coil
   - chromoplasts: appear as clumps and contain the carotenoid pigments that give the carrot its orange color
   Record your observations in Question 1 in the Evaluation.

8. Examine the sediment in the tubes from the centrifugation. The suspended particles will have settled to the bottom of the tube. The orange-tinted liquid that remains on top is the **supernatant.**

9. Decant (pour off) the supernatant from one of the tubes. Remove a *very small* amount of the sediment with a spatula and place it on a slide. Prepare a wet mount but use iodine-potassium iodide stain in place of water. Examine the slide carefully under low and high power.

10. Select a field that is representative of your slide and record your observations in Question 2 in the Evaluation.

Pea Seeds

1. Obtain 10–15 g of soaked pea seeds. Decant the water and add 100 mL of sucrose-buffer. Homogenize in the blender for 1 minute.

2. Filter the homogenate through four layers of cheesecloth and collect the filtrate in a beaker.

3. Centrifuge the filtrate in two centrifuge tubes as you did for the carrot separation.

4. While this separation is occurring, make a wet mount of the residue from the cheesecloth, using only a very small portion of this material.

5. Examine your slide under low and high power for unbroken cells, cell wall fragments, and starch grains. Stain as you did for the carrot and continue your observations. Record your observations in Question 4 in the Evaluation.

Centrifugation of the filtrate from the pea homogenate should result in several layers forming in the sediment of both tubes. The white layer consists of starch grains, the green layer contains chloroplasts, and the light brown layer contains nuclei. The top layer of liquid is the supernatant.

1. Observe both of these tubes and note how these layers are arranged. Record your observations in Question 5 in the Evaluation.

2. Reserve one of these tubes for the activity with mitochondria.

3. Decant the supernatant from the other tube. With a spatula remove a *very small* amount of sediment from the tube and make a wet mount. Use a drop of iodine-potassium iodide stain in place of water. Examine under low and high power of your microscope. Record your observations in Question 6 in the Evaluation.

Chemical Test for Mitochondria

Mitochondria are minute organelles that carry out cellular respiration. They are so small (about 0.5 μm $\times$ 3.0 μm) that they are not readily visible with the light microscope (see Figure 24.2). Because of their size, it is often easier to detect them by chemical means than by direct observation. During cellular respiration, glucose is degraded to pyruvate in the cytoplasm. After the pyruvate is produced, it enters the mitochondria, where it is broken down into carbon dioxide and water. Hydrogen ions are released during this process. If the compound tetrazolium chloride (TTZ) is added to a suspension and mitochondria are present, the hydrogen ions will combine with the TTZ and a pink color will develop. If the mitochondria are inactive or are not present in the suspension, no color will develop. The following test will determine whether mitochondria are in the suspension or in the sediment of the pea seed filtrate.

1. Follow Figure 24.3 closely as you do this activity.

2. Obtain three test tubes (10 mm $\times$ 100 mm). Mark them #1, #2, and #3.

3. Take the centrifuge tube saved from the previous activity and add the supernatant from it to tube #1 until it is about half full. Discard the remaining supernatant but *retain* the sediment.

Figure 24.2 Electron micrograph of mitochondria

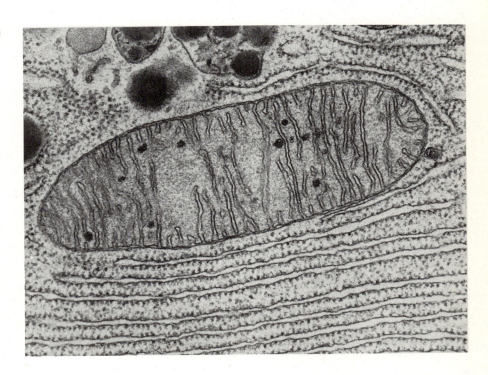

4. Carefully add 2 mL of TTZ solution to the top of the supernatant in tube #1 by trickling it down the side of the tube so that it forms a layer on the surface. Place this tube in a test tube rack.

5. Add 1 mL of sucrose buffer to the sediment in the centrifuge tube. Resuspend the sediment by placing the tip of a Pasteur pipette in the bottom of the tube and then squeezing gently until the sediment is in suspension. Add this suspension to test tube #2 until it is half full. Add TTZ solution as you did in the tube with the supernatant. Place this tube in a test tube rack.

6. Fill test tube #3 half full with sucrose buffer and add TTZ solution as you did before.

7. Incubate the three tubes in a warm water bath (37°C) for approximately 30 minutes or until a pink color develops in one or more of the tubes. Record your results in Question 9 in the Evaluation.

Figure 24.3   Test for mitochondria

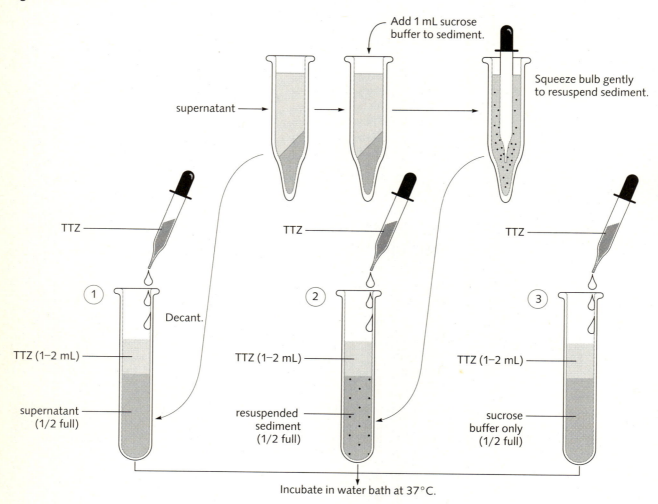

Add 1 mL sucrose buffer to sediment.

supernatant

Squeeze bulb gently to resuspend sediment.

TTZ

TTZ

TTZ

① Decant.

②

③

TTZ (1–2 mL)

TTZ (1–2 mL)

TTZ (1–2 mL)

supernatant (1/2 full)

resuspended sediment (1/2 full)

sucrose buffer only (1/2 full)

Incubate in water bath at 37°C.

EVALUATION 24

Cell Organelles: Their Separation and Observation

1. Diagram a representative field of the residue from the carrot homogenate. Label all of the organelles that you can identify.

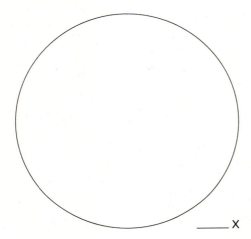

_____ X

2. Diagram a representative field from the carrot sediment. Again, label all of the parts that you can identifiy.

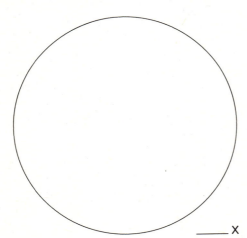

_____ X

3. (a) Compare the cellular material in the sediment with the residue.
   Explain any differences that you see.

   (b) The chromoplasts from the carrot cells have been retained in the
   residue or have settled in the sediment of the centrifuge tube, yet the
   supernatant remains orange. Explain why this occurred.

4. Diagram a representative field from the slide made with the residue from
   the pea homogenate. Label all of the organelles that you can identify.

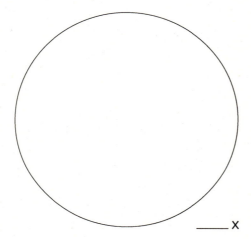

_____ X

5. (a) Which physical property of the nuclei, chloroplasts, and starch grains
   causes them to form layers in the sediment?

   (b) Rank each of these organelles in order of decreasing mass.

   _____ _____ _____

   (c) Where would cell organelles of less mass be found?

6. Diagram a representative field of the pea sediment from the centrifuge tube. Label all of the parts that you can identify.

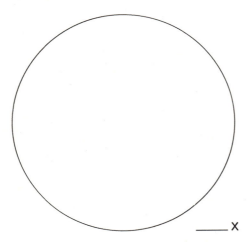

_____ X

7. How does the material on this slide differ from that found on the previous slide of the residue?

8. Biologists continuously look for the relationship between structure and function in organisms. As you observed cells from the carrot and pea, you should have noticed that the vessel elements supporting the cell walls were more prevalent in the carrot than in the pea and that the concentration of starch was higher in the pea. How can you account for these differences?

9. (a) Which one of the test tubes used in the test for mitochondria turned pink?

(b) Why was a third tube with only the sucrose-buffer included in this activity?

Transport Across the Cell Membrane

OBJECTIVES

At the end of this laboratory activity, you should be able to:

- define and give at least one example of diffusion.
- define the processes of dialysis and osmosis.
- explain the effect of solute concentration on the rate of osmosis.
- explain how the changing relationship between surface area and volume affects the ability of a cell to transport materials.

INTRODUCTION

Of the multitude of activities that must go on to maintain a cell in the proper **homeostatic condition** (stable internal environment), none is more important than the control of the movement of materials in and out of the cell by the cell or plasma membrane. Cell membranes are delicate boundaries that not only separate the cell contents from its environment but also have the remarkable ability to distinguish between different substances, allowing some to move through easily while others pass slowly or not at all. A membrane behaving in this manner is called **selectively permeable.**

There are many types of transport that occur between the cell content and its environment. Your textbook describes them in detail, and you should review this section before doing this laboratory activity. This activity investigates the type of transport generally known as **passive transport** and will

deal specifically with diffusion and osmosis. This type of transport is called passive because the source of energy for the movement of particles comes from the thermal energy of the cell's environment and not from energy stored in the chemical bonds of molecules such as adenosine triphosphate.

These activities are a type of modeling that is frequently done in science laboratories to demonstrate a process, since it is difficult to bring many real-life situations into the laboratory. Models are only representations of nature, and their performance must be interpreted with some degree of caution. The models of diffusion and osmosis you will use today, however, closely parallel the real world of the cell, and your results should present a very accurate representation of what happens as certain substances pass through the cell membrane.

Diffusion

Diffusion is the random movement of particles such as molecules or ions from a region of initial high concentration to a region of initial low concentration. Movement is brought about by the existing kinetic energy of the particles. The rate at which particles diffuse is a function of their size, shape, mass, electric charge, and the temperature of the environment. As diffusion occurs, the particles move in a straight line until they are deflected by another particle or by the side of the container. The particles then rebound and start moving in another direction. This movement continues until the particles have dispersed from their point of origin (high concentration) and finally distribute themselves uniformly throughout the available space.

Osmosis and Dialysis

Osmosis is a special case of diffusion involving the movement of water through a selectively permeable membrane to an area of lower water concentration. For example, pure water obviously has a higher water concentration than a solution of salt water. It may seem unusual to think of water as being concentrated since we usually use this concept only with reference to solutes (dissolved substances), but it can legitimately be applied to water as we are doing here. If the two liquids are separated by a selectively permeable membrane, the net movement of water is from the area of high water concentration (pure water) to the area of low water concentration (the saline solution). Although water molecules can move in either direction across the membrane, the net movement is to the area of lower water concentration. As water moves by osmosis into the area of lower water concentration, the pressure that builds up is called **osmotic pressure.** Eventually the osmotic pressure is sufficient to force water back across the membrane so that the system reaches an equilibrium.

The importance of osmosis to cells is difficult to overestimate. Whenever a tissue is exposed to a solution in which the solute concentration is higher than that in the cell's cytoplasm, water will leave the cells and the tissue will become soft and spongy. A piece of potato or carrot placed in a dish of salt water loses its stiffness. Conversely, when vegetables begin to wilt, spraying or soaking with fresh water will usually restore the cells to their former crisp state.

Solute particles may also move from an area of high concentration to an area of low concentration across a selectively permeable membrane. This process is called **dialysis** and will be studied along with diffusion and osmosis in this laboratory activity.

Cell Size and Rate of Diffusion

Although it may seem a contradiction in terms, one of the most noticeable characteristics of cells is their small size. Most eukaryotic cells range from 10–100 μm; prokaryotic cells may be considerably smaller. Few cells are visible without a microscope, and you never need fear being engulfed by a ten-ton ameba in the campus parking lot! There are exceptions to this generalization about small size: the yolk of a bird egg is technically a cell, but it is mostly concentrated nutrient, not cytoplasm; nerve cells may grow to a meter or more in length, but you still need a microscope to see them because they are so thin. Organisms grow due to the *addition* of cells, not because their individual cells grow larger.

Cells thrive on the exchange of materials with their environment, and this exchange must occur across the cell membrane. The membrane is a surface that allows materials to pass in or out of the cell's interior. It is believed that the relationship between surface area and volume in a cell is crucial to its survival. You will explore this relationship at the end of this activity.

MATERIALS

Culture bowls (4 in.)
Thermometers
Ice cubes (4° C)
Water (50° C)
Petri dishes
Potassium permanganate crystals ($KMnO_2$)
Dialysis tubing
Dental floss
Polypropylene closure
Pipettes (5 mL)
Tygon or rubber tubing
Ring stands and clamps
Sodium chloride solutions (5%, 10%, 20%, 30%)
Distilled water
Beakers (400 mL)
Marking pencil or pen
Starch solution (1%)
Dextrose solution (20%)
Iodine-potassium iodide (I_2KI) solution
Dropping bottles of Benedict's reagent
Agar blocks
Sodium hydroxide (NaOH) solution
Paper towels
Scalpel or spatula

PROCEDURE

Effect of Temperature on Rate of Diffusion

1. Take 2 4-inch culture bowls and fill one with ice and water (4°C) and the other with water at a temperature of about 50°C. Fill each bowl to the very top.

2. Place a petri dish top on each culture bowl and fill with water. Wait several minutes for the water in each dish to reach the temperature of the water in the culture bowl (see Figure 25.1).

Figure 25.1 Temperature and rate of diffusion

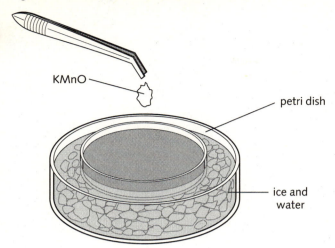

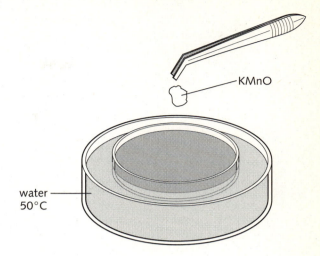

3. With a pair of tweezers, add a crystal of potassium permanganate ($KMnO_2$) to the center of each dish and observe carefully. Record your observations in Question 1 in the Evaluation.

➤ **CAUTION: Potassium permanganate is a strong oxidizing agent. Keep away from combustible materials. Handle only with tweezers, not your fingers.**

Rate of Osmosis and Solute Concentration

This activity demonstrates the effect of solute concentration on the rate of osmosis. Based on the previous discussion and on what you have read in your textbook about osmosis, you should be able to state a hypothesis predicting the outcome of the following experiment. After reading the directions below, turn to the Evaluation and complete Question 2 *before* setting up your experiment.

1. Obtain 5 pieces of dialysis tubing about 15 cm long. Hold the tubing under running water and rub between your fingers until the tubing separates. One end should be tied with dental floss or clamped with a polypropylene closure to make a bag.

2. Take a 5-mL pipette that has a piece of tygon or rubber tubing about 2 cm long attached to the delivery end of the tube. Mount the pipette in a clamp on a stand, delivery end down. Repeat this procedure for 4 more pipettes (see Figure 25.2).

3. Fill one bag until it overflows with 5% NaCl solution. This is bag #1.

4. Tie this bag with dental floss to the rubber or tygon tubing on one of the pipettes. Be sure it is tight, but don't tear the tubing.

5. Repeat this process with each of the following:

    Bag #2: 10% NaCl   Bag #4: 30% NaCl
    Bag #3: 20% NaCl   Bag #5: distilled H_2O

6. Rinse off each bag and then immerse each in a beaker of distilled water up to the area where it is tied.

7. It is important that the water level is high enough in the pipette to take an initial reading. If this is not the case, tie a piece of dental floss around the bag tightly enough to force water into the pipette to a level that can be read easily. You will not need to read the numbers on the pipette, just the graduations (marks). Record the value of each graduation for your pipette:

_____ mL.

Figure 25.2 Solute concentration and rate of osmosis

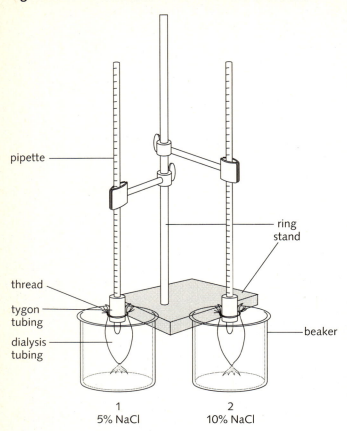

pipette

ring stand

thread

tygon tubing

dialysis tubing

beaker

1
5% NaCl

2
10% NaCl

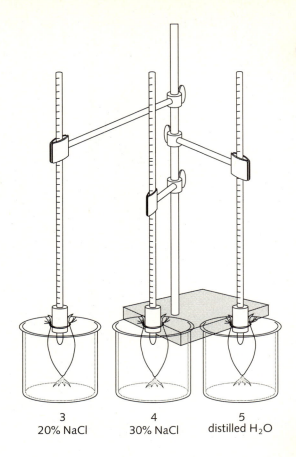

3
20% NaCl

4
30% NaCl

5
distilled H₂O

8. Mark the water level on each pipette with a marking pencil or pen and lower the bags into the beakers. Record the time: _____. Read the level of the solution in the pipette every 4 minutes for 5–6 readings and record your data in Question 3 in the Evaluation.

Osmosis and Dialysis

1. Take a 10-cm piece of dialysis tubing, wet with water, and rub between your fingers to separate. Tie one end securely with dental floss or clamp with a polyproplene closure.

2. Add enough 1% starch suspension to fill the bag one-quarter full. Add 20% dextrose solution until the bag is filled. Tie or clamp securely.

3. Fill a 400-mL beaker with distilled water and enough I₂KI solution to make it a distinct yellow color.

4. Immerse the bag in the beaker and observe any changes that occur in the bag or in the iodine-potassium iodide solution.

While observing this setup, keep in mind the following information:

• When iodine-potassium iodide solution comes into contact with starch it forms a complex and turns a purple-black color.
• Dextrose cannot be observed directly but may be detected by a chemical test. This is done by adding 3–5 mL of Benedict's reagent to an equal quantity of the unknown solution in a test tube and heating in a beaker of boiling water for

exactly 3 minutes. A positive test for dextrose is a color change from blue to yellow to orange and finally to red (highest concentration of sugar).

5. Observe this setup for approximateley 1 hour and test for the presence of dextrose in the beaker water at 30, 45, and 60 minutes. Record your observations in Question 5a in the Evaluation. While you are waiting for the results, continue with the next activity.

Rate of Diffusion and Surface Area/Volume

Why cells are so small is a question that does not have a complete answer yet. In this activity you will explore one hypothesis about the smallness of cells. It will be necessary to begin the activity by doing several calculations concerning the size of cells. To make these calculations simple, we will assume that cells have the approximate shape of a (six-sided) cube and will calculate the ratio of surface area to volume as our cube (cell) increases in size.

The surface area of a cube increases as the square of the length of a side.

$$Surface\ Area = 6 \times (length\ of\ side)^2$$
$$SA = 6S^2$$

The volume increases as the cube of this same dimension.

$$Volume = (length\ of\ side)^3$$
$$V = S^3$$

Use these formulas to calculate the surface area and volume of a cube or cell 4 cm on a side.

$$SA = 6(4)^2 = 96\ cm^2$$
$$V = 4^3 = 64\ cm^3$$

If you express these values as a ratio of surface area to volume you get

$$\frac{96\ cm^2}{64\ cm^3}$$

or in its simplest form

$$\frac{1.5\ cm^2}{1\ cm^3}.$$

This means that for a cell 4 cm on a side there is 1.5 times as much surface area as volume.

1. Use these formulas to complete the following chart, assuming that each cube represents a different size cell.

Cube size (cell) cm/side	0.1	1.0	2.0	3.0
Surface area $SA = 6S^2$				
Volume $V = S^3$				
Simplest ratio SA/V				

What happens to the relationship between surface area and volume as a cube increases in size?

Your instructor will have several trays of agar cubes available. These cubes are of different sizes— one much larger than the other. The agar used to make these cubes was mixed with phenolphthalein, a pH indicator that changes from colorless to pink when it comes into contact with a solution of sodium hyroxide.

► CAUTION: The sodium hydroxide used is irritating to the skin. If any is spilled on your skin or clothing, flush immediately with water and notify your instructor.

2. Take one cube of each size and place them in a beaker or culture bowl. Carefully cover them with the sodium hydroxide solution as directed by your instructor. The solution will slowly diffuse into the agar cubes and, as this happens, a pink color will develop. This color change will allow you to measure the rate of diffusion into each of the cubes.

3. Allow the cubes to remain in the solution for about 15 minutes and then remove them from the solution by pouring off the sodium hydroxide and rinsing with tap water. Place each cube on a paper towel, blot dry and slice in half with a spatula or scalpel. Record your observations in Question 6 in the Evaluation.

EVALUATION 25

Transport Across the Cell Membrane

1. Record your observations of the diffusion demonstration with the potassium permanganate by measuring and comparing the diameter of the ring of dye in each dish at selected intervals. What is the effect of temperature on the rate of diffusion?

2. State a hypothesis that predicts the changes you expect to occur in the experiment involving solute concentration and rate of osmosis.

3. Record your data in the following table.

Time (minutes)	Height of water column (mL)				
	Bag 1 5%	Bag 2 10%	Bag 3 20%	Bag 4 30%	Bag 5 0%
Initial reading	0	0	0	0	0

4. Interpret your data and indicate whether or not your hypothesis was substantiated. If your hypothesis is not substantiated, attempt to explain why.

5. (a) What conclusions can you reach concerning the movement (or lack of movement) of the starch, iodine-potassium iodide, dextrose, and water? Support your conclusions with any data that you collected.

   Starch _____

   Iodine-potassium iodide _____

   Dextrose _____

   Water _____

   (b) What does this tell you about the permeability of the dialysis tubing?

   (c) If this setup is allowed to remain undisturbed overnight the water in the beaker will become clear. Why?

6. Which agar cube showed the greatest amount of diffusion? Why?

7. Using the information in the table and the results of the diffusion demonstration, offer an explanation for the small size of cells.

Cellular Respiration

OBJECTIVES

At the end of this laboratory activity, you should be able to:

- explain the operation of a respirometer, including why it is necessary to use a carbon dioxide absorber and how changes in gas pressure affect the movement of fluid in the manometer.
- determine the rate of oxygen consumption per minute per gram of animal or plant tissue.
- construct and interpret a graph showing the relationship between body mass and rate of oxygen consumption in an animal.

INTRODUCTION

The existence of life, whether it be at the cellular or organismic level, requires continuous work. Organisms reproduce, grow, synthesize new materials, move, generate electricity, and, in some cases, light up. All of these activities require a continuous supply of energy that must come from the organic molecules that supply energy to the cell.

Cells release the energy stored in their fuel molecules through cellular respiration and store it temporarily in a more readily available form, **adenosine triphosphate,** or **ATP.** In aerobic organisms (those requiring oxygen), the release of this energy proceeds in three steps (see Figure 26.1). The first step begins with the conversion of a molecule such as glucose to **pyruvate,** which is also called **pyruvic acid** (Figure 26.1(a)). This initial set of reactions transfers a small amount of energy to ADP to make ATP. Energy is also transferred to a hydrogen/electron carrier, **nicotinamide adenine dinucleotide (NAD$^+$),** to make **NADH.** The pyruvate molecule is then converted into the compound **acetyl**

CoA, which enters a series of reactions known as **Kreb's cycle** (Figure 26.1(b)). During Kreb's cycle considerable energy is removed from the acetyl CoA molecule. Some of this energy is used to convert ADP to ATP. Additional energy is stored in the hydrogen/electron carriers nicotinamide adenine dinucleotide (NAD$^+$) and **flavin adenine dinucleotide (FAD).** These hydrogen/electron carriers are directed to a group of molecules called **cytochromes.** The cytochromes are embedded in the inner membrane of the mitochondria and are collectively known as the **electron transport system (ETS)** (Figure 26.1(c)). In the electron transport system, the NADH is dissociated into NAD$^+$, electrons, and hydrogen ions. As the electrons are transferred through the cytochromes, energy is released to produce ATP. The end products of this series of reactions are hydrogen ions, energy-poor electrons, and the carrier molecules NAD$^+$ and FAD (which are now available for another round). The hydrogen ions and electrons combine with molecular oxygen to make water, which is actually a waste product. Carbon dioxide, the other waste product, is eliminated into the water or air.

You should note that in the absence of oxygen (**anaerobic respiration**), some organisms pursue alternative pathways and produce ethyl alcohol or lactate (Figure 26.1(d)). Both of these reactions result in the production of only a small quantity of ATP.

In today's laboratory activity you will measure the metabolic rate of a mouse by determining the amount of oxygen used in a given period of time. This metabolic rate will then be compared with the mouse's mass in an effort to determine whether there is a relationship between these two variables. In addition, an optional activity involving respiration in plant seeds may be done, allowing you to compare respiration rates in animals and plants.

Figure 26.1 A summary of cellular respiration

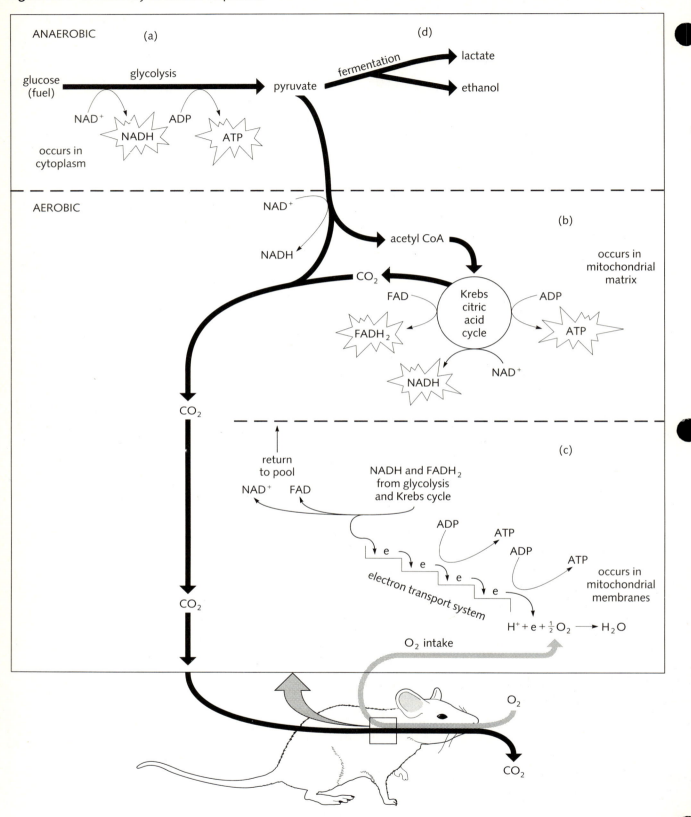

MATERIALS

Two-hole rubber stopper (#7) fitted with:
  Hypodermic syringe (5 cc)
  Manometer tube with fluid
  Glass tubing with rubber vent tube and clamp
Erlenmeyer flask (500 mL)
Mice (assorted sizes)
Filter paper
Test tube (150 mm × 25 mm)
Dropping bottle of potassium hydroxide (KOH)
  (20%)
Ring stand with clamp
Beaker/battery jar (2 L)
Marking pencil or pen

PROCEDURE

Operation of the Respirometer

A respirometer measures changes in gas pressure produced as an organism carries out cellular respiration. The respirometer used in this activity will measure oxygen consumption in the mouse over a given period of time (see Figure 26.2).

The operation of the respirometer is relatively simple and depends upon the relative changes in pressure that occur between the gas inside the flask and the atmosphere. When the vent at the top of the respirometer is closed by a clamp, the air inside

Figure 26.2 Respirometer apparatus

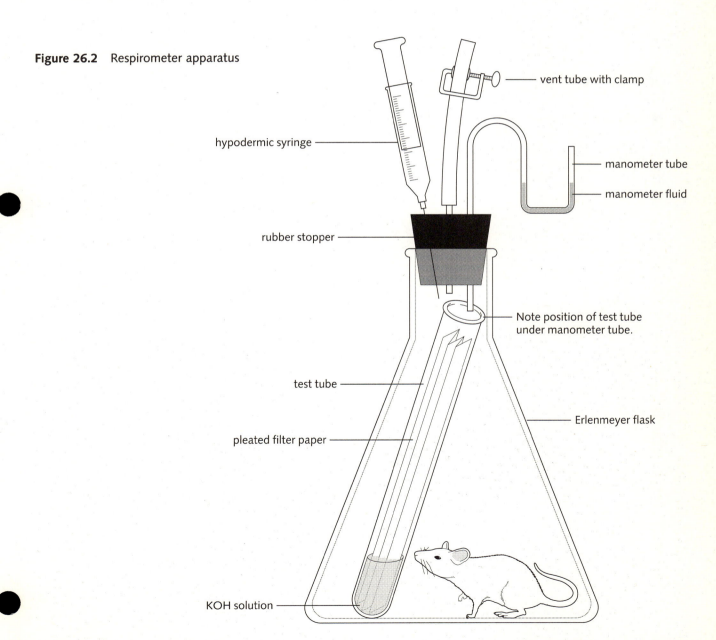

vent tube with clamp

hypodermic syringe

manometer tube

manometer fluid

rubber stopper

Note position of test tube under manometer tube.

test tube

Erlenmeyer flask

pleated filter paper

KOH solution

the flask cannot be exchanged with the atmosphere, and any pressure changes that occur in the flask will be detected by the movement of the manometer fluid. As the mouse respires it will consume oxygen and produce carbon dioxide. If the carbon dioxide is chemically removed as soon as it is produced, the pressure in the flask is reduced in proportion to the amount of oxygen consumed. A potassium hydroxide absorber in the flask removes this carbon dioxide.

Examine the respirometer on your laboratory table and locate the following four essential parts: a U-shaped manometer tube that contains kerosene and a red dye, a 5-cc hypodermic syringe inserted completely through a rubber stopper so that the needle is exposed on the underside, a piece of glass tubing topped by a flexible piece of rubber or tygon tubing and a clamp, and a test tube (150 mm × 25 mm).

Before completing the activity with a mouse, assemble the respirometer and test it for leaks, following these directions.

1. Making sure the rubber vent tube is *not* clamped shut, place the stopper firmly into the neck of the flask. Seat the stopper firmly in the neck of the flask by pushing down with your thumbs. This is the most common source of leaks and should be checked first if your apparatus does not operate correctly.

2. Close the clamp on the tubing. Hold your hands around the flask for several seconds and note how the manometer fluid moves. Which way does the fluid move?
   How can you account for this movement?
   If the manometer fluid does not move, check for leaks and try again.

Determination of Respiration Rate

The mice you will use should be of varying sizes and, if possible, of the same sex. Females are more docile than males and, thus, are preferable. Laboratory mice are easy to handle and normally do not bite. If you are nipped, inform your instructor immediately. Be careful to handle your mouse gently; you do not want to excite it unduly, because your data should reflect, as much as possible, its oxygen consumption while at rest.

1. Pick up your mouse by the tail and hold it in the palm of your hand. The mouse is an inquisitive creature and will usually crawl into the neck of the flask if the flask is held at a 45° angle and placed over the top of its head.

2. Fold a piece of filter paper that extends the length of the test tube and saturate it with KOH solution. There should also be about 3–5 mL of KOH solution in the bottom of the test tube. Insert this test tube into the flask under the end of the manometer tube. (The test tube will also serve as a reservoir in case you accidently force manometer fluid back into the flask by pulling the syringe barrel up with the vent tube closed.)

3. With the vent tube *open*, place the stopper in the neck of the flask and seat firmly.

4. Clamp the entire apparatus in a ring stand and place the flask in a beaker or battery jar of room-temperature water (see Figure 26.3). The room-temperature water is essential for keeping temperature fluctuations outside the flask from changing the pressure inside it. Allow the mouse to acclimate for about 5 minutes before you begin to take readings. While the mouse acclimates itself, look over the data tables in Question 1 of the Evaluation. Make sure you understand how to record data during the activity.

Figure 26.3   Respirometer apparatus

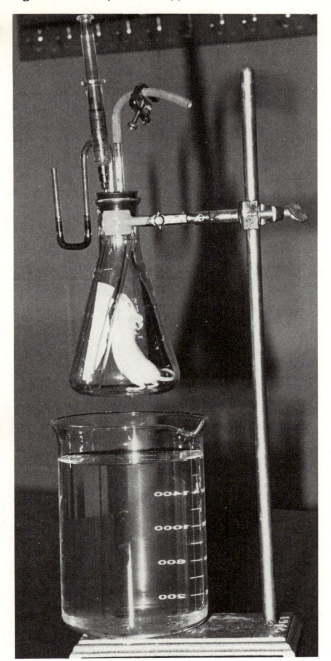

Follow these steps to obtain oxygen consumption readings for your mouse.

1. Set the plunger of the syringe on the mark at the top of the barrel.

2. Mark the level of the manometer fluid in the tube with a marking pencil or pen.

3. Close the clamp on the vent tube and record the time. As the mouse respires, the manometer fluid will move toward the flask.

4. Allow 1 minute to elapse and then return the manometer fluid to its original level by pushing the barrel of the syringe down. Read the volume (cc) of oxygen used and record this data in Question 1 in the Evaluation. Do *not* return the barrel of the syringe to its original position.

5. Continue to take readings at 1-minute intervals for a total of 3 minutes.

▶ **CAUTION: If your mouse shows signs of stress, shorten the total 3 minutes to 1–2 minutes. You may also have to ventilate your mouse more frequently if it depletes a volume of oxygen greater than the syringe capacity before 3 minutes have passed.**

6. At the end of 3 minutes, release the clamp on the vent tube and pump fresh air into the flask through the syringe. Pump the syringe about 20 times to ensure a good exchange of air. Close the clamp.

7. Repeat readings every minute, venting the flask every 3 minutes. Repeat the 3-minute cycle once more.

8. Remove your mouse from the flask and weigh to the nearest 0.1 g. Return it to its cage. Record the mass in Question 1 in the Evaluation.

9. Repeat this procedure for as many mice as your instructor indicates.

10. When you have finished, complete Questions 2–4 in the Evaluation.

OPTIONAL ACTIVITY: Respiration in germinating seeds

INTRODUCTION

The energy used by plants for their metabolism comes from the degradation of organic molecules such as starch that are produced during photosynthesis. The cellular respiration that occurs in a mouse is the same process that occurs in a plant as it releases energy to carry out the metabolic processes necessary for life. This activity provides you with the opportunity to study cellular respiration in germinating seeds.

MATERIALS

Potassium hydroxide (KOH) pellets
Cheesecloth
Test tube (200 mm × 25 mm)
Nonabsorbent cotton
Germinating seeds
Paper towels
Two-hole rubber stopper (#4) fitted with:
    Glass tubing with rubber vent tube and clamp
    Capillary tube
Ring stand with clamp
Beaker (2000 mL)
Metric ruler
Tape
Dye for manometer
Pipette

PROCEDURE

The apparatus used in this activity (see Figure 26.4) is very similar to the respirometer used for the mouse. As the seeds respire, oxygen is consumed and the volume of gas in the test tube decreases. Potassium hydroxide absorbs the carbon dioxide.

Preparation of Materials

1. Place 10–12 pellets of KOH in a small piece of cheesecloth. Tie the ends and drop the bag into the bottom of a test tube. Place a small loose wad of nonabsorbent cotton above the cheesecloth.

2. Gently roll about 20–25 germinating seeds between paper towels to remove excess water. Add a sufficient number of seeds to fill the test tube to about 3 cm from the top.

3. Make sure that the clamp on the vent tube is open and tightly insert the stopper with the manometer apparatus into the top of the test tube.

4. Clamp the test tube firmly in a ring stand and immerse the test tube in a beaker of room-temperature water. Allow the setup to equilibrate for 5 minutes.

Determination of Respiration Rate

1. Attach a millimeter ruler to the capillary tubing with tape so that the scale can be read easily.

2. Add a drop of dye to the end of the capillary tube with a pipette. Read the leading edge of the dye and record this measurement in millimeters. Clamp the vent tube and take readings at 1–5-minute intervals, depending on the amount of activity. Obtain at least 6 readings. Record your data in Question 1 under Optional Activity in the Evaluation.

3. If you need to repeat your readings, you can return the dye column to the end of the capillary tube by opening the vent and tilting the tube.

4. After the last reading, remove the seeds from the test tube and weigh to the nearest 0.1 g. Complete Questions 2 and 3 under Optional Activity in the Evaluation.

Figure 26.4 Seed respirometer apparatus

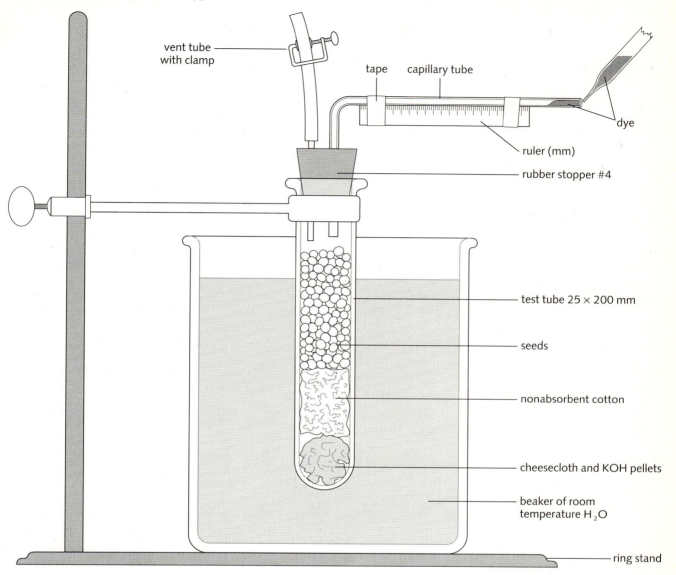

vent tube with clamp

tape   capillary tube

dye

ruler (mm)

rubber stopper #4

test tube 25 × 200 mm

seeds

nonabsorbent cotton

cheesecloth and KOH pellets

beaker of room temperature H_2O

ring stand

EVALUATION 26

Cellular Respiration

1. Record your data in the table below. If more than four mice are
   measured, make an additional table on a separate sheet of paper.

Time (min.)	Mouse # ___ Syringe reading (cc)	Mass ___ g Volume O_2 consumed (cc)	Mouse # ___ Syringe reading (cc)	Mass ___ g Volume O_2 consumed (cc)	Mouse # ___ Syringe reading (cc)	Mass ___ g Volume O_2 consumed (cc)	Mouse # ___ Syringe reading (cc)	Mass ___ g Volume O_2 consumed (cc)
0 1 2 3	5	0	5	0	5	0	5	0
Ventilation								
0 1 2 3	5	0	5	0	5	0	5	0
Ventilation								
0 1 2 3	5	0	5	0	5	0	5	0
Totals		cc		cc		cc		cc

2. (a) Calculate the rate of oxygen consumption per minute for each mouse by dividing the total amount of oxygen consumed by the total time. Next divide this result by the mouse's mass to calculate the rate of oxygen consumption per minute per gram of tissue. Show your calculations in the spaces below.

Mouse #	Mass (g)	O$_2$ cc/min.	O$_2$ cc/min./g

(b) Combine the data in your summary table with data from other students in a summary table for the entire class. Use this class data to plot a graph on the graph paper provided at the end of the Evaluation, using oxygen consumption as the dependent variable and mass as the independent variable.

Class data

Mass (g)	O$_2$ cc/min./g	Mass (g)	O$_2$ cc/min./g

Class data (*continued*)

Mass (g)	O₂ cc/min./g	Mass (g)	O₂ cc/min./g

3. The following graph (prepared by students in a general biology class) shows the relationship between respiration rate and body mass for 18 female mice ranging in size from 20–65 grams. Draw a line of best fit through these points.

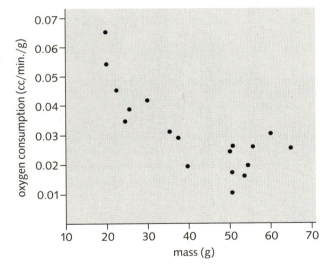

(a) What type of correlation—positive or negative—is indicated by this line?

(b) Does your data produce a similar line? _____ If so, return to Laboratory Activity 1 and review the procedure for calculating a Pearson r correlation coefficient. Use the class data to calculate a correlation coefficient to determine the strength of the relationship between respiration rate and mass. Complete the following table.

Variable X: _____   Variable Y: _____

Observation (N)	(X)	Deviation from mean (d_x)	Deviation squared $(d_x)^2$	(Y)	Deviation from mean (d_y)	Deviation squared $(d_y)^2$	Product of deviations $(d_x)(d_y)$
	$\Sigma =$		$\Sigma =$	$\Sigma =$		$\Sigma =$	$\Sigma =$
	$\overline{X} =$			$\overline{Y} =$			

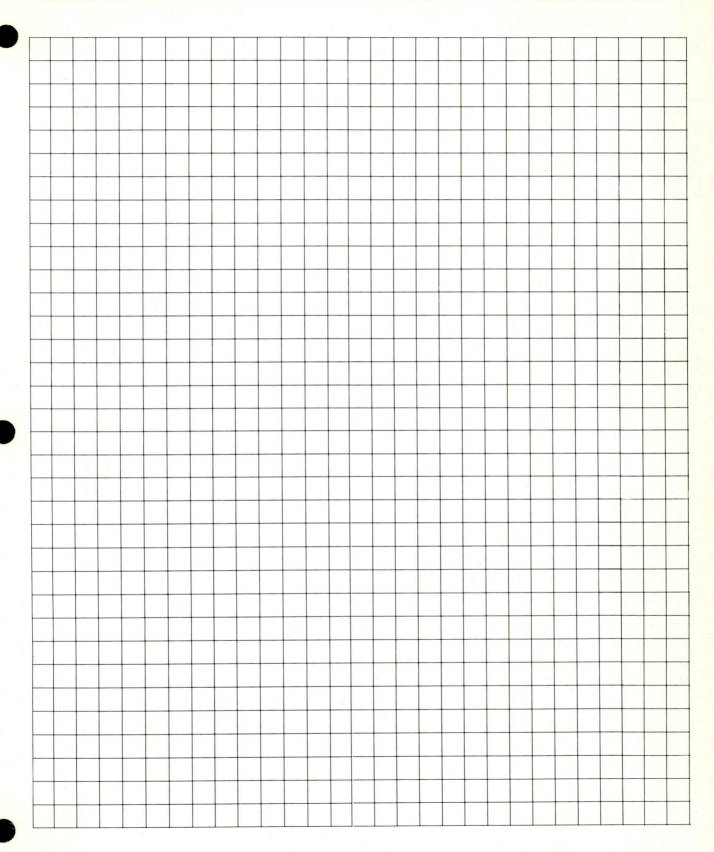

OPTIONAL ACTIVITY: Respiration in Germinating Seeds

1. Record the data from the seed experiment in the table below.

Time	Manometer reading (mm)
0	

2. Calculate the rate of oxygen consumption of the seeds per minute per gram of tissue (the rate will be in millimeters per unit of time). Show your calculations in the space below.

   Mass of seeds _____ g

   O_2 mL/min./g _____

3. How does the rate of oxygen consumption for the germinating seeds compare to the rate for the mice? Can you explain any difference?

Chromatography and the Absorption Spectra of Plant Pigments

OBJECTIVES

At the end of this laboratory activity, you should be able to:

- prepare a chromatogram of pigments from a green leaf extract.
- isolate chlorophyll a and b and the carotenoids.
- plot a graph showing percentage of light transmittance as a function of wavelength for the isolated pigments.

INTRODUCTION

Adsorption chromatography is a method used to separate a chemical mixture by passing it over a material that adsorbs different compounds at different rates. (**Adsorption** is the surface retention of compounds; **absorbtion** is the penetration of compounds into the absorbing substance.)

This technique was first developed in 1906 by Michael Tswett, a Russian botanist. Tswett used the new technique to separate leaf pigments, just as you are about to do. He filled a glass column with calcium carbonate, packed it tightly, and poured a petroleum ether solution of plant pigments into the column. The pigments separated into definite bands along the length of the column according to their adsorption characteristics. Tswett's work with partition chromatography enabled others to develop the technique of paper chromatography. In paper chromatography the materials separate out on a strip of paper instead of on an adsorbent material in a glass column. The resulting **chromatogram** is the strip of paper with bands or spots distributed along its length. This technique has proven most useful in separating small quantities of materials present in solution. You will use this technique to isolate plant pigments involved in photosynthesis.

The two important considerations in paper chromatography are the paper and the solvents. A small quantity of the liquid mixture to be separated is placed on a strip of paper and allowed to dry. The paper is then placed into the proper solvents, which begin to ascend the paper due to capillary action. As the solvents move up the paper, the components of the mixture on the spot move with them at different rates. Ultimately the different molecules in the mixture will distribute themselves over the length of the strip. This distribution is determined by the varying solubilities of the molecules in the solvents used and by the molecular weights of the molecules being separated.

In this activity you will separate the pigments present in the leaves of green plants.

MATERIALS

Acetone
Petroleum ether
Ventilated hood
Test tube (25 mm × 150 mm)
Cork with attached hook
Test tube rack
Mortar and pestle
Fresh spinach leaves
Beaker (50 mL)
Whatman #1 filter paper
Scissors
Dissecting needle or pin
Capillary tubing
Cuvettes for spectrophotometer
Spectrophotometers
Marking pencil or pen

PROCEDURE

The solvent chamber to be used in this chromatography activity consists of a large test tube sealed with a cork that is fitted with an attached hook. The solvent you will use is a mixture of 8% acetone and 92% petroleum ether. Your instructor will prepare this mixture for you just prior to this activity and store it under a ventilated hood.

1. Under the ventilated hood add 5 mL of this solvent mixture to the large test tube (25 mm × 150 mm) provided. This test tube will serve as your solvent chamber. Place a cork with an attached hook into the mouth of the test tube.

2. Return to your laboratory table and place the solvent chamber in a test tube rack until it is needed. This will allow the air in the solvent chamber to become saturated with the fumes from the solvent mixture.

3. Pull the shank of the hook through the cork until the hook is snug against the bottom of the cork.

The pigments you will separate are mixed with a multitude of other materials within the fluids of the leaf tissue. The following steps will permit you to extract these leaf pigments:

4. Add 5 mL of acetone to a mortar. Remove the large midvein from a spinach leaf and then tear the remaining leaf tissue into small pieces and place them into the acetone.

5. Using a pestle, grind the leaf tissue into a pulp. The leaf pigments will dissolve into the acetone at this time.

6. Examine the leaf extract carefully. It should be *very dark green* in color. If it is not, continue to grind and add more leaf tissue if needed.

7. Carefully pour off the extract into a small beaker, being extremely careful not to obtain any of the pulp. Set aside until needed.

Special care must be taken in the preparation of the chromatography paper strip to produce a strip of the correct size.

8. Cut a section of Whatman #1 filter paper approximately 15 cm in length and narrow enough so that it can be inserted in the solvent chamber without touching the sides.

9. Punch a hole in one end of the paper strip with a dissecting needle or pin.

10. Cut the paper strip to the correct length. This length should allow the cork with the hook and attached paper strip to be placed into the test tube (chamber) without letting the paper strip touch the solvent at the bottom of the tube. Do *not* open the solvent chamber at this time (see Figure 27.1(a)).

11. Cut indentations in the sides of the strip as shown in Figure 27.1(b).

12. Place the end of the glass capillary tubing into your leaf extract solution and allow the extract to rise up the tube. Place your index finger over the end of the capillary tube to hold the extract in the tube.

13. Touch the tube to the strip of paper at the point marked with an X on the sample shown in Figure 27.1(b). Be careful to make only a small spot. Hand dry the spot on the paper by waving the strip in the air.

14. Add another droplet on top of the first and repeat the process. Continue adding droplets and drying them until you have added 10–12 droplets.

15. Remove the cork from the solvent chamber. Pass the hook through the hole in the top of the strip.

Figure 27.1 Preparation of chromatograph paper

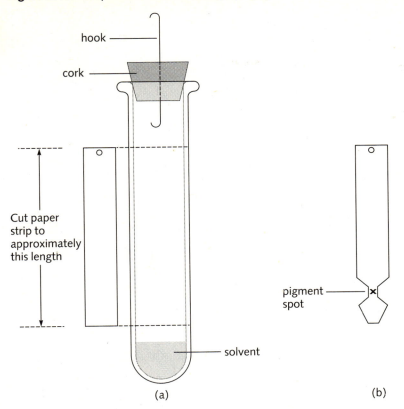

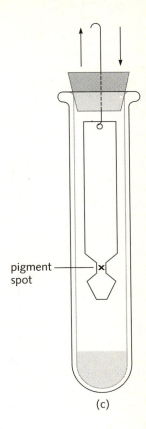

hook

cork

Cut paper strip to approximately this length

solvent

(a)

pigment spot

(b)

pigment spot

(c)

16. Pull the hook with the attached paper up snug with the bottom of the cork and place the cork into the test tube. Make sure the paper does not touch the solvent in the bottom of the solvent chamber. If necessary, remove the paper strip and trim additional paper from the top of it.

17. Replace the solvent chamber in the test tube rack. Allow the paper to hang undisturbed, over but not in the solvent, for 3–5 minutes so that the paper attains equilibrium with the atmosphere within the chamber (see Figure 27.1(c)). This will prevent evaporation of the solvent from the paper during the development of the chromatogram.

18. After 3–5 minutes gently push the shank end of the hook deeper into the test tube, permitting the end of the paper to drop 3–4 mm into the solvent. It is critical that the solvent does not splash over the paper at any time.

19. As the solvent mixture moves up the paper, colored bands will begin to appear. Each of these bands contains a different plant pigment. Allow the system to run until the bands have spread out over most of the length of the paper. Do not let the solvent reach the attached hook.

20. Remove your chromatogram from the chamber and allow it to dry.

The components of the bands can be identified from their Rf values when compared with standards. An Rf value can be calculated in the following manner.

$$Rf = \frac{\text{distance of the band from the origin}}{\text{distance of the leading edge of the solvent from the origin}}$$

The colored bands of your chromatogram contain the pigments involved in photosynthesis. Look closely at the chromatogram and identify the following bands. The bluish-green band is composed of **chlorophyll a,** which is found in all photosynthetic eukaryotes and in the cyanobacteria. The light green band composed of **chlorophyll b** is an accessory pigment to chlorophyll a in the process of photosynthesis. The assortment of bands in the yellow to orange range includes the carotenoid **carotene.** The carotenoids are also accessory pigments involved in photosynthesis. Carotene can be oxidized into a variety of additional yellow pigments, some of which you should be able to find on your chromatogram.

In this second part of the laboratory activity you will be using the spectrophotometer to determine which wavelengths of light are absorbed by these isolated plant pigments. Before proceeding, read Appendix E, which contains a description of this instrument and its operation.

1. Carefully cut your chromatogram to separate the chlorophyll a, chlorophyll b, and the carotenoids.

2. Collect all the chlorophyll a bands from the students at your laboratory table and place them into a small beaker containing 5 mL of acetone. Repeat this process, collecting chlorophyll b and the carotenoids in separate beakers.

3. Pour the acetone containing the dissolved chlorophyll a into a cuvette and label with a glass marking pen at the extreme top of the cuvette.

4. Repeat this procedure with the acetone solutions of chlorophyll b and the carotenoids. At this point you will have three cuvettes: one containing chlorophyll a, another containing chlorophyll b, and a third containing the carotenoids.

The following instructions concerning the operation of the spectrophotometer are pertinent to this particular activity. Note that there are two spectrophotometers available for use during this activity. They are identical except that the light source of instrument A produces light over a wavelength range of 350 nm–600 nm and instrument B produces light over a wavelength range of 600 nm–750 nm. To get the full range of wavelength measurements required, your sample must be read by both instruments. Remember that you are interested in the absorption of specific wavelengths by these pigments. You are reading percentage of light transmittance because this measurement is easier to read and to graph. Keep in mind that if a pigment is absorbing a large amount of a specific wavelength, it is transmitting little of that wavelength.

5. Turn on the spectrophotometer by rotating the power knob clockwise. Allow at least 5 minutes for warm up.

6. Set the transmittance at 0% by turning the zero-adjust knob.

7. Set the spectrophotometer at the desired wavelength.

8. Fill a cuvette with 5 mL of pure acetone. Insert it in the sample holder and close the cover. Adjust the light control knob so that 100% transmittance is read on the scale. Remove the sample.

9. Put the cuvette with pigment sample into the sample holder. Read the percentage of transmittance directly from the scale. Record this value on your data sheet in Question 1 in the Evaluation. Remove the sample.

10. Before changing to the next wavelength, turn the light control knob counterclockwise approximately a quarter turn to prevent the indicator needle from pinning.

11. Set the next desired wavelength.

12. Standardize the instrument for this new wavelength, using the same tube of acetone used before.

13. Measure the percentage of transmittance of the sample at the new wavelength.

14. Repeat this procedure of standardization and measurement until the percentage of transmittance at all required wavelengths listed in Question 1 has been measured. (You will have to switch spectrophotometers at 600 nm.)

After the data have been collected, plot your information on the graph paper in Question 2 in the Evaluation. You should have 3 lines on your graph that show the absorption characteristics for chlorophyll a, chlorophyll b, and the carotenoids.

EVALUATION 27

Chromatography and the Absorption Spectra of Plant Pigments

1. Fill in the data sheets.

Chlorophyll a

Instrument A				Instrument B	
Wavelength nm	% Transmit.	Wavelength nm	% Transmit.	Wavelength nm	% Transmit.
350		500		625	
375		525		650	
400		550		675	
425		575		700	
450		600		725	
475				750	
Visible light range (350–600 nm)				Infrared range (625–750 nm)	

Chlorophyll b

Instrument A				Instrument B	
Wavelength nm	% Transmit.	Wavelength nm	% Transmit.	Wavelength nm	% Transmit.
350		500		625	
375		525		650	
400		550		675	
425		575		700	
450		600		725	
475				750	
Visible light range (350–600 nm)				Infrared range (625–750 nm)	

Carotenoids

Instrument A				Instrument B	
Wavelength nm	% Transmit.	Wavelength nm	% Transmit.	Wavelength nm	% Transmit.
350		500		625	
375		525		650	
400		550		675	
425		575		700	
450		600		725	
475				750	
Visible light range (350–600 nm)				Infrared range (625–750 nm)	

2. On the graph paper below, plot your percentage of transmittance readings from your data sheet for chlorophyll a, chlorophyll b, and the carotenoids.

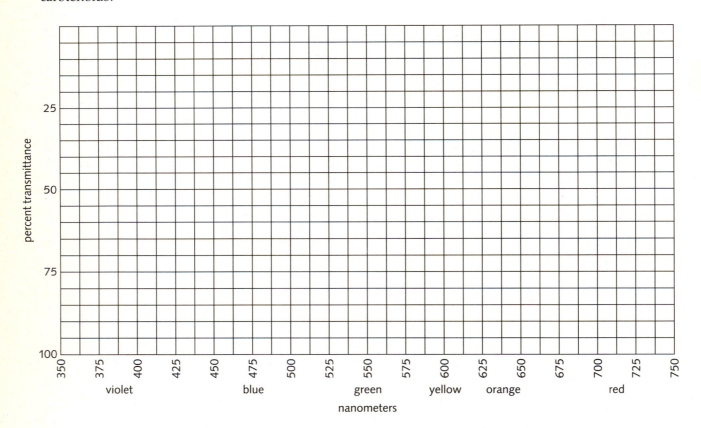

3. Why do the percentage of transmittance values start at 100% and decrease as you move up the *y*-axis of your graph?

4. Write a brief interpretation of the graph you prepared in Question 2.

Photosynthesis

OBJECTIVES

At the end of this laboratory activity, you should be able to:

- describe how chloroplasts may be isolated from green leaves.
- calculate the quantity of chlorophyll in a suspension of chloroplasts.
- state the effect of light intensity on photoreduction.
- state why the chloroplast membrane must be intact for photoreduction to occur.

INTRODUCTION

Of the multitude of processes and structures that are necessary for life to exist, none is more fundamentally important than the energy-storing process of photosynthesis. By capturing the energy of sunlight, the autotrophs of our world are the first link in a series of chemical reactions providing the energy that drives the engine of life. The details of the entire photosynthetic process are provided in your textbook, and you should be familiar with them before you begin. For now it will be helpful to review the steps of photosynthesis that apply to this activity.

Photosynthesis is divided into two stages, the **light reactions** and the **dark reactions.** During the light reactions water is **oxidized** (electrons are removed) and molecular oxygen is produced as a by-product. In addition, adenosine triphosphate (ATP) and nicotinamide adenine dinucleotide (NADPH) are produced. The light reactions occur in the membranes of the **thylakoids,** flat saclike structures located in the chloroplasts (see Figure 28.1). The dark

reactions, which will not be studied in this activity, involve the **reduction** (addition of electrons) of carbon dioxide to carbohydrate.

The initial step in the light reaction occurs when photons of light cause the chlorophyll to be oxidized in **photosystem II** (see Figure 28.2a). At this same time water is also oxidized, resulting in the production of hydrogen ions and molecular oxygen (see Figure 28.2b). The chlorophyll molecules are then reduced, using electrons donated from the

Figure 28.1 Electron micrograph of a chloroplast with thylakoid membranes

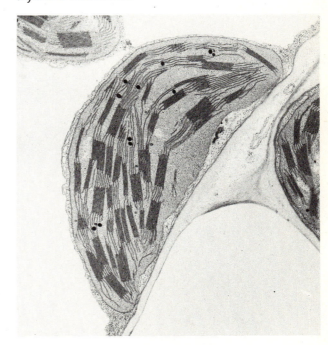

Figure 28.2 Simplified diagram of electron transport chain in thylakoid membrane of a chloroplast

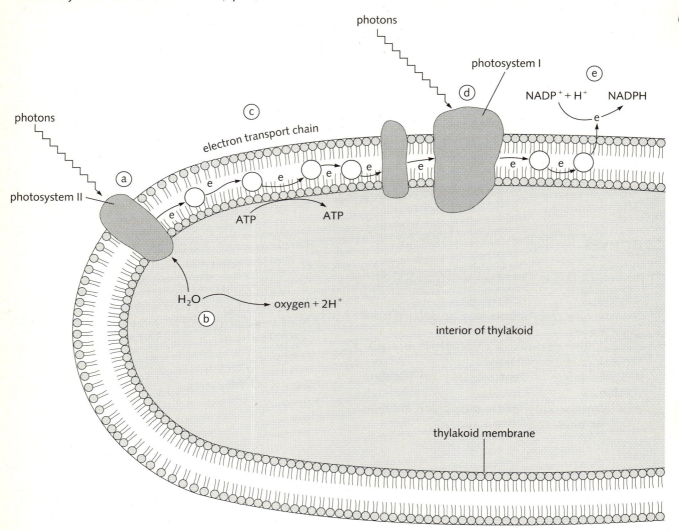

oxidation of water. The electrons released from the original oxidation of chlorophyll are passed through a series of carriers in the electron transport chain to the final acceptors in **photosystem I** (see Figure 28.2c). As this electron transport is occurring, some energy is diverted to synthesize ATP from ADP and inorganic phosphate (P_i). In photosystem I, photons of light also oxidize chlorophyll, and the resulting electrons are replaced with the electrons from the chlorophyll in photosystem II (see Figure 28.2d). The electrons from photosystem I are passed to the final electron acceptor, **nicotinamide adenine dinucleotide phosphate (NADF$^+$),** which is reduced to NADPH (Figure 28.2e).

Let's summarize several of the important points of the light reactions.

1. The final electron carrier NADP$^+$ is reduced to NADPH by the addition of electrons and hydrogen from the chlorophyll in photosystem I.

2. The electrons lost from photosystem I are replaced with electrons from photosystem II.

3. Electrons are removed from water and transferred to the chlorophyll in photosystem II. Hydrogen and molecular oxygen are also produced during this reaction.

4. Photosystems I and II are linked by an electron transport chain. As electrons move along this chain, ATP is produced.

5. The final products of the light reactions are ATP, NADPH, and O_2. These reactions may also be summarized in the following equation:

$$ADP + P_i + NADP^+ + H_2O \longrightarrow$$
$$ATP + NADPH + 2H^+ + \tfrac{1}{2}O_2$$

In today's activity you will study only the light reactions of photosynthesis. First you will isolate chloroplasts from green leaves. During this process the final electron acceptor, NADPH, will be washed out. Then you will add a synthetic acceptor, dichlorophenol indophenol (DCPIP) to take the place of the NADP. DCPIP is a blue dye that turns colorless when it is reduced. This characteristic will allow you to follow the light reaction by measuring the change in the DCPIP from blue to colorless with a spectrophotometer. Because the electrons used to reduce the DCPIP are ejected by photons of light, the reaction may be referred to as **photoreduction.** The reaction with DCPIP is as follows:

$$\text{DCPIP} + \text{H}_2\text{O} \longrightarrow \tfrac{1}{2}O_2 + \text{DCPIP} + 2H^+$$

oxidized (blue) hydrogen/electron source reduced (colorless)

MATERIALS

Fresh spinach leaves
Phosphate buffer (0.1 M)
Blender
Beakers (250 mL, 400 mL)
Container with ice
Centrifuge tubes
Centrifuge
Tris-magnesium buffer (7.0 mM)
Acetone
Graduated cylinder
Large test tubes
Pipette (1 mL)
Vortex stirrer
Funnel with filter paper
Cuvettes for spectrophotometer
Spectrophotometer
150-watt light
Marking pencil or pen
Dichlorophenol indophenol (DCPIP) (0.3 mM)
Test tube racks
Dropping bottles of ethanol (95%)

PROCEDURE

Preparation of Chloroplast Suspension

First you must prepare a homogenate of spinach leaves in a buffered salt, as follows.

1. Add 100 g of fresh deveined spinach leaves to 250 mL of 0.1 M phosphate buffer and grind in a blender for 1 minute.

2. Pour this homogenate into a 400-mL beaker and keep in a container of ice. Remove from the ice only to carry out an activity.

3. To prepare a chloroplast suspension that is free of other cell organelles, centrifuge 50 mL of the homogenate at 300 g (gravity) for about 3 minutes. This will separate the larger cell organelles (nuclei, cell walls) and debris from the chloroplasts, which will remain in suspension in the **supernatant** or top layer.

4. Decant the supernatant into a second tube and discard the pellet of material that has settled to the bottom.

5. Recentrifuge the supernatant at 2000 g for 10 minutes. This will pellet the chloroplasts. Discard the supernatant from this spin.

6. Resuspend the chloroplast pellet in 15 mL of Tris-magnesium buffer. This is the chloroplast suspension that you will use in the remaining activity. Return the tube to the container of ice.

Determining Chlorophyll Concentration

It is important that you standardize the concentration of chlorophyll in your suspension for the remaining activities. Determine the concentration of chlorophyll (mg/mL) of your suspension by completing the following.

1. Measure 19 mL of acetone in a graduated cylinder and pour into a large test tube.

2. Using a 1-mL pipette, add 1 mL of your prepared chloroplast suspension to the acetone. Mix well by gently rolling the test tube between your hands or by using a vortex stirrer. Acetone is an organic solvent and will dissolve the chloroplast membranes, allowing the chlorophyll to go into solution.

3. Filter the mixture through filter paper to remove any insoluble material. Collect about 6 mL of clear green extract.

4. (You may need to turn to Appendix E to review the operation of the spectrophotometer before continuing.) Using a cuvette, prepare a blank tube of 100% acetone and set the spectrophotometer to zero transmittance at 645 nm (645 nm is the maximum absorbance peak for chlorophyll a).

5. Insert the cuvette with the chlorophyll extract into the well and record the absorbance at this wavelength in the space below. Repeat this entire process with the wavelength setting at 663 nm (663 nm is the maximum absorbance peak for chlorophyll b). Record your results below.

Percent absorbance		
Chlorophyll a (645 nm)	_____	%
Chlorophyll b (663 nm)	_____	%

6. Use the following formula to calculate the concentration of chlorophyll in the suspension.

$$\underset{\substack{\text{percentage}\\\text{absorbance}\\\text{CHL a}}}{\underline{\hspace{1cm}}} (20.2) + \underset{\substack{\text{percentage}\\\text{absorbance}\\\text{CHL b}}}{\underline{\hspace{1cm}}} (8.02) = \underset{\substack{\text{chlorophyll}\\\text{in suspension}}}{\underline{\hspace{1cm}}} \text{mg}$$

Your answer is the number of milligrams of chlorophyll in 50 mL of homogenate.

7. Divide this value by 50 mL to determine the milligrams of chlorophyll per milliliter.

$$\frac{\underline{\hspace{0.5cm}} \text{ mg}}{50 \text{ mL}} = \underline{\hspace{1cm}} \text{mg/mL}$$

During the remaining activities you will need to add enough chlorophyll suspension to each tube to equal 30 μg of chlorophyll. Calculate this quantity for your suspension as follows.

8. Convert the number of milligrams of chlorophyll/mL to μg/mL by multiplying by 1000.

$$\underline{\hspace{1cm}} \text{mg/mL CHL} \times 1000 = \underline{\hspace{1cm}} \mu\text{g/mL}$$

9. Dividing 30 μg by the result in the previous step will give you the number of milliliters of suspension needed in each test tube.

$$\frac{30 \ \mu\text{g}}{\underline{\hspace{0.5cm}} \ \mu\text{g/mL}} = \underline{\hspace{1cm}} \text{mL of chloroplast suspension}$$

Therefore each _____ mL of chloroplast suspension contains 30 μg of chlorophyll.

The Photoreduction of an Electron Acceptor

As indicated in the introduction, the synthetic electron acceptor DCPIP accepts electrons during the light reaction of photosynthesis and changes from a blue color to nearly colorless. As the DCPIP is reduced, an increasing amount of light is allowed to pass through the solution. The spectrophotometer records this as an increase in percentage of light transmittance. The following activity will demonstrate how to use this dye to measure the process of photoreduction during the light reaction to photosynthesis.

1. Place in a cuvette the amount of chloroplast suspension that contains 30 μg. Add 5 mL of Tris-magnesium buffer. This is the blank tube that sets the spectrophotometer for 100% light transmittance by the solvent (see Appendix E).

2. Set the wavelength on the spectrophotometer to 600 nm, insert the blank tube in the well, and adjust the light transmittance to zero.

3. Remove the blank tube.

4. Place in a second cuvette the amount of chloroplast suspension that contains 30 μg. Add 4.5 mL of Tris-magnesium buffer and 0.5 mL DCPIP solution. This is the reaction tube.

▶ **CAUTION: To prevent the photoreduction of DCPIP, do not add the DCPIP solution until the room lights have been turned off.**

5. After adding the DCPIP, insert the reaction tube in the well and record the light transmittance at zero time ($T = 0$) in the table below.

6. Place the reaction tube 30 cm from a 150-watt light and record the light transmittance at 1-minute intervals until no more change occurs. Record your data in the following table.

Time	Transmittance

7. Plot this data in the graph provided.

Does the slope of the line agree with your expectations? _____ If not, check with your instructor to see if your experiment was set up and carried out properly.

At this point you should understand how to measure photoreduction.

Investigating Photoreduction

The following experiments are designed to show how various factors may alter the rate of the light reaction in photosynthesis.

➤ **CAUTION: Remember you must work with the lights out until it is time to run the experiment.**

Effect of Light Intensity on Photoreduction
Be sure to read the entire activity before beginning.

1. With a marking pencil, number 5 cuvettes near the top of the tube.

2. Prepare each cuvette with the following:
   • amount of chlorophyll suspension that contains 30 μg of chlorophyll
   • 4.5 mL Tris-magnesium buffer
   • 0.5 mL DCPIP solution

3. Place each cuvette in its own test tube rack and then station them at distances of 20, 40, 60, 80, and 100 cm from the light source.

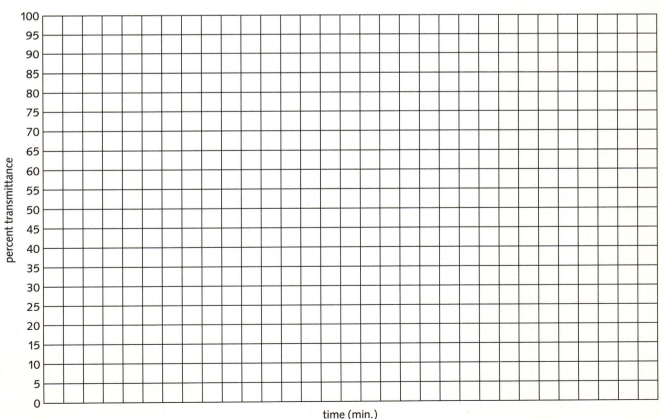

time (min.)

4. Expose the suspensions to light for a length of time that will give you the most reduction (color change) in the least amount of time (see the previous activity for the best time). Measure the percentage of transmittance of each cuvette with the spectrophotometer.

5. Record your data in Question 1 in the Evaluation.

► CAUTION: You must work quickly and not expose the chlorophyll suspension to light other than during the time of the experiment.

Dissolving the Membrane of the Chloroplast
Electron transport in the chloroplast depends on many factors for its success. Chief among these is the integrity of the membrane of the chloroplast and, in particular, the membrane of the thylakoid that contains the pigments and electron carriers of photosystems I and II (see Figure 28.2). This membrane is composed partly of lipids that are soluble in organic solvents such as alcohol. This experiment will demonstrate the importance of an intact membrane system to electron transport.

1. Prepare 4 cuvettes with the amount of chlorophyll suspension that contains 30 μg of chlorophyll. Add 0.5 mL of DCPIP and 4.5 mL Tris-magnesium buffer. Number each tube.

2. To tube #1 add five drops of 95% ethanol, to tube #2 add 10 drops of alcohol, to tube #3 add 15 drops, and to tube #4 add 20 drops.

3. Expose each cuvette to the light source for the appropriate length of time determined previously. Measure the percentage of transmittance with a spectrophotometer and record the results in Question 2 in the Evaluation.

EVALUATION 28

Photosynthesis

1. (a) Record your data from the light intensity experiment in the following
        table.

Exposure time	Distance (cm)	% Transmittance
	20	
	40	
	60	
	80	
	100	

   (b) Prepare a graph of this data on the graph paper at the end of this
        activity. Plot percentage of transmittance as a function of light
        intensity (measured as distance in cm).

   (c) What type of curve did you get? Explain the shape of this curve.

2. (a) Record the data from the chloroplast membrane experiment in the table below.

Exposure time	Drops of alcohol	% Transmittance
	5	
	10	
	15	
	20	

(b) What effect does an organic solvent such as alcohol have on photoreduction?

(c) At what concentration does this effect occur?

(d) Why is the membrane of the chloroplast important to the light reactions of photosynthesis?

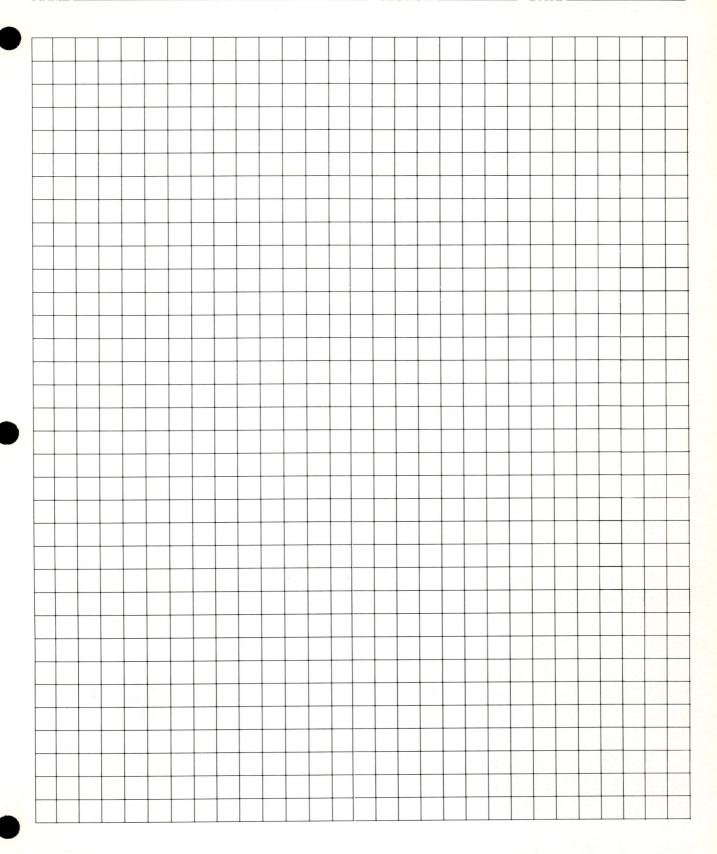

Plant Reproduction

Students of biology are usually well informed about the reproductive cycle of most animals. The mature animal is diploid (each cell having homologous pairs of chromosomes) and possesses reproductive organs capable of meiosis (reduction division), which produces haploid (*n*) gametes. These gametes fuse (syngamy) and produce a diploid (2*n*) zygote, which can grow by means of mitosis into a multicellular diploid organism. This reproductive cycle appears to have some evolutionary advantage in that the organism has pairs of genes, and a large variety of gametes are produced by meiosis.

These general reproductive characteristics have also evolved in plants, but in a very different manner. The terminology needed to explain the haploid and diploid stages of plants in this cycle reads like nothing you have studied before in biology. In most plants the generalized reproductive cycle shown in Figure 1 is in operation.

Figure 1   Generalized plant cycle

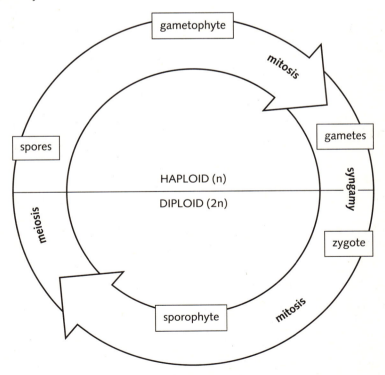

The gametophyte stage is composed of cells with an (n) number of chromosomes; that is, the chromosomes appear singly. This generation of the plant produces **gametes** by mitosis. Gametes from such gametophytes fuse to produce diploid ($2n$) **zygotes.** The zygote grows into a multicellular $2n$ **sporophyte** by mitotic division. The sporophyte stage produces haploid (n) **spores** by meiosis. These spores can germinate individually into another gametophyte stage. Plants demonstrate this "alternation of generations" in many different forms. With a little effort it is possible to identify these sexual and asexual generations by studying the life cycles of selected plants.

Your task in the next three activities is to identify the generalized plant cycle as it occurs in mosses, ferns, and flowering plants.

You may wish to review the general reproductive process in your textbook before continuing with these plant reproduction activities.

The Life Cycle of the Moss

OBJECTIVES

At the end of this laboratory activity, you should be able to:

- diagram the life cycle of the moss, showing the significant stages.
- define alternation of generations.
- identify each stage of the moss cycle as (*n*) or (2*n*).

INTRODUCTION

The term "moss" is misused considerably. For example, Spanish "moss" is really a flowering plant of the pineapple family (see Figure 29.1(a)), and reindeer "mosses" are lichens. In this activity we are concerned with the true mosses belonging to the class Musci of the division Bryophyta (see Figure 29.1(b)).

Figure 29.1 Mosses
(a) Spanish moss

(b) True moss

The true mosses elegantly demonstrate alternation of generations. The specific generations of mosses and their development are quite easy to observe, making them the logical start for a study of plant reproduction.

During this activity you will study the stages of the moss life cycle shown in Figure 29.2.

The gametophyte of the true moss plant lacks vascular tissue to transport water throughout the plant; however, it has organs that superficially resemble leaves, roots, and stems. The gametophyte is the conspicuous and independent stage of the moss cycle. This is the plant that you visualize when you think of a moss and usually is not more than several centimeters in height. The gametophyte plant can be unisexual or have sperm and egg producing organs on the same plant. In either event eggs are produced in the **archegonium** and sperm are produced in the **antheridium** by mitosis. Each cell of the gametophyte is in the haploid (*n*) condition. This is significant because any mutation will manifest itself in the cell containing it because the cell contains single genes, not pairs.

The archegonium is vase-shaped, swollen at the base, and contains a single egg. The neck of the archegonium contains canal cells that disintegrate as the archegonium matures, leaving a canal through which the sperm can swim to reach the mature ovum.

The antheridium is oval in shape and is outlined with a nonreproductive jacket of cells. Large numbers of the immature sperm are held within the antheridium. The antheridia are surrounded by sterile hairs, long filamentous organs called **paraphyses.**

When both the archegonium and the antheridium are mature, the flagellated sperm swim to the egg by way of surface water on the plant. This results in fertilization and the production of a diploid (2*n*) zygote, which grows by mitotic division to produce a sporophyte **embryo** within the archegonium. This embryo will grow into a mature sporophyte plant while remaining within the archegonium. This sporohyte plant is a single stalk with a terminal **capsule.**

Figure 29.2 Moss life cycle

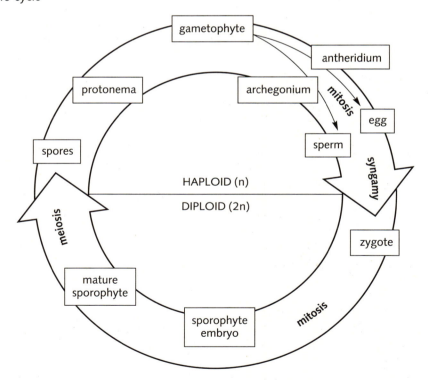

The sporophyte plant is capable of photosynthesis although it is somewhat dependent upon the gametophyte for the duration of its short life. When the sporophyte matures, meiosis occurs within the capsule, producing many haploid (n) spores that are released into the surrounding area. In the proper environment these spores will germinate and grow mitotically into a **protonema.** The cells that are produced are arranged in unicellular rows and at first glance resemble filamentous algae. Further development will produce a small leafy gametophyte.

MATERIALS

Mature living gametophyte moss plant with a
  sporophyte
Prepared slides of various stages of the moss
  cycle
Compound microscope

PROCEDURE

1. Obtain a mature living gametophyte moss plant with a sporophyte growing from its archegonium. Examine it carefully as you review the life cycle of the mosses described in the introduction. Complete Question 1 in the Evaluation.

As you study each of the following slides, complete the appropriate diagrams in Question 2 in the Evaluation.

2. Obtain a prepared slide of moss antheridium. Observe under low power to locate the elongated antheridium within the cup-shaped male head. The numerous minute cells within the antheridium are the developing sperm. Diagram the entire head, showing the antheridium and the enclosed sperm.

3. Obtain a prepared slide of moss archegonium. Observe under low power to locate the vase-shaped archegonia. If the slide has been properly prepared you should be able to see at least one developing egg in an archegonium. Diagram the archegonium.

4. Obtain a prepared slide of a moss sporophyte embryo. Observe under low power. Diagram and label the gametophyte supporting the sporophyte.

5. Obtain a prepared slide of a sporophyte capsule. Observe under low power and diagram the capsule and the enclosed spores.

6. Obtain a prepared slide of moss protonema. Observe under low power. Diagram and label.

EVALUATION 29

The Life Cycle of the Moss

1. Diagram a mature moss gametophyte plant with an attached sporophyte. Label each generation.

2. Complete the moss life cycle below by drawing the appropriate diagrams in the circles.

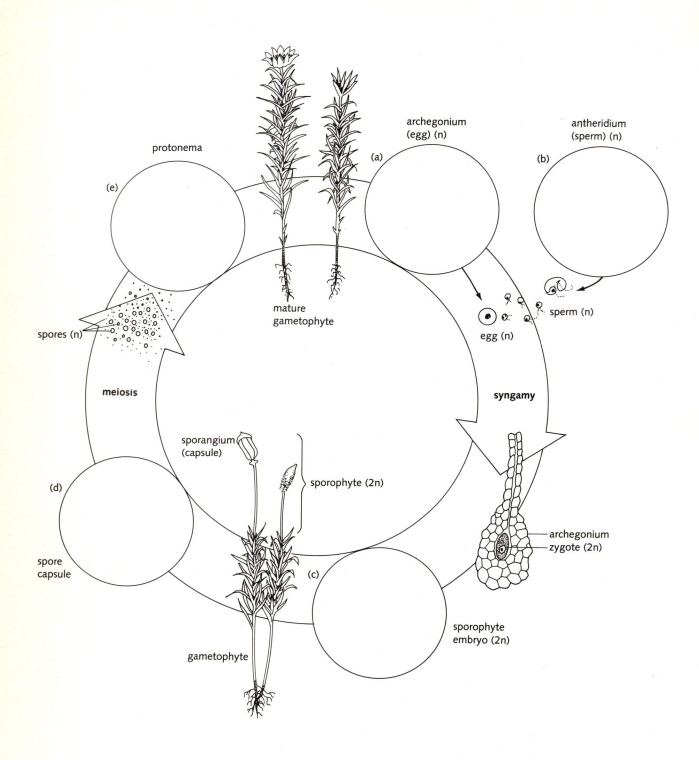

protonema

archegonium (egg) (n)

antheridium (sperm) (n)

(e)

(a)

(b)

mature gametophyte

sperm (n)

spores (n)

egg (n)

meiosis

syngamy

sporangium (capsule)

sporophyte (2n)

(d)

archegonium
zygote (2n)

spore capsule

gametophyte

(c)

sporophyte embryo (2n)

The Life Cycle of the Fern

OBJECTIVES

At the end of this laboratory activity, you should be able to:

- diagram the life cycle of the fern, showing the significant stages.
- identify each stage of the fern cycle as haploid (*n*) or diploid (*2n*).

INTRODUCTION

Ferns are among the most interesting and enchanting of the vascular plants (see Figure 30.1). They originated about 400 million years ago, and today about 12,000 different species cover our earth. About two-thirds of these grow in tropical regions. The stage of the fern reproductive cycle that most people are familiar with is the sporophyte, a large leaf with numerous subdivisions called **leaflets.** The main leaf grows from an underground stem called a **rhizome,** from which grow many fine roots. When the leaf first emerges from the ground it is wrapped in a tight coil known as a **fiddlehead** (see Figure 30.2). Ferns have well-developed vascular tissue, which allows many species to grow several meters high.

The life cycle of the fern shows the typical alternation of generations you studied in the moss.

Figure 30.1 Ferns
(a) Bird's nest fern

(b) Ostrich fern

(c) Brittle maidenhair fern

Figure 30.2 Fern fiddlehead

The major differences are that the sporophyte generation is more dominant and that the gametophyte generation, while still reproductively important, is physically less conspicuous.

The life cycle of the fern is summarized in Figure 30.3. The leaves of the fern have structures called **sori,** each of which is actually a group of microscopic reproductive structures known as **sporangia.** The diploid (2n) sori are located on the under surface of the leaf and appear as little black, red, brown, or white dots arranged in a rather regular pattern (see Figure 30.4). The formation of haploid (n) spores by meiosis occurs in the sporangia. After the spores mature, specialized cells in the sporangium contract, causing it to spring open and eject the spores into the air. If the environment is suitable, each spore will germinate into a small, flat, heart-shaped plant called the **prothallus,** with numerous rootlike rhizoids that extend into the ground. On its ventral (bottom) surface, each prothallus bears a sperm-producing antheridium and an egg-producing archegonium. The multiflagellate sperm and the eggs are produced by mitosis.

Ferns may self-fertilize or cross-fertilize. Cross-fertilization occurs when the mature sperm are released from one prothallium to fertilize the egg in

Figure 30.3 Fern life cycle

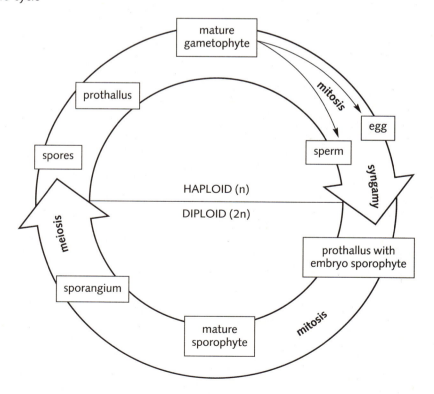

Figure 30.4 Fern sori

a different one. Fertilization takes place in the lower portion of archegonium and is followed immediately by mitosis and the development of a young embryo plant. This young sporophyte is nurtured by the prothallium for a short time and eventually differentiates into a mature sporophyte that roots itself in the soil and becomes an independent photosynthetic structure. Then subsequent disintegration of the gametophyte completes the cycle.

MATERIALS

Prepared slides of various stages of the fern life
 cycle
Living fern plant
Glass slides and coverslips
Dissecting needles
Pasteur pipettes
Compound microscope

PROCEDURE

As you study the mature fern plant and each of the prepared slides, complete the appropriate diagrams in Question 1 in the Evaluation.

1. Diagram a fern leaflet and show the arrangement of the sori in (a).

2. Obtain a prepared slide of fern sporangia or make your own slide by scraping the contents of a sorus into a drop of water on a glass slide. Examine under low power and diagram a sporangium with enclosed spores in (b).

3. Obtain a prepared slide of a mature gametophyte and examine it under low power. Make an outline diagram of this structure and label the antheridia, archegonia, and rhizoids in (e). The sex organs are located on the ventral surface of the prothallium. Why is this an advantage?

4. Obtain a prepared slide of a prothallium (gametophyte) with a young sporophyte. Make an outline diagram of this stage and label each of these structures in (i). Both the moss and fern show an alternation of generations in their life cycles. What advantages does this type of reproduction have for the organism?

EVALUATION 30

The Life Cycle of the Fern

1. Complete the fern life cycle below by drawing the appropriate diagrams in the circles.

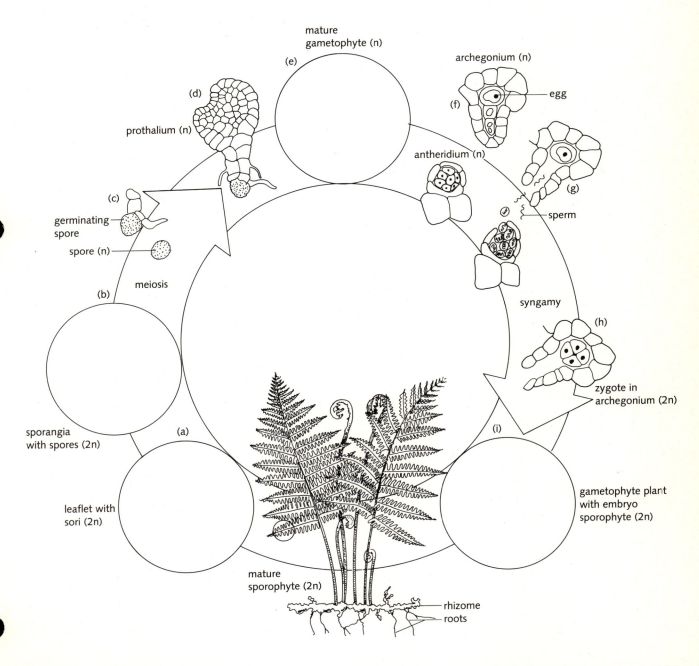

The Life Cycle of Flowering Plants (Angiosperms)

OBJECTIVES

At the end of this laboratory activity, you should be able to:

- diagram the life cycle of the flowering plants, showing the significant stages.
- identify each stage of the flowering plant cycle as haploid (*n*) or diploid (2*n*).

INTRODUCTION

Flowering plants (also known as **angiosperms**) are economically important to humans for many reasons. The number of products we use each day that come from flowering plants is staggering. They produce much of our food and large amounts of fiber for our housing and clothing. These products include your morning paper, the wooden chair you may be sitting on, or the watermelon you had for lunch.

The life cycle of flowering plants may be confusing because sexual qualities are often mistakenly attributed to the sporohyte generation and because the gametophyte is usually not known at all. The life cycle of the flowering plants is similar to the life cycle of the mosses and the ferns, but with some major modifications. During this activity you will study the stages of the flowering plant cycle described in Figure 31.1. You may want to review the parts of a typical flower shown in your textbook.

There are many different types of flowers that can appear on sporophyte plants. Some flowers contain both **stamens** and a **pistil,** while others have just one or the other. Both the stamen and pistil contain tissue where spores are produced by meiosis. Angiosperms can produce two types of spores within the flower. The smaller of the two kinds of spores (microspores) develop into the male gametophyte and the larger spores (megaspores) develop into the female gametophyte.

Special cells within the **anther** portion of the stamen produce four cells called **microspores** from one microspore mother cell, or **microsporocyte.** The single nucleus of each microspore divides mitotically and produces a cell with a generative nucleus and a tube nucleus. This microscopic single cell with two nuclei, a **pollen grain,** is the male gametophyte plant, which will ultimately produce two sperm mitotically from the generative nucleus.

How do these sperm reach the egg? The entire male gametophyte plant (pollen grain) is carried by the wind, insects, birds, or bats to the pistil containing the egg. This process is called **pollination.**

The base of the pistil of the flower contains the ovary. Within the ovary one or more ovules are found, containing a specialized cell called a megaspore mother cell, or **megasporocyte.** This diploid cell produces four haploid cells by meiosis. Three of these atrophy and the remaining one develops into a megaspore. As you have learned previously, spores germinate and grow, but in this instance the megaspore develops within the ovule, which is within the ovary. The development proceeds as follows: The single nucleus produces two, then four, then eight nuclei by mitosis. One nucleus from each set of four migrate to the center of the **embryo sac** (female gametophyte) and become the two polar nuclei. Cytokinesis then occurs, walling off three cells at one end of the developing female gametophyte and three at the opposite end. Thus the microscopic female gametophyte contains eight nuclei in seven cells.

Huge quantities of pollen must be produced to ensure that pollen arrives at a pistil of another flower

Figure 31.1 The life cycle of flowering plants

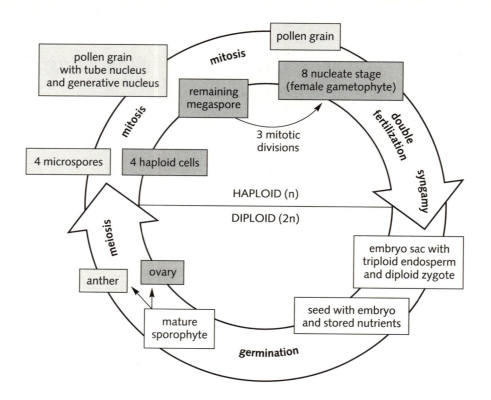

of the same species. Once the pollen reaches the pistil, the tube nucleus of the pollen grain causes a pollen tube to grow through the tissue of the pistil and enter through the **micropyle** (an opening in the ovule) to reach the egg within the female gametophyte. One sperm fertilizes the egg and produces a diploid (2*n*) zygote, which grows into an embryo sporophyte. The second sperm fertilizes the two already-fused polar nuclei, producing a

triploid (3*n*) endosperm nucleus. This endosperm cell will produce a large number of cells by mitosis to serve as food for the sporophyte embryo. Portions of the ovule, the embryo, and the endosperm tissue make up the resulting seed. When the seed is planted it germinates, and the embryo sporophyte grows rapidly into a new mature sporophyte.

MATERIALS

Living flower or flower model
Prepared slides of the flowering plant cycle
   Anther cross section
   Ovary cross section
   Pollen grains
   Four-nucleate stage of female gametophyte
   Eight-nucleate stage of female gametophyte
Compound microscope

PROCEDURE

1. Study a living flower or a model of a typical flower and complete a drawing of this flower or model in Question 1 in the Evaluation.

As you study each of the prepared slides complete the appropriate diagrams in Question 2 in the Evaluation.

Development of the Male Gametophyte

1. Obtain a slide of a transverse section of an anther and examine it under the scanning power of your microscope. Make an outline diagram of the anther and label the four pollen sacs and microsporocytes in (a). What is the function of the microsporocytes?

2. Obtain a slide of pollen grains and diagram several of these grains, showing the tube nucleus and generative nucleus in (b). What kind of cell division has produced these nuclei?

3. Examine a slide of pollen grains and diagram a germinating pollen grain, showing a pollen tube in (c).

Development of the Female Gametophyte

1. Obtain a slide showing a cross section of an ovary. Using the scanning power of your microscope, diagram the ovary, showing the position of the ovules in (d). The ovules may or may not be visible on your slide depending on how the ovary was sliced during preparation. Note how the outer covering of the ovules does not fuse but leaves an opening, the micropyle, through which the pollen will enter just before fertilization. Label the ovary, ovule, megasporangium, integuments, and micropyle.

2. Of the four haploid cells that are produced from the megasporocyte, three disintegrate, leaving a megaspore. What is the advantage to the plant in having these three cells disintegrate? What type of cell division produces the two-, four-, and eight-nucleate stage?

3. Obtain a slide of the four-nucleate stage in the development of the female gametophyte. Obtain another slide showing the eight-nucleate stage of development. Diagram these nuclei and label the fusion nuclei and egg nucleus in (e) and (f). The chromosome condition of the zygote is _____ (n) and the triple fusion nucleus (endosperm) is _____ (n). What will each of these develop into in the seed?

EVALUATION 31

The Life Cycle of Flowering Plants (Angiosperms)

1. Study the living flower or flower model provided by your instructor. What is the relative position of the stamens to the pistil? Label the anther, filament, stigma, style, and ovary in the diagram below.

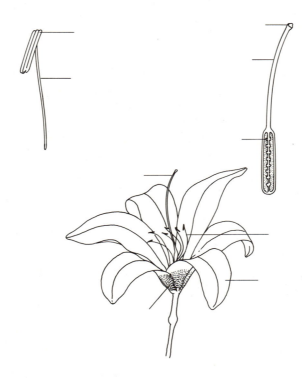

2. Complete the flowering plant life cycle below by drawing the appropriate diagrams in the circles.

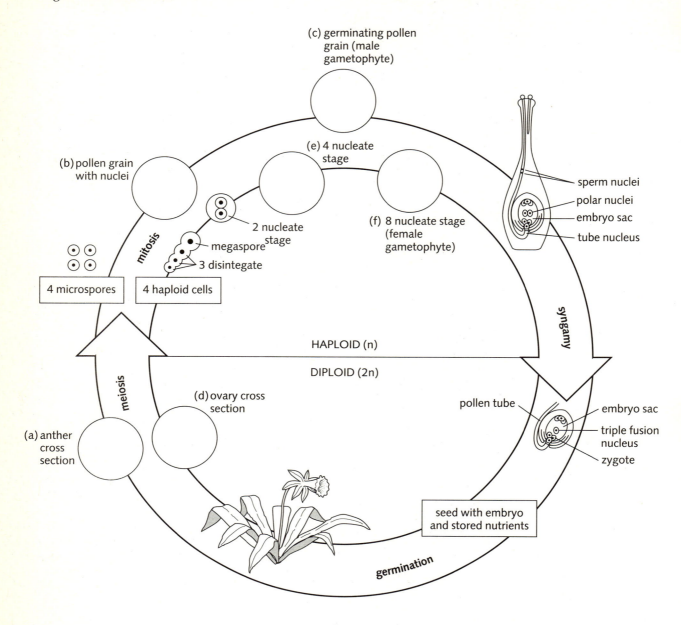

(c) germinating pollen grain (male gametophyte)

(e) 4 nucleate stage

(b) pollen grain with nuclei

2 nucleate stage

megaspore

3 disintegate

4 microspores

4 haploid cells

mitosis

(f) 8 nucleate stage (female gametophyte)

sperm nuclei
polar nuclei
embryo sac
tube nucleus

syngamy

HAPLOID (n)

DIPLOID (2n)

meiosis

(d) ovary cross section

(a) anther cross section

pollen tube

embryo sac
triple fusion nucleus
zygote

seed with embryo and stored nutrients

germination

The Water Vascular System of Plants

OBJECTIVES

At the end of this laboratory activity, you should be able to:

- trace the movement of water through a vascular plant.
- illustrate the arrangement of xylem tissue in roots.
- section, stain, and identify selected tissues in a herbaceous plant stem.

INTRODUCTION

Photosynthetic organisms need only water, certain minerals, carbon dioxide, and light to produce energy-rich organic compounds. In addition, such organisms require oxygen for respiration. Water plants can quite easily meet these relatively simple requirements, but as plants evolved and invaded the land, relatively little water was available to them. Without direct contact with water on all surfaces, land plants had to obtain it from the soil and distribute it throughout the plant in some efficient manner.

This problem was solved in many land plants by the evolution of a **water vascular system,** which allows for the movement of water from the soil to the roots and on to the stems and leaves. This system made water available for all cellular activity in the plant, including photosynthesis.

Xylem tissue is the main component of the water vascular system. The cells composing xylem tissue include **tracheids, vessels, parenchymal** cells for storage, and **sclerenchymal** cells (fibers) for support. In this activity you are concerned only with those cells that conduct water, the tracheids and vessels. Both kinds of cells are elongated and at maturity lack the living portion of the cell. The dead cells line up end to end to form water-conducting tubes. Tracheids have overlapping tapered ends and pits that line up, allowing water to flow from one cell to another. Vessels have end walls containing actual perforations or holes through which water flows.

Conducting tubes composed of xylem cells are found in the roots, stems, and leaves of vascular plants. In this activity you will investigate root hairs and the xylem cells of roots, stems, and leaves. You will also examine how the arrangement of xylem cells differs in each location.

Almost all of the water absorbed by vascular plants occurs through the younger part of the roots. This intake of water takes place specifically through the epidermal covering. Certain cells of the epidermis elongate to produce **root hairs,** through which most of the absorption of water and minerals by vascular plants occurs. Water enters these cells by osmosis, then passes from cell to cell across the root tissue, again by osmosis, until it reaches the water-conducting xylem cells. Water may also travel along the cell walls by a process called **apoplastic movement.** Eventually the water reaches the central core of the root, the **stele.**

Keep in mind that the cells you are about to examine generally appear circular in shape. The tissue was cut in cross section; thus, an elongated cell appears as a circle. As you view this dicot root cross section you will be able to view the tortured route that water must make to cross the root and enter the xylem cells. The dead elongated xylem cells begin to conduct water in the root of plants by capillary action. In roots you will see these cells arranged in the form of an X, but as the xylem tissue continues into the stem, the cells will take on another arrangement.

Water will ultimately move from the xylem cells in the vascular bundles within the stem to xylem cells within the leaves. The water is then available to the **mesophyll** cells in the leaf, which are employed in photosynthesis.

Stomata (tiny openings), mostly on the lower surface of the leaf, permit air to enter the intercellular spaces within the mesophyll (see Figure 32.1). Water and carbon dioxide are now available for photosynthesis. The stomata are really double-edged swords. Although the stomata allow air containing carbon dioxide to enter, they also promote **transpiration,** the loss of water from intercellular spaces (see Figure 32.2). In the event that too much water is lost in this manner, the plant will suffer from dehydration and wilt.

The specialized epidermal cells of the stomata are the **guard** cells. They are the only cells of the epidermis containing chloroplasts and thus are readily seen under the microscope. The guard cells surround the opening into the intercellular spaces. Collectively, this entire stomatal apparatus is usually referred to as a **stoma** (plural **stomata**).

MATERIALS

Radish (*Raphanus sativus*) seedlings
Microscope slides
Cover slips
Compound microscope
Prepared slide of typical dicot root
Boiled geranium (*Geraniaceae*) stem tissue
Tweezers
Prepared slide of sunflower (*Helianthus*) stem
Dicot stem tissue
Cork
Razor blade
Petri dishes
Ethanol 50%
Dissecting microscope
Toluidine blue stain
Distilled water
Glycerine 50%
Prepared slide of leaf cross section
Geranium (*Geraniaceae*) leaf tissue

Figure 32.1 Stomata from underside of leaf

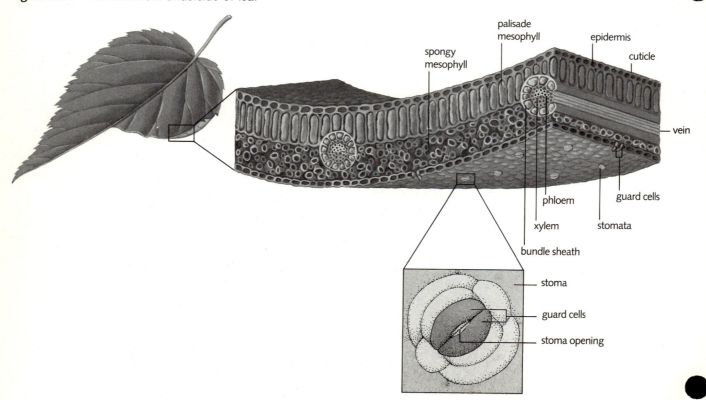

Figure 32.2 Water droplets on leaf caused by transpiration

PROCEDURE

The microscopic examinations you will be making during this activity should help you in visualizing the movement of water through a typical vascular plant. After completing each observation prepare a labeled drawing in the appropriate section of the flowchart in Question 1 in the Evaluation.

Root Hairs

1. Obtain a radish seedling (*Raphanus sativus*) from the petri dish provided for you. Place it on a slide in a drop of water. Observe under scanning or low power.

2. Sketch several of these root hairs of the epidermis in the appropriate area of the flowchart in the Evaluation.

Root Cross Section

1. Obtain a prepared slide of a typical dicot root cross section.

2. Study the slide under low power, noticing the single cell layer of epidermis covering the outside of the root. Also notice the stele, which is surrounded by a single layer of cells called the **endodermis.** Within the stele you will find the xylem and associated tissue of the vascular system. The xylem cells will appear as large thick-walled cells, which when viewed collectively, appear as the arms of an X (see Figure 32.3).

3. Make a sketch of the root cross section on the flowchart in the Evaluation and label the stele, xylem, and cortex.

Figure 32.3 Dicot root cross section

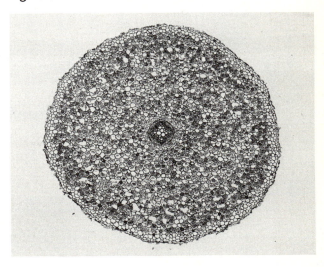

Stem: Whole Xylem Cells

You have seen the cells of xylem tissue in cross section, but how do they appear when they are whole? Your instructor will provide you with some geranium plant tissue that has been gently boiled in nitric acid to break down the stem tissue. This procedure separated the cells of the stem.

1. Using tweezers, place a small quantity of this plant material into a drop of water on a microscope slide and cover with a cover slip. Examine under low and high power to observe the xylem cells.

2. Review the description given of these cells in the introduction. Diagram several different types of xylem cells in the flowchart in the Evaluation.

Stem: Cross Section

1. Select a prepared slide of the sunflower (*Helianthus*) stem and observe under low power. Starting your observation from the outside edge, locate the epidermis. Continue inward, locating the familiar cortex composed of large, oval, thin-walled cells. The vascular tissue is arranged in bundles (resembling acorns) around the periphery of the stem. Notice the large thick-walled xylem cells toward the inside of the bundle. A sheath of thick-walled sclerenchyma cells surrounds each vascular bundle. The central portion of the stem is composed of pith (see Figure 32.4).

Figure 32.4 *Helianthus* stem cross section

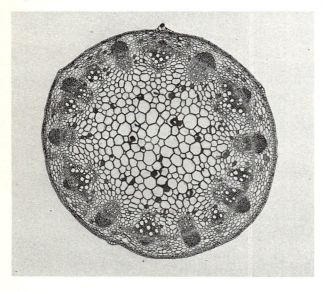

2. In the Evaluation, make a drawing of a wedge section of your field of view, showing one or more vascular bundles.

Preparing Your Own Stem Cross Section

Before you continue to trace the water from the stems to the leaves, take a brief detour to make your own cross sections. Your instructor will provide herbaceous dicot stem specimens for your use.

1. Obtain a 2-cm length of stem, a cork, a razor blade, and a petri dish containing 50% ethanol.

2. Place a section of stem in the groove cut into the side of the cork. Allow a small portion of the stem to extend over the end of the cork. Slice, do not chop, the thinnest possible sections of stem tissue. Pretend you are slicing yourself a piece of angel food cake! See Figure 32.5.

3. Place 10 or more slices (sections) into the 50% ethanol in the petri dish. The sections should be left in the ethanol for at least 5 minutes to kill the cells and fix the tissue.

4. Examine the sections under a dissecting microscope. Place the entire petri dish on the microscope stage for examination. Select the thinnest sections for further study.

5. Place the selected sections in a second petri dish containing toluidine blue stain. Toluidine is a differential stain, which means that it will combine with different tissues to varying degrees. Use tweezers to gently transfer the sections from one petri dish to another.

6. Allow the sections to sit for 5 minutes in the stain. Transfer them to a third petri dish containing distilled water to remove any excess stain.

7. Place a section into a drop of 50% glycerine solution on a microscope slide and cover with a cover slip. Observe under low and high power of your compound microscope. In all likelihood you will be pleased with your results; if not try again, and again . . . and maybe again, if necessary.

8. Draw a portion of your stem cross section in the Evaluation and label the stele, xylem, and cortex.

Figure 32.5 Preparing stem cross section

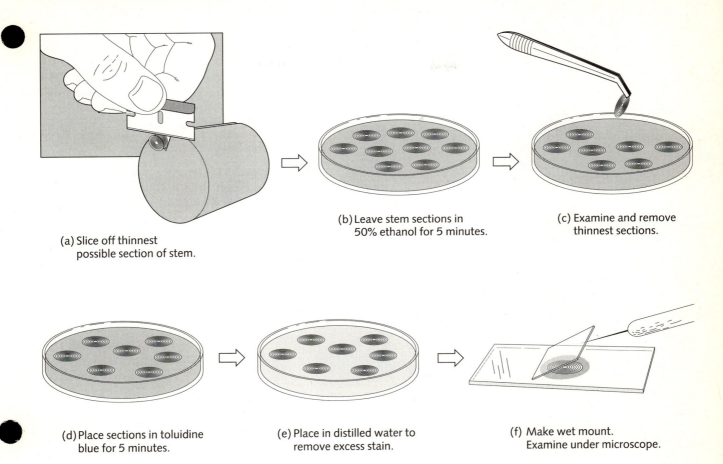

(a) Slice off thinnest possible section of stem.

(b) Leave stem sections in 50% ethanol for 5 minutes.

(c) Examine and remove thinnest sections.

(d) Place sections in toluidine blue for 5 minutes.

(e) Place in distilled water to remove excess stain.

(f) Make wet mount. Examine under microscope.

Leaf

1. Obtain a prepared slide of a leaf cross section and study it under low power. Notice a single layer of epidermal cells covering both the upper and lower surfaces of the leaf. Also notice the vascular bundles containing the xylem cells. The remaining cells are leaf ground tissue or mesophyll.

2. Diagram and label a section of this leaf cross section in the Evaluation.

Stomata

1. Obtain a geranium leaf and tear it to obtain a ragged edge (Figure 32.6). Examine the edge carefully to see an extended portion of the clear epidermis. Pull off a section of this epidermal layer with tweezers and prepare a wet mount slide.

2. Observe under the compound microscope. You will find several stomata, appearing as oval openings between the two guard cells. The guard cells are the only cells of the epidermis that contain chloroplasts.

3. Diagram and label several stomata on the flow-chart in the Evaluation.

As a water molecule escapes into the air through a stoma it is replaced by another water molecule. Each water molecule is attracted to an adjacent water molecule throughout the water vascular system by cohesion. Once water enters the vascular system it is pulled along in this fashion.

Figure 32.6 Preparing geranium leaf for observation of stomata

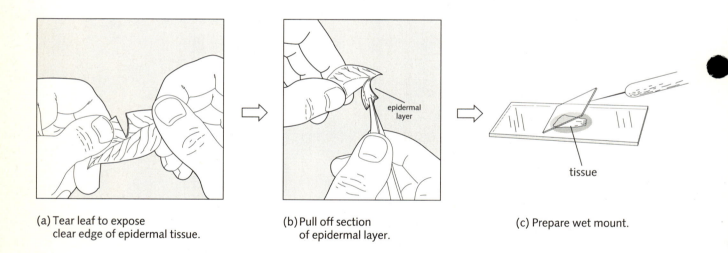

(a) Tear leaf to expose clear edge of epidermal tissue.

(b) Pull off section of epidermal layer.

(c) Prepare wet mount.

EVALUATION 32

The Water Vascular System of Plants

1. Complete the diagrams in the flowchart below.

leaf stomata

_____ X

leaf cross section

_____ X

hand-sectioned stem

_____ X

helianthus stem

_____ X

whole xylem cells

stele of dicot root

_____ X

root hairs

_____ X

Water and Mineral Transport in Plants

OBJECTIVES

At the end of this laboratory activity, you should be able to:

- describe the passage of water through the plant body from root to leaf.
- name and describe the physical characteristics of water that are involved in the process of transpiration.
- explain how a potometer works.
- explain how various environmental factors may influence transpiration rate.

INTRODUCTION

During Laboratory Activity 32 you traced the anatomy of the xylem tissue, which transports water through the plant body from root to leaf. In this laboratory activity you will work with a plant to measure the rate at which water moves out of the plant in the process botanists call **transpiration.**

If you have ever had the responsibility for watering the garden around your house you are sure to have noticed that most plants require a considerable amount of water to survive. One early botanist, Stephen Hale, calculated that his sunflower plant "imbibed" and "perspired" 17 times as much water as he himself did, pound for pound, every 24 hours. One of the reasons for this is that most of the water absorbed by the plant root is transported through the plant and evaporates from the leaves. A plant transpires about 90% of the water that it absorbs. (An animal, although it loses water through evaporation from its body surface, recirculates most of its water in its blood plasma and other body fluids.) In vascular plants, xylem

tissue and the many unique properties of water allow this remarkable conducting process to occur. Without it vascular plants would never have reached the heights so admirably demonstrated by the oaks, spruces, and redwoods of our forests.

Water absorption begins with root hairs, which are an extension of epidermal cells lining the outer surface of roots. Water passes into these epidermal cells primarily by osmosis and then along the walls of the cells in the cortex to a central area called the **stele.** In the stele, water enters the xylem tissue and begins its journey upward into the stem and finally to the leaf (see Figure 33.1). The question of how large volumes of water can be transported through a plant has intrigued botanists for years. The most accepted explanation draws on several of the physical characteristics of water you studied in Laboratory Activity 20 and on the fact that water is able to evaporate from the leaf as a result of being heated by the radiant energy from the sun. The idea may be summarized as follows.

1. Water enters the root hairs by osmosis and then travels to the xylem to begin its journey upward.

2. As the water column moves up, it is held together by the cohesiveness of the water molecules. This attraction of water molecules for each other gives water its high tensile strength and enables it to move in a long column without breaking.

3. Water molcules are also attracted to the walls of the xylem cells (adhesion) and this also helps the water to move upward. (Remember the exercise in Activity 20 on capillary action as a function of the diameter of the capillary tubing.)

Figure 33.1 Water absorption by root hairs

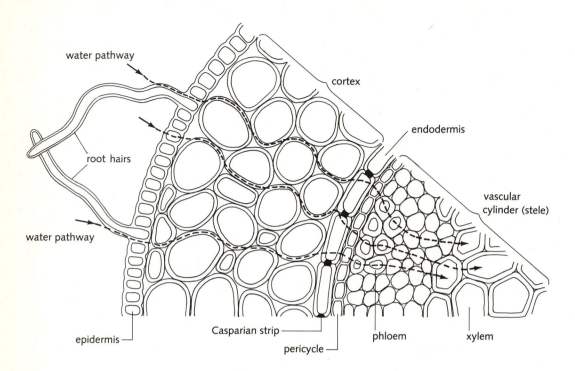

water pathway

cortex

endodermis

root hairs

vascular
cylinder (stele)

water pathway

epidermis

Casparian strip

pericycle

phloem

xylem

4. Finally, when the water reaches the leaf it moves by osmosis to the mesophyll cells where it evaporates into the air spaces in the leaf, through the stomata, and into the atmosphere. As the water evaporates from the mesophyll cells it pulls the entire column behind it, replacing what is being lost. The energy that causes the water to evaporate comes from the sun and is the driving force behind the whole process of transpiration (see Figure 33.2).

This very brief explanation describes the transpiration-cohesion-tension theory of water movement in vascular plants. Consult your textbook for a more detailed account of this process and for a discussion of the evidence that supports this hypothesis.

In today's laboratory activity you will examine the rate at which transpiration occurs and several of the environmental factors that may influence the process.

MATERIALS

Cobalt chloride filter paper
Geranium plant
Tweezers
Plastic wrap
Paper clips or tape
Celery stalk
Beaker
Dropping bottle of methylene blue stain
Glass slide
Compound microscope
Thoday potometer apparatus
Cuttings of herbaceous dicot plants
Razor blade
Petroleum jelly or silicone sealant
Paper towel
White paper
Metric ruler

Figure 33.2 The pulling force of transpiration

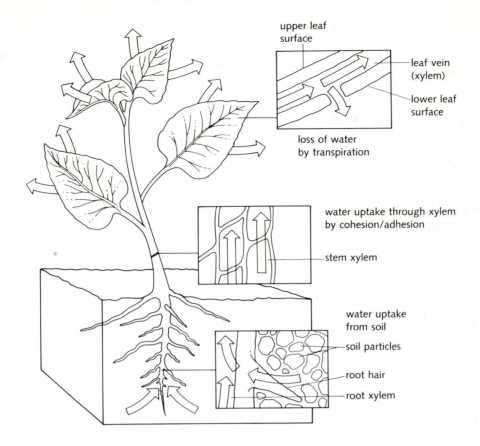

PROCEDURE

Observing Water Transport in Celery and Geranium

1. Obtain several pieces of filter paper (about 2 cm square) that have been soaked in a saturated solution of cobalt chloride ($CoCl_2$). The paper should be dry and a distinct blue color. If it is not, ask your instructor for directions on how to dry it. Take a well-watered geranium plant and, with a pair of tweezers, place one piece of cobalt chloride paper on the top of a leaf and the other on the bottom.

2. Completely cover the leaf with plastic wrap and fasten with paper clips or tape to exclude moisture from the atmosphere. Prepare several leaves in this manner. The cobalt chloride paper should turn light blue and then pink as it absorbs moisture.

3. Observe each piece of paper over a period of about 1 hour and record your observations in Question 1(a) in the Evaluation section.

4. Take a fresh stalk of celery with the leaves intact and cut about 1 or 2 cm from the end. Place the stalk in a beaker of water to which methylene blue stain has been added. The solution should be extremely dark. Allow the celery stalk to remain in the solution until the dye becomes visible in the stem and leaves.

5. Make a thin cross section of a piece of the stem that is not too saturated with the dye (near the top) and mount on a glass slide. Observe under low power of your microscope and record your observations in Question 2 in the Evaluation.

Measuring the Rate of Transpiration

The device you will use to measure the rate of transpiration is called a Thoday potometer (*poto* = drink). During assembly, you must be careful not to allow any air to enter the tubing or the cut plant stem.

1. Use a cut branch from a plant such as a geranium or some small woody plant and hold the stem under water while cutting about 1 or 2 cm from the end. This will remove any dead tissue and air bubbles that may have entered the xylem. Cut on an angle as shown in Figure 33.3. Do not remove the plant from the water after cutting.

2. Submerge the entire potometer apparatus and allow it to fill completely with water. There should be no air bubbles in the tubing. Carefully insert the cut stem about 2 cm into the rubber tubing. Be sure that the rubber fits snugly around the stem but not so tightly that it may restrict the flow of water. At this point the entire apparatus may be removed from the water and mounted on the ring stands. Check to be sure there are no air bubbles and that the apparatus does not leak. It may be necessary to seal the tubing around the plant stem with petroleum jelly or silicone sealant. The plant should be held gently in the clamp with a folded paper towel. The capillary tubing should rest on the table on a piece of white paper with a metric ruler parallel to the tubing. The clamp at the bottom of the thistle tube must be closed.

3. As water leaves the plant by transpiration it is replaced by water from the tubing. The rate of water loss may be determined by measuring how far the meniscus moves in the capillary tubing over a specified period of time. Read the meniscus at 1-minute intervals for 10 minutes. You may need to repeat a 10-minute run several times and average the results. Water should be replaced in the tubing by loosening the clamp at the bottom of the thistle tube.

4. Establish the normal rate of water loss for your plant and record the data in Question 3 in the Evaluation.

Environmental Conditions and Transpiration Rate

The rate at which plants transpire may be altered by changing environmental conditions or by changing some structure or function of the plant itself. Your task in this exercise is to develop a hypothesis concerning how changes in air movement, light, temperature, humidity, and the opening and closing of stomata may influence transpiration, and then design an experiment to test your hypothesis. Remember, scientists do experiments only if they seem reasonable on the basis of previous experience and information, so your reasons for doing this experiment should be based on your knowledge of transpiration and plant structure. In Question 4 in the Evaluation write a clear, succinct statement giving the background information that leads to your experiment and then state the hypothesis you wish to test. The data collected in the first activity should be used as the control.

Figure 33.3 Preparing stem for measuring transpiration

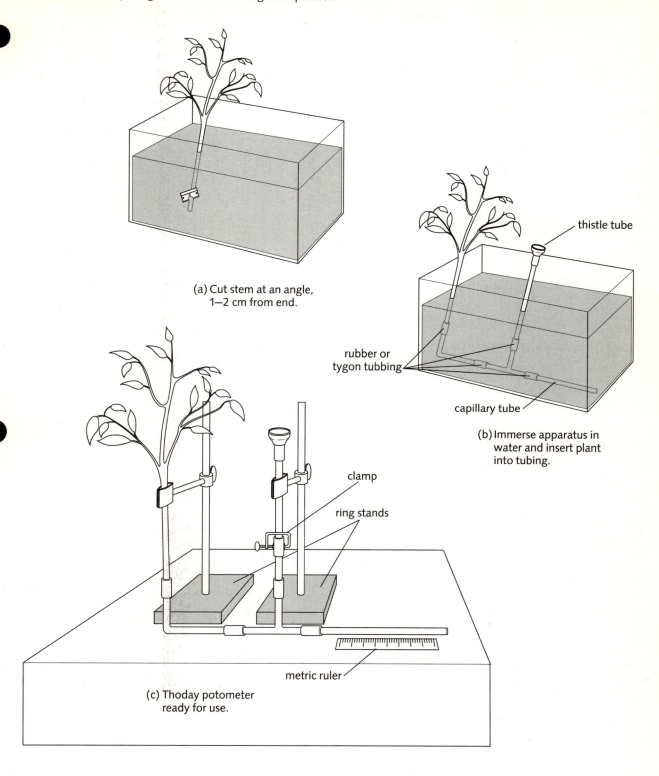

(a) Cut stem at an angle,
1–2 cm from end.

thistle tube

rubber or
tygon tubbing

capillary tube

(b) Immerse apparatus in
water and insert plant
into tubing.

clamp

ring stands

metric ruler

(c) Thoday potometer
ready for use.

EVALUATION 33

Water and Mineral Transport in Plants

1. (a) Record your observations of the color changes of the cobalt chloride paper.

   (b) It is common for deciduous terrestrial plants to have the majority of their stomata on the underside of the leaf. Do your results confirm this? Why would this be an advantage to the plant?

2. Record your observations of the distribution of the dye in the stem and leaves of the celery.

3. Record the data on transpiration rate in the following table. Record water loss in millimeters.

	Time (minutes)									
Run	1	2	3	4	5	6	7	8	9	10
1										
2										
3										
Average										

4. State the background information for your experiment on how changing an environmental condition may effect the rate of transpiration in a plant.

5. State your hypothesis for this experiment.

6. Record the data from your experiment in the following table.

Run	Time (minutes)									
	1	2	3	4	5	6	7	8	9	10
1										
2										
3										
Average										

7. On the graph paper provided, prepare a graph of the data from the control (first activity) and experimental setups. Be sure to plot the independent variable (which is _____) on the ordinate (*x*-axis) and the dependent variable (which is _____) on the abscissa (*y*-axis).

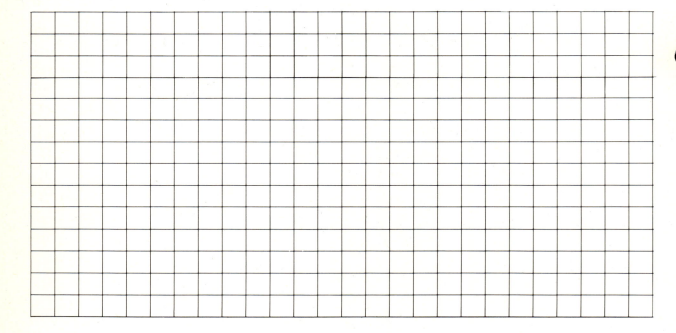

8. Interpret your results from this experiment. Was your hypothesis substantiated? If your hypothesis was not substantiated, explain what steps you would take; e.g., restate the hypothesis for a better one, redesign the experiment.

The Dissection of the Fetal Pig

OBJECTIVES

At the end of this laboratory activity, you should be able to:

- locate and identify each of the major organs of the fetal pig.
- place each organ in its proper organ-system.

INTRODUCTION

People have been dissecting animals for a very long time in order to learn more about animal structure and function and also to learn about human anatomy. One of the earliest dissectors was Galen (A.D. 129–199), a physician to the Roman Emperor Marcus Aurelius. He made many important discoveries through his dissections of the pig, the dog, and the rhesus monkey. Galen showed us that much can be learned about human anatomy by dissecting animals like the pig. For example, most people believed that the arteries contained air and Galen demonstrated that they carried blood instead. Most, if not all, of his dissections were made of animals other than humans. He then extrapolated the information obtained and applied it to humans. Galen's work was considered infallible for the next 1200 years!

The fetal (unborn) pigs you will dissect were found in the uterine horns of a mature sow as it was being processed for meat. These fetal pigs would simply have been discarded if they were not used for dissection. The gestation period for the pig is 112–115 days. Just before birth it measures about 30 cm from the tip of the snout to the base of the tail. Your specimen will be approximately 22 cm in length and about 100 days old.

The pig and human are quite similar in their internal anatomy; they both belong to the same subphylum, Vertebrata, and to the same class, Mammalia. However, the pig (*Sus scrofa*) belongs to the order Artiodactyla (even-toed hoofed mammals) while humans (*Homo sapiens*) belong to the order Primates (upright animals).

This dissection is by no means a complete one, and you will be specifically concerned only with locating and identifying the major organs of the pig's body. You will not attempt a dissection of the brain because this tissue is particularly fragile and difficult to work with. In addition, only the larger blood-carrying vessels will be studied. The function of each organ is not discussed in depth, and you should consult the appropriate chapters in your textbook for additional information.

MATERIALS

Latex gloves
Petroleum jelly
Goggles
Fetal pig, double-injected
Dissecting pan
Scissors
String
Dissecting pins
Scalpel or razor blade
Glass rod
Dissecting needles or probe
Tweezers

PROCEDURE

You and your laboratory partner will be given a fetal pig by your instructor. As you dissect the pig, be careful to follow the instructions in your manual as well as any additional directions given by the laboratory instructor. As you work on the pig, you will need to know the anatomical terms, listed in Table 34.1. Familiarize yourself with them before you begin your dissection.

Your pig is preserved in a chemical preservative that may dehydrate your fingertips. To prevent this you should wear gloves or cover your fingers with a thin film of petroleum jelly as often as needed.

► **CAUTION: If you wear contact lenses, the chemical used to preserve the pig may present a hazard. Your instructor will provide special goggles for you to wear during the dissection.**

Table 34.1 Anatomical terms

Term	Region
anterior	toward the head
posterior	toward the hind end
longitudinal	long axis from head to tail
transverse	perpendicular to the long axis
pectoral	the chest and shoulder region
pelvic	the hip region
dorsal	the back region
ventral	the belly region
lateral	toward the side
median	toward the middle
left and right	refer to the pig's left and right

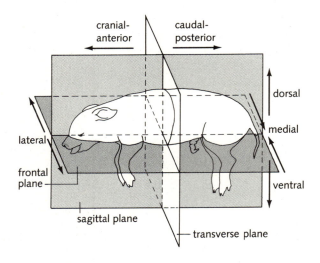

External Anatomy

1. Place your animal in the dissecting pan and examine the following external body parts. Refer to Figure 34.1 while examining your pig.
   - **Limbs:** The pig's limbs are adapted for walking on its toes (digitigrade locomotion) and varies from the human foot, which is adapted for walking on the sole (plantigrade locomotion).
   - **Digits:** Carefully examine the pig's foot and find the four digits. Note that the middle two are flattened and have hooves that are homologus to human toenails. Hold the pig in a walking position and notice that only the middle two digits touch the pan.
   - **Hock:** Find the first joint up from the toes; this is called the **hock** and is homologous to the human ankle.
   - **Snout:** The snout of the pig is flattened on the end and is well adapted for rooting in the soil.
   - **Umbilicus:** On the ventral surface is the umbilical cord, which carries nutrients and wastes to and from the fetus. This cord was originally attached to the fetal portion of the placenta, the organ responsible for exchanging nutrients and wastes between the fetus and mother.

2. Make a fresh cut with scissors to remove the ragged end of the umbilicus. Examine this cut end and locate two small **umbilical arteries** with thick walls and one **umbilical vein,** which is larger and has a thinner wall. If the pig was injected with latex, the larger arteries will be red and the larger veins blue.

Male or Female

3. Examine the region just posterior to the umbilicus to determine the sex of your pig. If you find a small opening, the **urogenital** opening,

you have a male. The **penis** lies under the skin posterior to the urogenital opening. You can feel it by pinching the skin in this area. The **scrotal sacs** are located ventral to the anus. Squeezing the hind legs together will cause the scrotal sacs to bulge slightly.

4. If the urogenital opening is not present, examine the area just ventral to the anus. A urogenital opening here indicates a female. Complete Question 1 in the Evaluation.

Internal Anatomy

In preparation for studying the internal anatomy, you will have to reposition your pig in the dissecting pan and tie it firmly before making any abdominal incisions. Follow these directions carefully.

1. Place the pig in the dissecting pan, dorsal side down. Tie a piece of string to one ankle, pass the string under the pan and tie again to the opposite ankle. The legs should be spread apart under some degree of tension. Repeat this procedure with the other pair of legs.

2. With your scissors, make a mid-ventral longitudinal incision beginning just anterior to the umbilicus. Cut in an anterior direction up to and completely through the **sternum** (breastbone), stopping at the **papilla** (hair on chin). Make the remaining incision by following the dotted lines in Figure 34.1.

➤ **CAUTION: Be careful to cut through only the skin and muscle, making certain you do not penetrate too deeply since this will damage some of the internal organs.**

3. Retie the strings to take up any slack that may have developed during cutting. Use dissecting pins to hold the body well against the pan.

Figure 34.1 External anatomy of the fetal pig
(a) Male

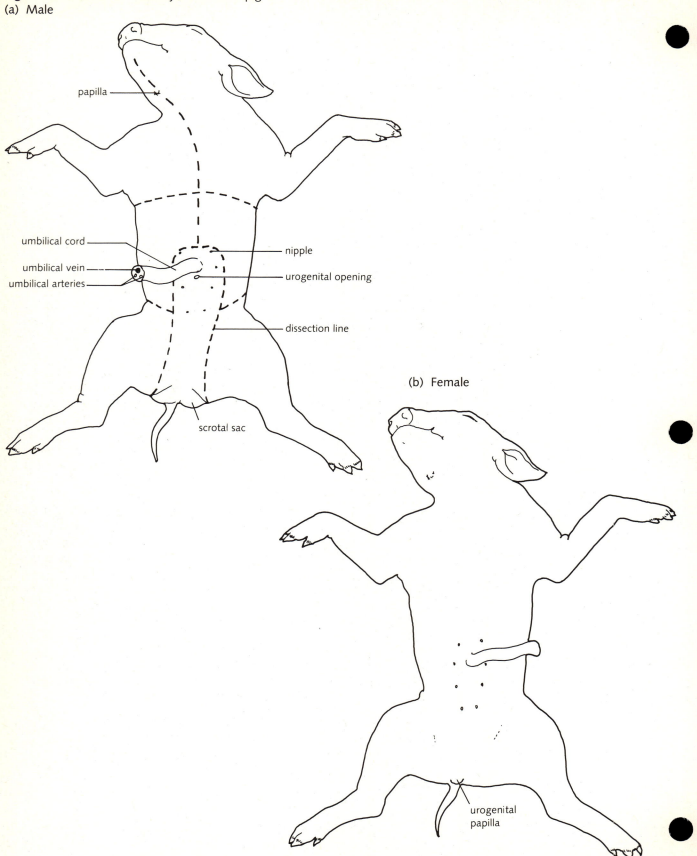

papilla

umbilical cord
umbilical vein
umbilical arteries

nipple

urogenital opening

dissection line

scrotal sac

(b) Female

urogenital
papilla

Digestive System

As you examine the two body cavities refer to Figures 34.2 and 34.3. The anterior **thoracic cavity** contains the lungs and heart. Posterior to the thoracic cavity is the **abdominal cavity.** These cavities are separated by the **diaphragm,** a very thin muscular sheet of tissue attached to the body wall.

1. Using a razor or scalpel, cut through the angle of the jaws so that the lower jaw can be pulled down. Locate the **hard palate** in the roof of the mouth and the **soft palate** posterior to it. Find your own hard and soft palate with your tongue.

2. Feel along the outer edge of the mouth of the pig for the developing teeth. Expose them by cutting into the jaw and removing the covering of tissue. Remove as many of the teeth as possible. Try to identify the molars and premolars (for grinding) and the canines (for tearing).

Figure 34.2 Internal anatomy of the fetal pig

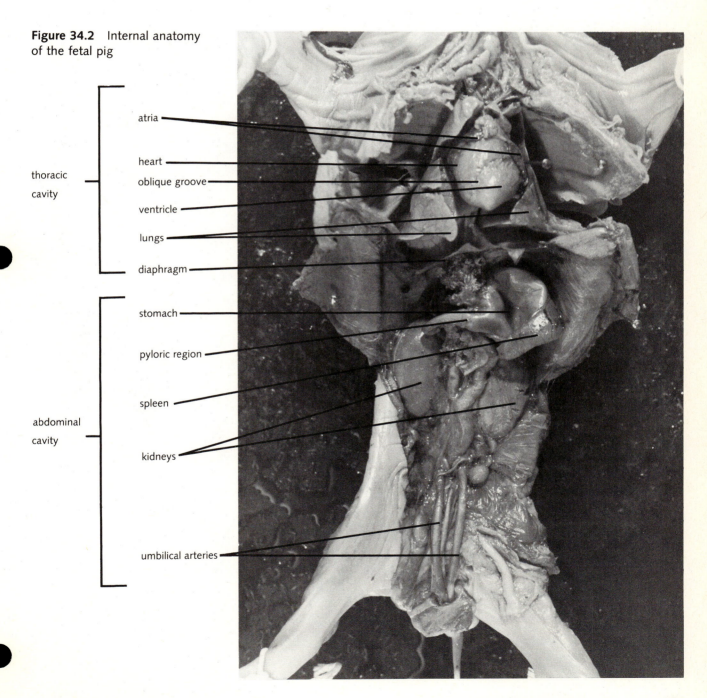

thoracic cavity

- atria
- heart
- oblique groove
- ventricle
- lungs
- diaphragm

abdominal cavity

- stomach
- pyloric region
- spleen
- kidneys
- umbilical arteries

Figure 34.3 Internal anatomy of the fetal pig

larynx

thyroid

thymus

intercostal artery
and nerve

liver

spleen

stomach

pancreas

small intestine

colon

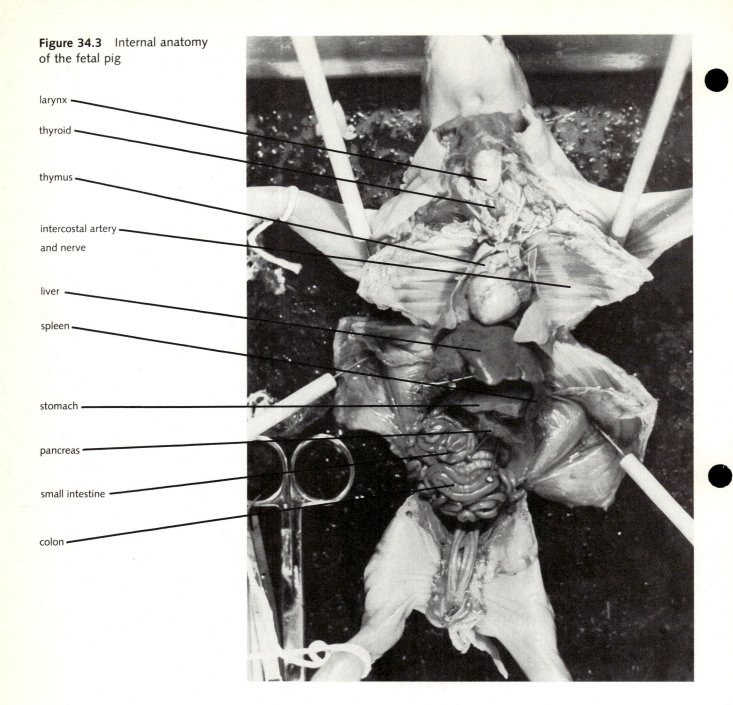

3. Carefully push a glass rod down the **gullet** or **esophagus** into the **stomach.** Leave the rod in place, to help you examine the esophagus later.

4. Locate the **liver** in the abdominal cavity. This is the large, brown, three-lobed organ posterior to the diaphragm. The liver has numerous functions, including the production of bile to aid in digestion, detoxification, production of blood proteins, and the processing and storage of digested food. Note the large vein that extends from the base of the umbilical cord to the middle of the liver. This is the **umbilical vein,** which originates in the **placenta** and carries nutrient-rich blood to the liver. Cut the umbilical vein and move this median strip of tissue out of the way.

5. Find the right (pig's right) lobe of the liver and under it the small saclike organ called the **gall bladder.** The gall bladder, which stores bile made in the liver, is sometimes difficult to find since it is embedded in the liver. Locate the **common bile duct,** which carries bile from the gall bladder to the **small intestine.** Bile is used as an emulsifying agent in the digestion of fats.

6. Blood leaves the liver by way of the **hepatic vein,** enters the **inferior vena cava,** and returns to the heart. These organs are located under the liver and are more easily seen after it is removed. Remove the liver by carefully cutting the hepatic vein where it joins the inferior vena cava and any connective tissues that attach it to the stomach and diaphragm. Do not remove any other structure at this time!

The upper part of the stomach joins the esophagus and is known as the **cardiac region.** The lower part of the stomach continues into the small intestine and is called the **pyloric region.** Refer to Figures 34.2 and 34.3 to help you identify structures in the abdominal cavity.

7. Locate the **pyloris,** a ring-shaped sphincter muscle marked by a slight constriction. By alternately opening and closing, the pyloris controls the movement of food from the stomach to the small intestine.

8. Cut the stomach open and view the openings of the esophagus and pyloris from inside. Also find the folds on the stomach wall, the **rugae.**

9. Lift up the stomach and gently pull down the intestinal mass to expose the **pancreas.** The pancreas is a spongelike organ that extends from the pyloric region of the stomach transversely to the left body wall. Like the liver and gall bladder, the pancreas is an accessory organ in the process of digestion. The pancreas also secretes the hormone insulin, which helps regulate glucose metabolism in the body.

10. On the left side of the stomach locate the **spleen,** a long, flattened, reddish-brown organ. The spleen functions in blood storage, production of white blood cells in adults, and blood filtration.

11. Examine the small intestine by first locating its point of attachment to the stomach. Using the end of the dissecting needle or probe, move the intestine around as you examine it. Note carefully the thin tissue that holds the coiled intestine together. This is the **mesentery** and is an extension of a thin membrane, the **peri-**

toneum, which lines the body wall. Note that the mesentery contains numerous blood vessels.

12. Remove the small intestine by first cutting it just below the pyloric region of the stomach and then by cutting the mesentery, which will allow you to extend it to its full length. Look carefully for the point where the small intestine joins the **colon** (large intestine). At this point there is a pouch, the **caecum.** In humans a fingerlike extension, the appendix, is attached to the end of the caecum. Does the pig have an appendix? _____

13. Locate the colon, a tightly coiled mass on the left side of the abdominal cavity. It merges with the **rectum,** which is embedded in the body wall, and exits by way of the **anus.** Remove both the small intestine and colon from the abdominal cavity. Be careful not to remove any other organs!

14. Complete Question 2 in the Evaluation.

Urinary System

1. Remove the peritoneum with tweezers and scissors to obtain a clear view of the **kidneys,** the two bean-shaped organs located against the dorsal body wall. On the inner side of each kidney locate a thin white tube, the **ureter.** The ureter of each kidney carries urine to the **urinary bladder.**

2. On the inside wall of the flap of tissue that attaches to the umbilicus, find two **umbilical arteries** and the **urinary bladder** located between them.

3. Locate the **renal artery** and **vein,** which leave the kidney on the inner or concave side.

The kidney is composed of distinct layers, each with a specific function. These layers are listed below, beginning with the outer covering or capsule and ending with a cavity, the pelvis.

- **Renal capsule:** outer layer surrounding and supporting the entire kidney
- **Cortex:** located directly under the renal capsule, the cortex is a thick brownish layer where the filtration of blood occurs
- **Medulla:** located under the cortex, the medulla is a smooth layer lighter in color than the cortex
- **Renal pelvis:** a funnel-shaped cavity where urine collects and the ureter joins the kidney

4. Remove one of the kidneys and make a longitudinal incision through it. Identify the renal capsule, cortex, medulla, and renal pelvis.

5. Complete Question 3 in the Evaluation.

6. Remove the second kidney and store according to your instructor's directions. It will be used in Laboratory Activity 39.

Reproductive Systems

During this activity you should be certain to examine pigs of both sexes.

Male (refer to Figure 34.4)

The **testes** of the male begin development in the abdominal cavity and later migrate to the **scrotal sacs** between the hind limbs. If the pig is in its

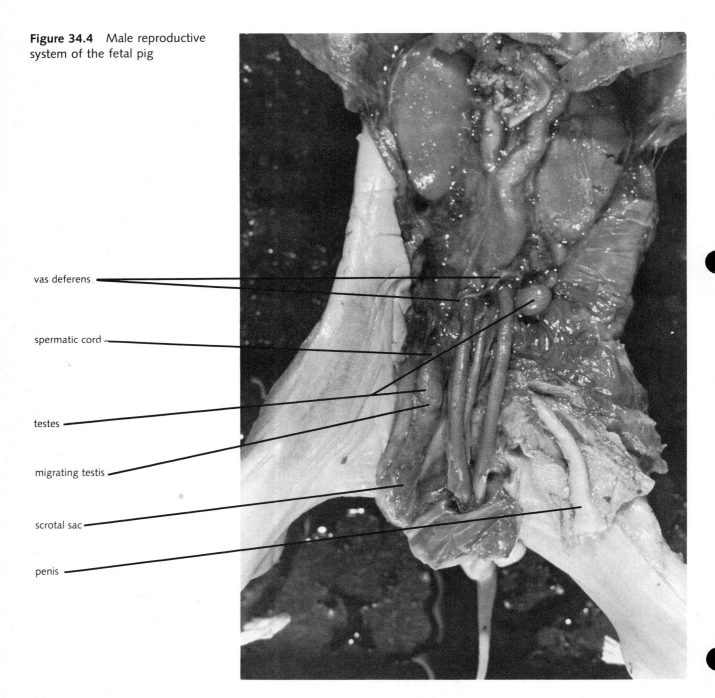

Figure 34.4 Male reproductive system of the fetal pig

vas deferens

spermatic cord

testes

migrating testis

scrotal sac

penis

early stage of development, the oval-shaped testes will be found just posterior to the kidneys. If the pig is close to term (near birth), the testes will have migrated to the scrotal sacs.

1. Examine the abdominal cavity for the testes. If the testes are not found here they will be located in the scrotal sacs. Carefully cut the skin near the juncture of a leg and the body; the scrotal sac will be white-to-gray in color.

If the testes have descended, each will be attached to a **spermatic cord** composed of the **vas deferens** (sperm duct), a **spermatic vein** and **spermatic artery,** and a **spermatic nerve.**

2. Use tweezers to pull gently on the spermatic cord. This should produce some movement of the testes within the scrotum.

3. The urethra and penis are found in the strip of tissue containing the urogenital orifice, located posterior to the umbilicus. By pinching this strip of tissue, you will be able to locate the muscular tube (penis) easily. On the inside of this strip of tissue, make a medial cut to expose the urethra and penis.

4. Complete Question 4 in the Evaluation.

Female (refer to Figure 34.5)

1. Locate the **ovaries** suspended from the dorsal wall of the abdominal cavity posterior to the kidneys. They are held loosely in the abdominal cavity by thin mesentary.

2. Adjacent to each ovary is the **oviduct,** a tightly coiled tube that attaches to the **uterine horns.** The uterine horns join to form the **uterus,** which leads to the **vagina.**

3. Locate the uterus by first locating the urinary bladder and then the urethra, which extends from the bladder. Dorsal to the urethra is the uterus. The vagina is attached to the distal end of the uterus. You may have to cut some of the muscle tissue in the body wall to find the uterus and vagina.

4. Complete Question 4 in the Evaluation.

Respiratory System

The cone-shaped rib cage forms the boundary of the **thoracic cavity.** The ribs are found along its inner surface. Refer to Figure 34.6 while dissecting the organs of the respiratory system.

1. Locate the small red **intercostal arteries** and **intercostal nerves** between the ribs. The nerve is normally white-to-gray in color.

2. Locate the thin **pleural membrane,** which bounds the outer surface of the **lungs** and the inner surface of the **rib cage.**

3. Extend the mid-ventral incision up through the neck region. Be very careful not to cut too deeply since you might damage underlying organs. In the neck region locate the **larynx.** It is easily recognized as a hard enlargement of the breathing tube and is composed of cartilage.

4. Cut into the larynx along the mid-ventral line and separate the two sides. Inside find the **vocal folds,** two shelflike membranes, which in humans are better developed as the vocal cords. Attached to the larynx is the **trachea.** The wall of the trachea is composed of rings of cartilage that prevent it from collapsing. The trachea leads from the larynx to the lungs where it branches to form the **bronchi.** Trace the trachea to this juncture.

5. Two glands of the endocrine system can be found. The **thyroid** is posterior to the larynx and is a small reddish-brown body. The thyroid secretes the hormone thyroxine, which increases the rate of metabolism in the body. The **thymus gland** will be seen as a mass of spongy white tissue on each side of the larynx and may extend as far as the heart region. During your dissection you may have cut this gland into several pieces. The thymus is the site for the development of one type of cell in the immune system.

6. Remove both lungs and store according to your instructor's directions for use in Laboratory Activity 37.

7. Complete Question 5 in the Evaluation.

Figure 34.5 Female reproductive system of the fetal pig

ovaries

uterine horns

oviduct

uterus

urinary bladder

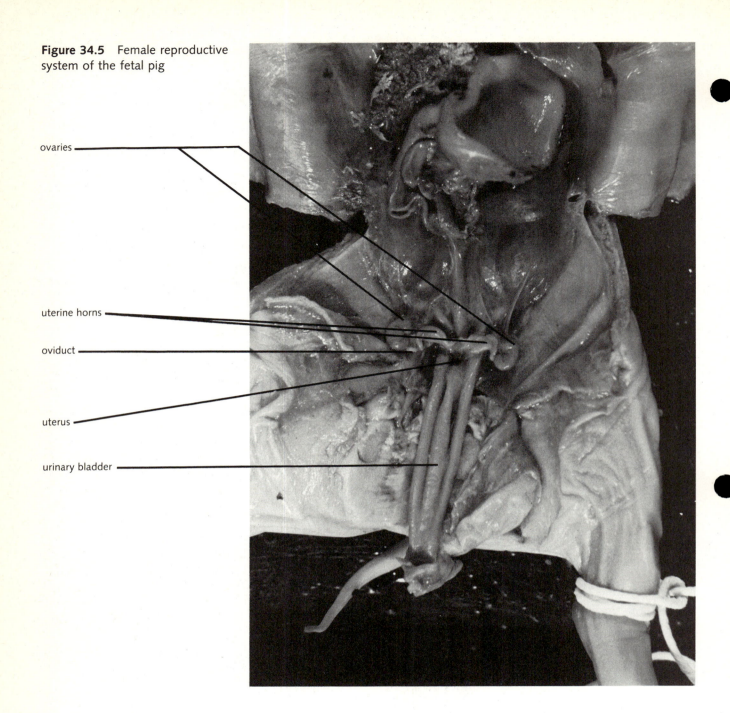

Circulatory System

Mammals, including the pig, have a four-chambered heart. The two lower chambers, the **ventricles,** are thick and muscular and make up the main body of the heart. The upper chambers, the **atria,** are thin-walled, darker in color, and appear to be draped over the ventricles. The atria collect blood from the lungs and other regions of the body and pass it to the ventricles. The ventricles then pump blood away from the heart.

1. Locate the **heart** between the lungs. It is slightly pointed at its posterior (ventricular) end and is covered by a thin membrane, the **pericardium** (see Figures 34.2 and 34.6).

Figure 34.6 The thoracic cavity of the fetal pig

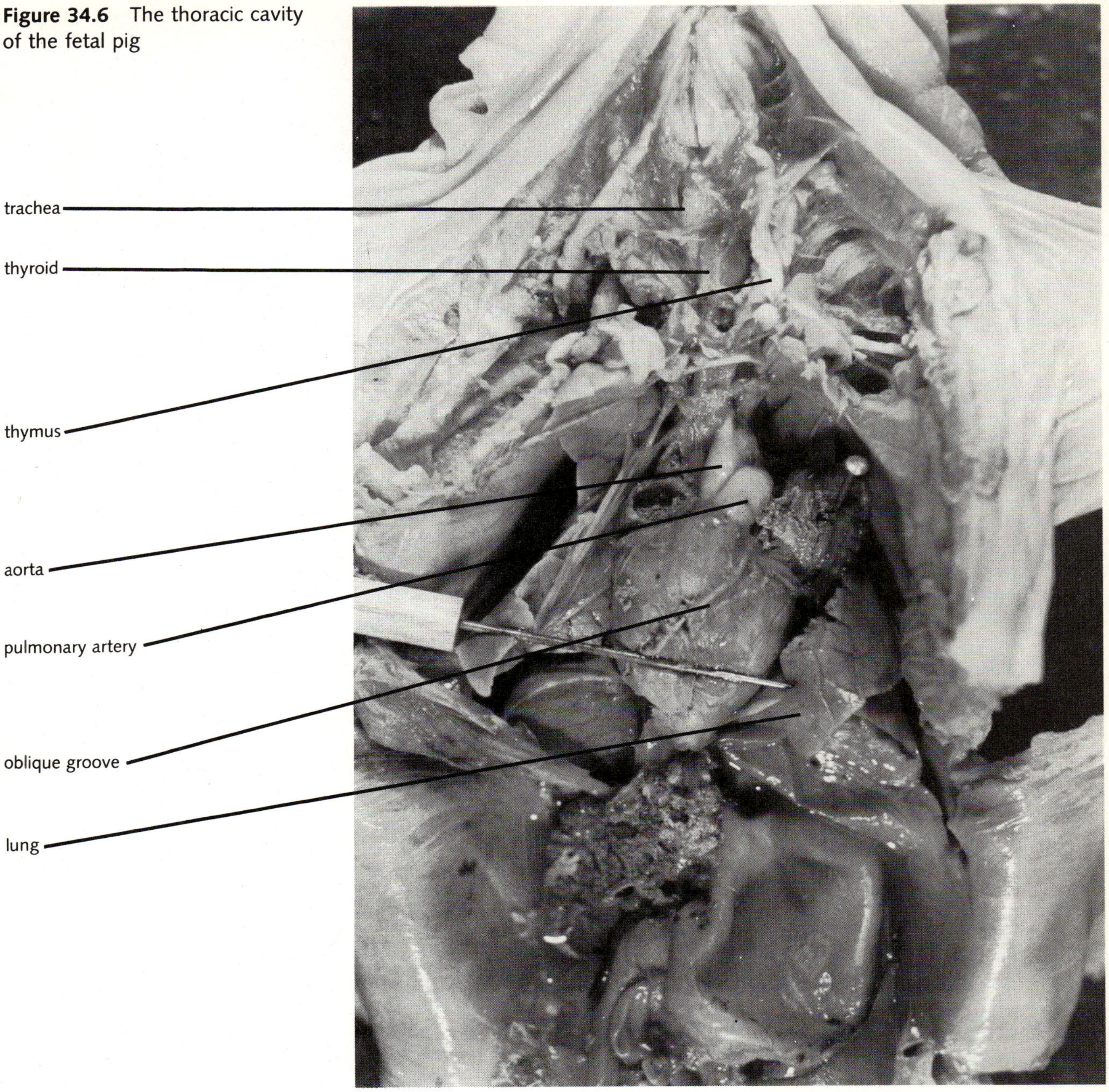

trachea —————————

thyroid —————————

thymus —————————

aorta —————————

pulmonary artery —————————

oblique groove —————————

lung —————————

2. Carefully remove the pericardium. Locate the **oblique groove,** which runs diagonally from left to right on the ventricles and contains a **coronary artery** and **coronary vein.** These vessels are a part of the blood vascular system of the heart muscle.

3. Gently push the heart to the pig's left to expose the two large veins that bring blood to the heart from the body. These veins, the **inferior vena cava** and **superior vena cava,** join at the right atrium and are injected with blue latex.

4. Examine the anterior portion of the heart and locate the two large arteries that join there. The one nearest the ventral side is the **pulmonary artery;** it curves to the left. This artery carries blood from the right ventricle and branches in the back of the heart to each lung.

5. Under the pulmonary artery, find the **aorta,** which carries blood from the left ventricle to all parts of the body.

6. Locate the **ductus arteriosus,** a very short, wide vessel connecting the aorta and pulmonary artery (see Figure 34.7). This vessel is found only during the fetal stage and serves to shunt blood from the pulmonary artery to the aorta for distribution throughout the entire body. This is necessary in order to short-circuit the blood around the nonfunctional lungs. At birth this duct will close, and blood will go to the lungs by way of the pulmonary artery to be oxygenated. In the fetus blood is oxygenated in the placenta.

Figure 34.7 The circulatory system of the fetal pig

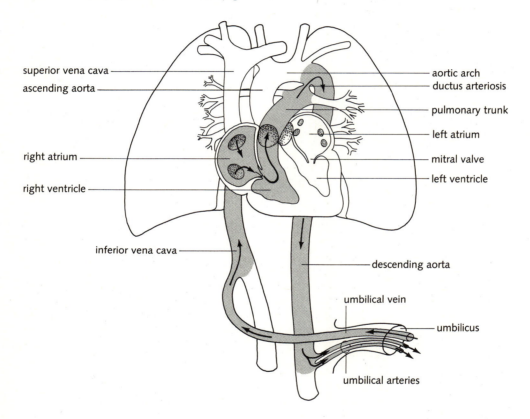

superior vena cava

ascending aorta

right atrium

right ventricle

inferior vena cava

aortic arch

ductus arteriosis

pulmonary trunk

left atrium

mitral valve

left ventricle

descending aorta

umbilical vein

umbilicus

umbilical arteries

7. Carefully dissect the tissue at the anterior end of the aorta to reveal two blood vessels, the **carotid arteries,** which carry blood to the head region.

8. Cut the pulmonary artery and aorta where they join the heart and remove the heart from the thoracic cavity.

9. Pull the inferior vena cava to the left to expose the pulmonary artery entering the right lung. Under the inferior vena cava locate the **dorsal aorta,** which carries blood to the lower part of the body. Note where it branches off by way of the renal arteries to each kidney.

10. Find the renal veins that branch from the inferior vena cava.

11. Trace the dorsal aorta to the point where it joins the umbilical arteries. Locate the **femoral arteries,** which carry blood to each leg. Complete Question 6 in the Evaluation.

You have finished the fetal pig dissection. Review the Evaluation section to be sure that it is complete. Dispose of your pig according to your instructor's directions.

EVALUATION 34

The Dissection of the Fetal Pig

1. Compare the pig's foot to a human foot with regard to structure and function.
   Label external features of the fetal pig in the following two diagrams.

External anatomy of the fetal pig

(a) Male

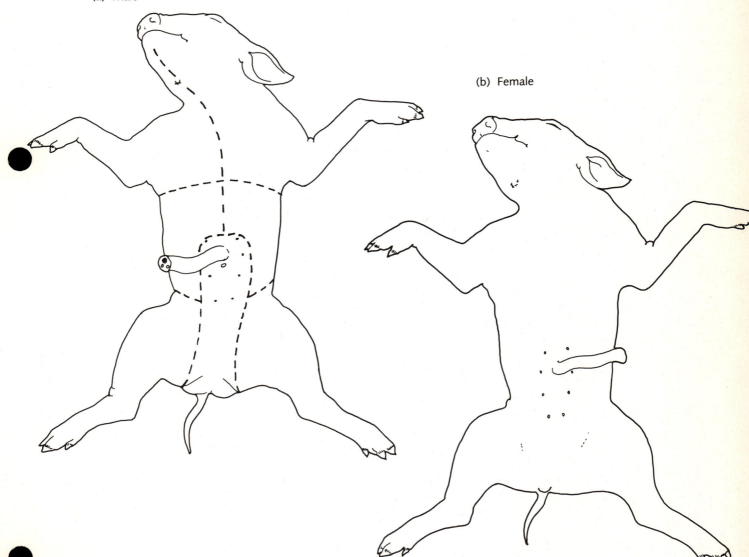

(b) Female

2. Using the following outline of the pig's body, diagram and label the organs of the digestive system: abdominal cavity, diaphragm, esophagus, liver, gall bladder, common bile duct, small intestine, pyloris, rugae, pancreas, spleen, mesentary, colon, caecum. If an organ is located under another one, diagram it using dotted lines.

The digestive system of the fetal pig

3. Using the following outline of the pig's body, diagram each of the organs of the urinary system: kidneys, ureter, urinary bladder, renal artery and vein, renal capsule, cortex, medulla, renal pelvis.

The urinary system of the fetal pig

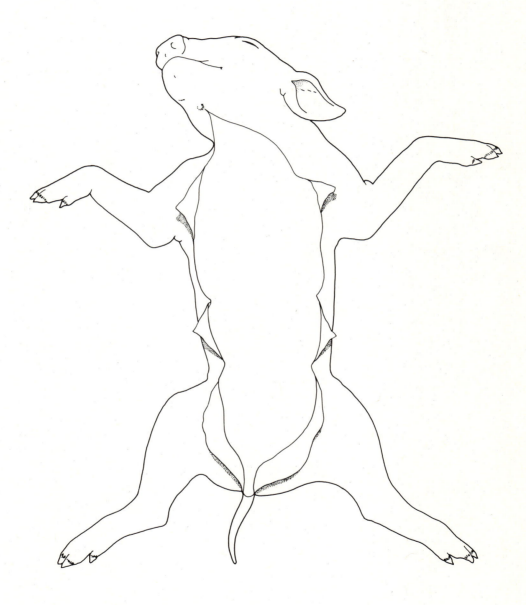

4. What is the sex of your pig? _____

Using the following outlines (male and female) of the pig's body, diagram and label the organs of the reproductive system: testes, scrotal sac, spermatic cord, penis, ovaries, oviducts, uterine horns, vagina.

The reproductive system of the fetal pig

(a) Male

(b) Female

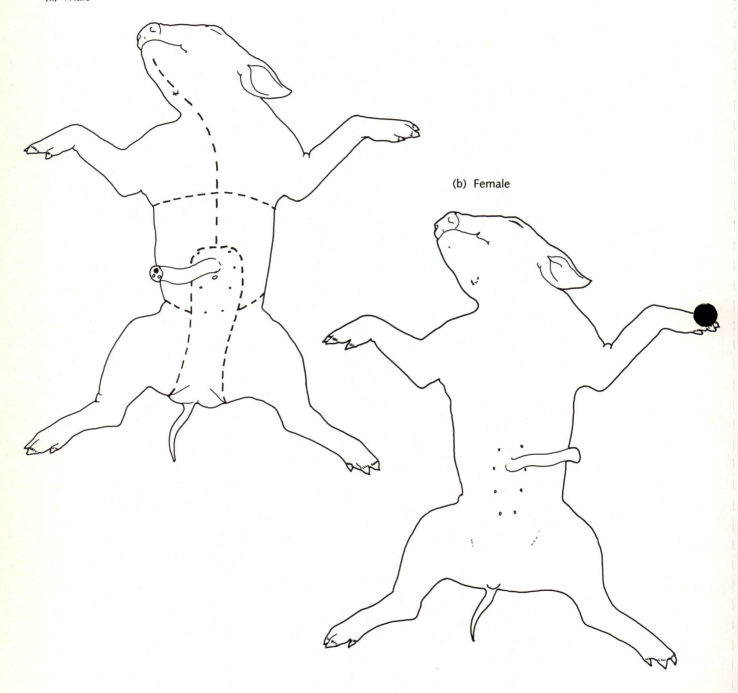

5. Using the following outline of the pig's body, diagram and label the organs of the respiratory system: diaphragm, thoracic cavity, intercostal artery and nerve, pleural membrane, lungs, rib cage, larynx, vocal folds, trachea. Also diagram the two endocrine glands, thyroid and thymus, observed in this area of the body.

The respiratory system of the fetal pig

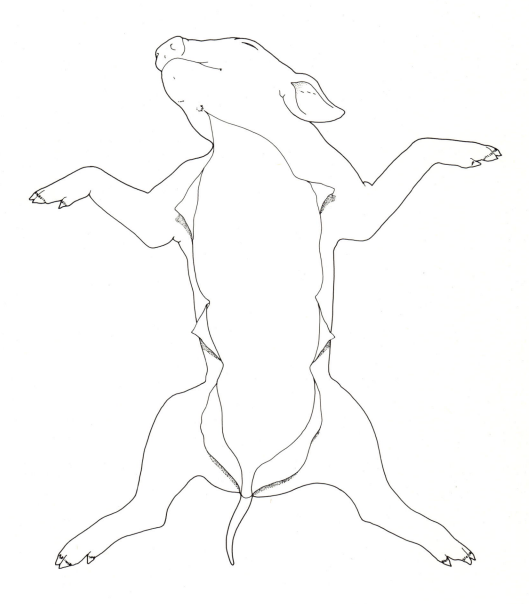

6. Draw a schematic diagram showing the flow of blood to and from the heart. Include the pulmonary artery, aorta, ductus arteriosis, and venae cavae. How will this circulation differ in a mature pig?

Animal Tissues

OBJECTIVES

At the end of this laboratory activity, you should be able to:

- state the major function of each tissue studied.
- identify the major parts of each tissue studied.
- identify a photograph or prepared slide of each tissue studied.

INTRODUCTION

This laboratory activity focuses on the area of biology known as **histology.** As you know from previous study in biology, all multicellular organisms are made of cells, and these cells combine to form distinct structures called **tissues.** A tissue is a group of closely associated cells that carry out a specific function. The histologist recognizes four main types of tissue in animals—**epithelial tissue, connective tissue, muscle tissue,** and **nervous tissue**—although many subgroups exist. These four tissue types are formed from the primary germ layers—ectoderm, mesoderm, and endoderm—during embryonic development. By forming tissues an organism can become specialized, leading to the development of systems such as the integumentary, digestive, nervous, and skeletal systems. This ability to specialize endows organisms with great versatility and is a major factor in the successful evolution of the diverse organisms we see in the world today.

Table 35.1 shows the classification scheme used by histologists. Each of these main tissue types will be discussed in this laboratory activity, and, in some cases, reference will be made to a particular subgroup.

MATERIALS

Compound microscope
Prepared slides of the following tissues:
   blood smear
   elastic connective tissue (artery wall)
   loose connective tissue
   cartilage (trachea)
   bone (ground)
   stratified squamous epithelium (skin)
   simple columnar epithelium (intestinal tract)
   pseudostratified ciliated columnar epithelium
    (trachea)
   skeletal muscle
   smooth muscle
   cardiac muscle
   spinal cord
   cross section of sciatic nerve

PROCEDURE

During this activity you will compare the photomicrographs on the following pages with slides given to you by your instructor. Read the accompanying text material as you study each photomicrograph and then identify the important elements in the prepared slide. As you examine each slide, make a diagram in the appropriate section in the Evaluation. Although each slide will not be an exact copy of the photomicrograph, you should be able to identify all the structures labeled in the photomicrograph. As you study each slide, attempt to relate the structure of the tissue to its function in the body.

Table 35.1 Animal tissue classification

Tissue	Subgroup	Typical location	Function
Connective	Blood	Vascular system	Transport, immunity
	Elastic connective tissue	Large arteries	Support, framework
	Loose connective tissue	Under skin, around muscle	Binds organs
	Dense connective tissue	Tendons, ligaments	Support
	Cartilage	Ends of bones, nose, larynx, trachea	Support, protection flexibility
	Bone	Skeleton	Support, protection
Epithelial	Stratified epithelium (two or more cell layers)	Skin, linings of body openings and ducts	Protective, strengthening
	Simple epithelium (one cell layer)	Covering of organs, linings of body cavities	Diffusion, filtration, secretion
Muscle	Skeletal	Skeleton	Voluntary movement
	Smooth	Internal organs	Involuntary movement
	Cardiac	Heart	Rhythmic contraction
Nervous		Brain, spinal cord, peripheral nerves	Communication, information storage and processing

Connective Tissue

Connective tissue is found in every organ of the body. Its primary function is to support, strengthen, and provide a structural framework for the organs in which it is found. Connective tissue is characterized by few cells and large amounts of extracellular material. This extracellular matrix is composed of various kinds of fibers and a mixture of tissue fluids in which these fibers are embedded. The function of connective tissue cells is to secrete these fibers and tissue fluids. There are almost a dozen different kinds of connective tissue in the body; you will study four of them in this activity.

Blood

Blood is classified as a connective tissue because it contains cells distributed in an extracellular fluid called **plasma** (see Figure 35.1). Unlike other connective tissues, blood does not have a fibrous or hard matrix, and the liquid plasma allows for the free movement of the white and red cells through the body. The most numerous cells in the blood are the **red blood cells,** or **erythrocytes,** which appear in the photomicrograph as small cells with a central depression. Their primary function is to carry oxygen to the tissues. Mature erythrocytes have no nuclei.

Figure 35.1 Blood smear showing red blood cells, white blood cells, and platelets

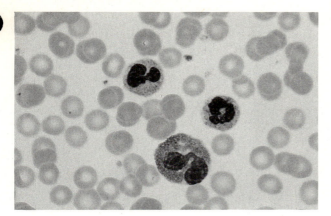

Figure 35.2 Cross section of artery wall showing elastic connective tissue

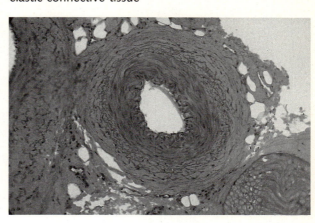

White blood cells, or **leucocytes,** are less numerous than red blood cells. They have a large nucleus and are motile, moving from the circulatory system into the surrounding tissues. There are five different types of white blood cells, and their primary function is to fight infection in the body. The white blood cell in the photomicrograph is a **lymphocyte.** Note that it has a nucleus that almost fills the cell.

1. Examine a prepared slide of a blood smear and identify the red and white blood cells. In Question 1(a) in the Evaluation, make a sketch of a smear of blood tissue and label several red and white cells.

Elastic Connective Tissue
Elastic connective tissue is composed of many elastic fibers that have an irregular appearance. Figure 35.2 shows a cross section of an artery wall, in which the elastic fibers can be seen as dark, thick, wavy lines. These fibers give considerable strength to the artery wall, and their elasticity allows the wall to expand and constrict with each surge of blood. The cells that produce the fibers are not visible. This artery is in a state of constriction; when relaxed, the fibers and the artery wall lose their wavy appearance.

1. Examine a prepared slide of a cross section of an artery wall and identify the elastic fibers. In Question 1(b) in the Evaluation, make a sketch of a section of the wall and label the elastic fibers.

Loose Connective Tissue
A type of connective tissue that is widely distributed in the body is **loose connective tissue.** This tissue is responsible for binding structures together and holding them in position while still providing flexibility. For example, this tissue permits your skin to move when rubbed and then return to its normal shape. It also serves as a padding and pathway for nerves and blood vessels.

Numerous types of cells are found in loose connective tissue. In Figure 35.3 the dark areas are the nuclei of **fibroblast** cells, which secrete the fibers into the extracellular matrix. The fibers give support to the tissue. The clear area is the extracellular matrix and, unlike that of bone or cartilage, is fluid and does not anchor the cells firmly.

1. Examine a prepared slide of loose connective tissue and identify the fibroblast cells, elastic fibers, and extracellular matrix. Sketch a section of your slide in Question 1(c) in the Evaluation and label these structures.

Cartilage
One of the most important connective tissues in the body is **cartilage.** Figure 35.4 shows **hyaline** cartilage, which is used to cover the ends of bones and to support the nose, larynx, and trachea. When covering the **distal** (end) portion of a bone it allows the smooth articulation of a joint during movement.

Cartilage cells are called **chondrocytes** and assume different shapes depending on their location.

Figure 35.3 Fibroblasts in loose connective tissue

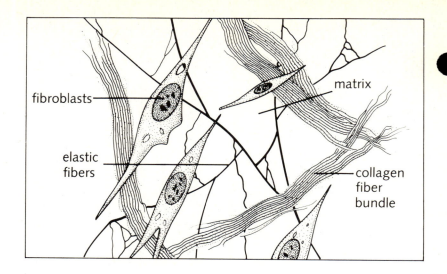

fibroblasts

elastic
fibers

matrix

collagen
fiber
bundle

Figure 35.4 Hyaline cartilage

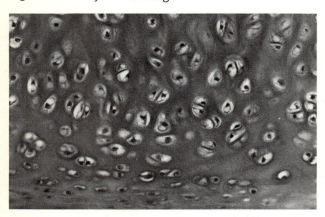

The flat cells at the top of the photomicrograph are near the cartilage surface. The cells in the center are more oval and are surrounded by a **capsule** that is derived from the matrix. The matrix is solid and holds the chondrocytes firmly in place.

1. Examine a prepared slide of cartilage and identify the chondrocytes, capsules, and matrix. In Question 1(d) in the Evaluation, make a sketch of a section of your slide and label these structures.

Bone
Bone tissue is the hardest of the connective tissues and forms the skeleton of vertebrate animals. The skeleton provides a rigid framework for the attachment of muscles, protects soft body parts, and provides support.

Compact bone tissue forms the outer layer of bone, while the inner portion is more porous or spongy. In compact bone the bone cells, or **osteocytes,** are arranged in concentric layers (see Figure 35.5) around a **Haversian canal** that contains blood vessels and nerves. Each osteocyte is contained in a space called the **lacuna.** The lacuna and osteocyte appear only as a dark area in this photomicrograph. Radiating from the osteocytes in the lacuna are fine cytoplasmic processes or **canaliculi** that extend through the matrix to adjacent osteocytes. Nutrients diffuse the canaliculi to nourish the bone cells. The light area between the lacunae is the hard matrix of calcium salts.

1. Examine a prepared slide of bone tissue and identify the lacunae with the osteocytes, Haversian canals, and canaliculi. In Question 1(e) in the Evaluation, sketch a section of your slide and label these structures.

Epithelial Tissue

Epithelial tissue covers various parts of the body, such as the skin and cavities of internal organs. The epithelium that covers the outside of the body forms the skin, or **integument.** Inside the body, epithelial cells line the alimentary tract, blood vessels, and respiratory system. Epithelial cells are also found in many glands of the body, where they secrete fluids. Anything entering or leaving the body must pass through a layer of epithelium. As a result, this tissue regulates the flow of materials between the body and the environment and between different parts of the body. During this activity you will examine epithelial tissue from the skin, digestive tract, and trachea (windpipe).

Figure 35.5 Osteocytes surrounding Haversian canal

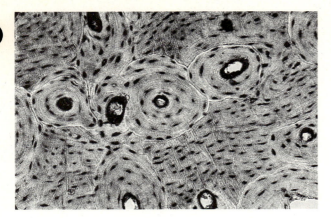

Figure 35.6 Stratified squamous epithelium

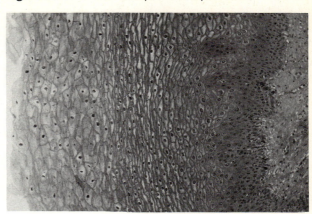

Stratified Squamous Epithelium

Human skin is composed of epithelial cells (see Figure 35.6) that vary in shape from round or oval to flat. "Squamous" refers to the flat shape of some of the individual cells; "stratified" means occurring in layers. The outer cells of skin contain keratin, a protein that forms a tough resilient covering and gives skin its resistance to the drying and abrasive effects of the environment. Stratified squamous epithelium is also found in the mouth, esophagus, nasal cavity, anus, vagina, and in the openings of the ears.

Figure 35.7 shows a cross section of skin. The outer layer is the **epidermis.** The keratinized cells in this layer are continually sloughed off and replaced by cells of the **stratum germinativum,** which are constantly dividing. Immediately beneath the epidermis is the second layer, or **dermis.** This is a layer of connective tissue that supports the epidermis and binds it to the underlying muscle.

1. Examine a prepared slide of epithelial tissue (skin) and identify the epidermis, dermis, keratinized layer of cells, and stratum germinativum. Sketch a section of this slide in Question 2(a) in the Evaluation and label these structures.

Figure 35.7 Cross section of skin

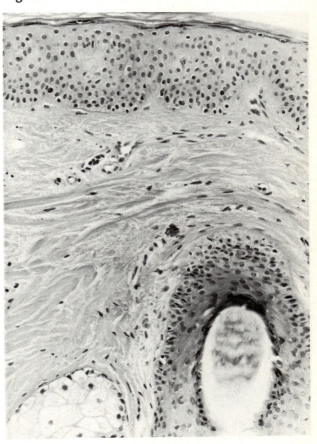

Simple Epithelium—Columnar Type

Figure 35.8 shows a cross section of epithelial tissue from an internal organ such as the stomach or intestine. The shape of these cells is quite different from those in the skin. This tissue lines the stomach and intestinal tract and secretes fluids.

The most abundant cells are the **epithelial cells.** Note their elongated shape and large, deeply stained nuclei at the base of the cell. The larger, rounder cells are **goblet cells,** which secrete mucus onto the tissue surface.

1. Examine a prepared slide of a cross section of tissue from the stomach or intestine. Make a sketch of a section of your slide in Question 2(b) in the Evaluation and label the epithelial cells, nuclei, and goblet cells.

Pseudostratified Ciliated Columnar Epithelium

The epithelial tissue shown in Figure 35.9 is from the **trachea** (windpipe) and is similar to the columnar epithelium. However, this type of epithelium has numerous hairlike projections, or **cilia,** extending from the surface. The cilia have a coordinated wavelike movement that transports a layer of mucus secreted by the goblet cells over the surface of the trachea, helping to remove foreign particles from the lungs. The name *pseudostratified* is used because the cells give the appearance of being in several layers even though they are only one layer thick.

1. Examine a prepared slide of a cross section of a trachea and identify the epithelial cells, goblet cells, and cilia. Make a sketch of a section of your slide in Question 2(c) in the Evaluation and label these structures.

Muscle Tissue

Three types of muscle tissue are structurally distinguishable: skeletal, smooth, and cardiac. **Skeletal muscle** is innervated by nerves from the central nervous system (brain and spinal cord) and is under voluntary control. The other two types are innervated by the autonomic nervous system and are not subject to voluntary control. **Cardiac muscle** is found only in the heart, while **smooth muscle** is located in the internal organs such as the stomach, blood vessels, and alimentary tract.

Muscle tissue is made up of long fibers (cells) that are encased in a sheath of connective tissue. This structure is particularly adapted for the process of contraction.

Skeletal Muscle

Skeletal muscle forms the largest portion of muscle tissue and is the most abundant tissue in the body. It is attached to and covers much of the skeleton and controls most of the voluntary types of movement—walking, writing, and lifting. The muscle fibers (cells) are long and thin and contain many nuclei (see Figure 35.10). Skeletal muscle is frequently referred to as striated muscle because of the numerous dark bands that cross at right angles to the long axis of the muscle. These striations are produced by alternating bands of several types of protein in the muscle cell. Muscle fibers are bound together by connective tissue for support. Each muscle fiber has many dark nuclei.

Figure 35.8 Cross section of simple columnar epithelial tissue

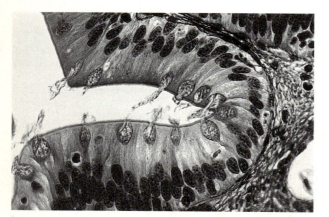

Figure 35.9 Cross section of pseudostratified ciliated columnar epithelial tissue

Figure 35.10 Muscle fibers from skeletal muscle

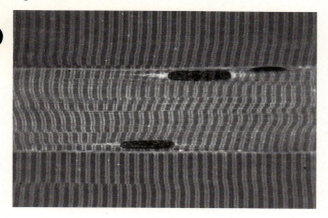

Figure 35.11 Cross section of smooth muscle

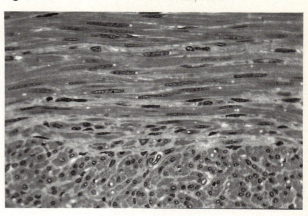

1. Examine a prepared slide of skeletal muscle and identify the muscle fibers, striations, connective tissue, and nuclei. In Question 3(a) in the Evaluation, sketch a portion of your slide and label these structures.

Smooth Muscle (Intestinal Wall)

Smooth muscle is found in the walls of most of the internal organs. Figure 35.11 shows a section of the human intestine. The smooth muscle of the intestinal tract undergoes rhythmic contractions in order to move the food along as it is digested. The cells of smooth muscle are long and spindle-shaped, and the nucleus lies near the center. Notice the lack of striations and that each cell has only one nucleus.

1. Examine a slide of smooth muscle and locate the muscle fibers and nuclei. In Question 3(b) in the Evaluation, make a sketch of a section of your slide and label the muscle fibers and nuclei.

Cardiac Muscle

Cardiac muscle is found only in the heart, and the fibers form a branching network (see Figure 35.12) that gives heart cells the unique ability to contract as if they were one fiber. Like skeletal muscle, car-

Figure 35.12 Cardiac muscle

diac muscle is striated but is not under voluntary control. The cells in cardiac muscle are separated by cross-bands called **intercalated disks.** These disks are formed by the junctions between adjacent cell membranes and help hold the cells together and transmit contractions between cells. Each fiber has one nucleus, which appears as a dark oval.

1. Examine a prepared slide of cardiac muscle and identify the muscle fibers, nuclei, and intercalated disks.

Nervous Tissue

The nervous system is primarily responsible for conducting nerve impulses throughout the body. The functional unit of this system is the nerve cell or **neuron** (see Figure 35.13). The neurons shown in this photomicrograph are from the spinal cord. Nerve cells in the spinal cord are a part of the central nervous system and consist of three types: motor, association, and sensory. The cells seen here are motor neurons and demonstrate the three main parts of a nerve cell: the cell body, the dendrites, and the axon. The cell body contains the organelles typically found in a cell and most of the cytoplasm. The neuron in the photomicrograph shows a darkly stained nucleus. Your specimen may also show small dark granules, the **Nissl bodies,** which are specialized layers of the endoplasmic reticulum (see Laboratory Activity 23). Extending from the cell body are numerous branched processes called **dendrites.** Dendrites carry impulses to the cell body. **Axons,** the second type of cytoplasmic extension, are long and cylindrical and conduct impulses away from the cell body. The length of an axon varies considerably from only a few millimeters in the brain and spinal cord to over a meter in nerves located in the extremities.

The term **nerve** is often confused with neuron or nerve cell. A nerve is a bundle of neurons. Nerves are usually macroscopic and are surrounded by a sheath of connective tissue for support and protection. The nerve in Figure 35.14 is a cross section of the sciatic nerve that extends from the base of the spinal cord to the lower leg. Each oval structure is an individual axon extending from a cell body. Some axons are covered with layers of fatty material collectively known as the **myelin sheath.** This sheath insulates each axon and allows a faster rate of conduction than occurs in fibers that are not covered.

1. Examine a prepared slide of neurons from the spinal cord and identify the cell body, dendrites, axon, and nucleus. In Question 4a, sketch several neurons and label these structures.

2. Examine a prepared slide of a nerve cross section and identify the individual axons and myelin sheath. In Question 4b in the Evaluation, sketch a portion of your slide and label these structures.

Figure 35.13 Neuron from spinal cord

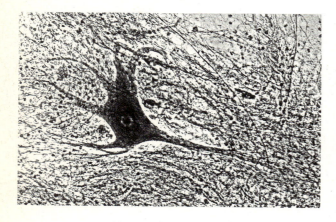

Figure 35.14 Cross section of sciatic nerve

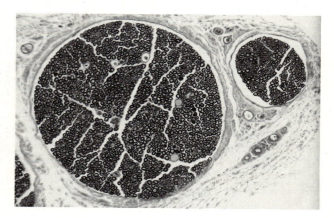

EVALUATION 35

Animal Tissues

When studying a specimen under the microscope field, it is often very helpful to make a simple but accurate diagram of what you are viewing. Drawing forces you to observe closely the object under study, and this attention to detail will help you to retain a mental image of the specimen. Drawings need not be artistic to be useful. Your drawing should include only those features that will help you to recognize the specimen; do not try to include every detail that you observe. Make your diagrams with a sharp pencil, use narrow lines, and make them large enough to fill the field of view provided. Label the major features of each diagram with a straight line and try to avoid crossing lines.

1. Connective tissue

   (a) Blood

   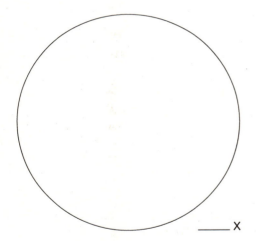
   _____ X

   (b) Elastic connective tissue

   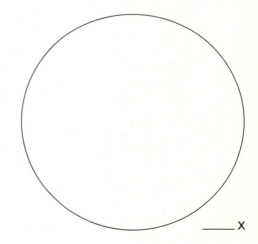
   _____ X

   (c) Loose connective tissue

   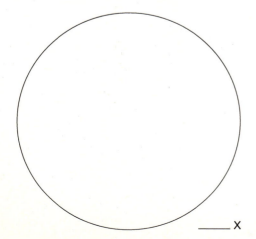
   _____ X

   (d) Cartilage

   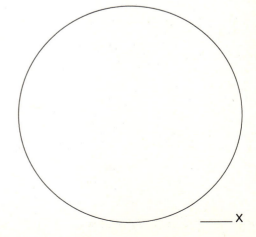
   _____ X

(e) Bone

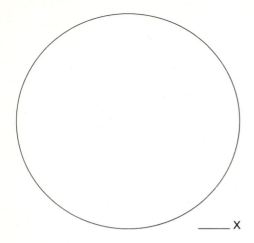

_____ X

2. Epithelial tissue

(a) Stratified squamous epithelium

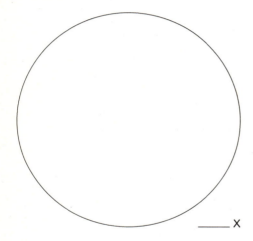

_____ X

(b) Simple columnar epithelium

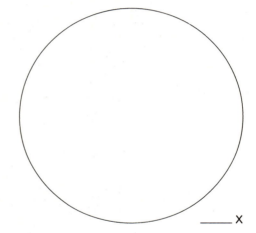

_____ X

(c) Pseudostratified ciliated epithelium

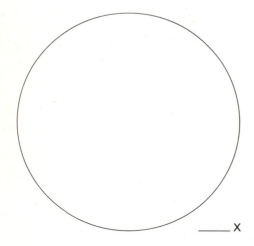

_____ X

3. Muscle

   (a) Skeletal muscle

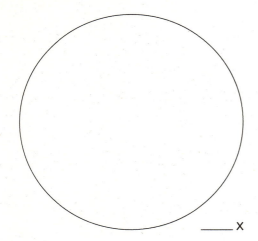

_____ X

   (b) Smooth muscle

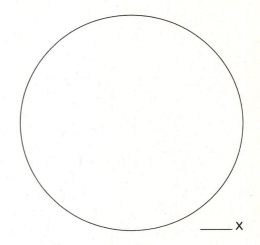

_____ X

   (c) Cardiac muscle

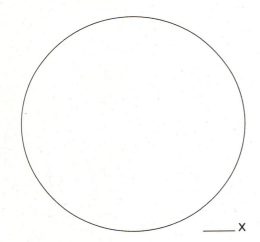

_____ X

4. Nervous tissue

   (a) Neurons from spinal cord

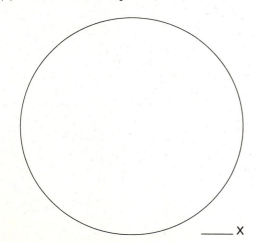

_____ X

   (b) Cross section of a nerve

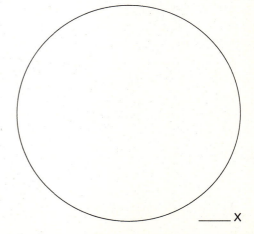

_____ X

5. Biologists are always concerned with the relationship between structure and function. Select one tissue from each group studied and attempt to relate its structure to the function it performs in the body. You may need to consult your textbook for additional information about the tissue you are studying.

Tissue	Structure-function relationship

Vertebrate Circulation

OBJECTIVES

At the end of this laboratory activity, you should be able to:

- describe what produces a pulse.
- use a sphygmomanometer to measure blood pressure.
- explain how arteries assist in the movement of blood.
- state the function of capillaries.
- describe how blood is returned to the heart in veins.

INTRODUCTION

The first living organisms evolved in primitive seas and remained in the aquatic environment for millions of years. Because of this evolutionary history, life on our planet is intimately associated with water. Water became the arena in which life performed and continues to perform. Much later, as organisms moved onto the land, they encountered a multitude of problems, the most obvious of which was the limited amount of water present. A well-developed circulatory system, which would enable animals to provide an aquatic environment for every cell of the body, was necessary if animals were to establish themselves on land. Circulatory systems already existed in many aquatic animals but were improved upon in terrestrial animals.

In present day vertebrates the complex circulatory system is an intricate part of every body system. In *homeothermic* (warm-blooded) vertebrates like humans, the system is powered by a double pump with four separate chambers (see Figure 36.1). In essence this heart is two pumps; one receives blood from the body and pumps it to the lungs; the other receives blood from the lungs and pumps it to all parts of the body through a vast network of arteries and capillaries. Blood low in oxygen returns from all parts of the body, collects, and enters the **right atrium**. This thin-walled muscular chamber has the small task of forcing the blood into the **right ventricle** through the **tricuspid valve**. Simultaneously, oxygenated blood from the lungs is returned to the **left atrium**. Upon constriction of this chamber blood is forced through the **mitral** or **biscuspid valve** into the **left ventricle**. The ventricles constrict simultaneously, pushing blood into major arteries. The right ventricle sends blood to the lungs via the pulmonary arteries and the left ventricle has the incredible task of forcing blood throughout the arteries and capillaries of the entire body. When these ventricles constrict, backflow of blood into the atria is prevented by the one-way valves mentioned previously.

This laboratory activity cannot cover all aspects of such a complex system. Instead you will study selected topics that deal with several of the more important functions of blood circulation. Blood is a complex tissue that was studied briefly in Laboratory Activity 35.

Figure 36.1 The major chambers and valves of the human heart

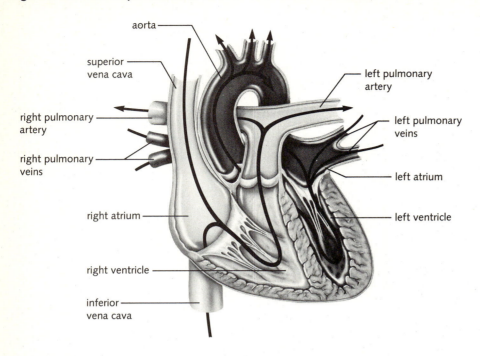

- aorta
- superior vena cava
- right pulmonary artery
- right pulmonary veins
- right atrium
- right ventricle
- inferior vena cava
- left pulmonary artery
- left pulmonary veins
- left atrium
- left ventricle

MATERIALS

Watch with second hand
Sphygmomanometer
Stethoscope
Pumping apparatus
Goldfish
Fish net
Petri dish
Cheesecloth
Compound microscope
Meter stick
Rubber tubing

PROCEDURE

Pulse

As the left ventricle contracts it pushes blood into the **aorta,** causing the vessel to expand. This expansion produces a pressure wave that continues along the arterial system. When you take a pulse you are feeling this pressure wave, not the actual flow of blood (see Figure 36.2). You can feel this pulse where an artery is close to the surface of the body, just under the skin. In the following activity

Figure 36.2 Measuring pulse

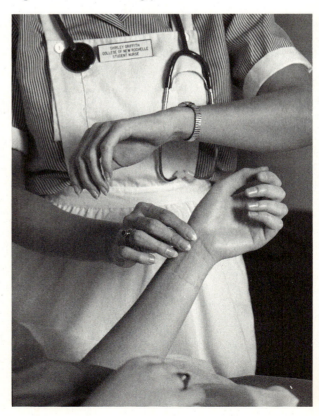

you will take your own pulse reading at the **carotid artery** on either side of the neck or at the **radial artery** on the radial side of the wrist at the base of the thumb.

1. Place your index and middle finger on the artery and detect a strong pulse.

2. Using a watch with a second hand count the pulse for one minute. Record your pulse.

Your pulse rate can be affected by many environmental factors such as exercise, physical health, and mental stress. Devise an experiment concerning the pulse. For example, you could collect data from individuals before and after exercise and compare the results of athletes with nonathletes. Or you could compare the pulse of smokers with that of nonsmokers. Design your experiment according to the following format. Record all information in Question 2 in the Evaluation.

3. Devise a hypothesis you wish to test.

4. Write out your intended procedures. Make certain to state all specific details.

5. Follow your procedures and record all collected data.

6. Write a conclusion based on your hypothesis and collected data.

Blood Pressure

When the left ventricle contracts, blood is forced into the aorta at considerable pressure and travels into arterial branches that carry blood to all regions of the body. One of these branches is the **brachial artery** on the inside of the upper arm. The blood pressure is usually taken from the brachial artery by using a **sphygmomanometer** (see Figure 36.3). Blood pressure is measured in millimeters and refers to the height to which a column of mercury (Hg) would rise under that amount of pressure. A reading for a healthy young person at rest might range between 100/60 mg Hg and 120/80 mg Hg. The upper figure of this reading is the pressure in the brachial artery just after **systole** (ventricular contraction); the lower reading is the pressure present in the artery during **diastole** (the period between heartbeats).

Figure 36.3 Measuring blood pressure

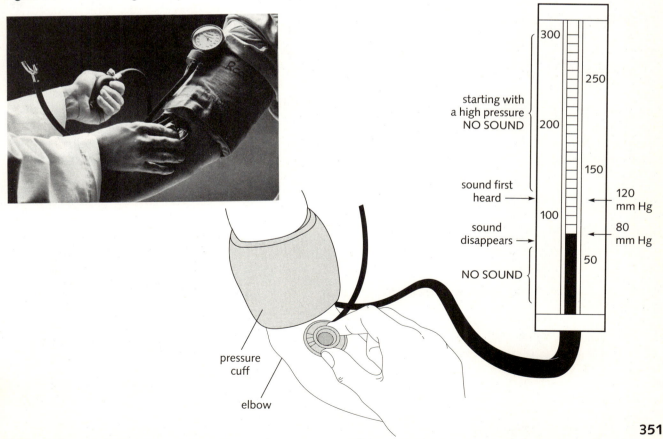

Blood pressure is a very useful diagnostic tool that tells us a great deal about the health of the heart and the circulatory system. Follow the procedure given below to take your laboratory partner's blood pressure with the sphygmomanometer.

1. Have your partner sit comfortably in a chair with both arms resting on a table top, level with the heart. Your partner should relax in this position for at least three minutes before having the blood pressure taken.

2. Wrap the cuff securely around the arm above the elbow. Place the stethoscope over the brachial artery just below the elbow.

3. Pump air into the bladder within the cuff until the gauge reads approximately 150 mm of Hg. At this point the pressure within the cuff should be sufficient to cut off the flow of the blood through the brachial artery.

4. Gradually release air from the bladder within the cuff by turning the valve on the bulb in a counterclockwise direction. Listen for a faint tapping sound at each heartbeat. When this sound is first noted check the pressure on the gauge. This is the systolic blood pressure.

5. Continue to release air from the cuff. As you do so the quality of the sound will change from loud to soft and will eventually disappear. Record the reading on the gauge when the sound disappears; this is the diastolic pressure.

6. Now have your partner take your blood pressure. Record your blood pressure in Question 2 in the Evaluation.

If difficulty is encountered in reading the pressure gauge, release the cuff and hold your hand over your head for thirty seconds. Wait three minutes before making another attempt.

➤ CAUTION: Do not attempt this procedure repeatedly without removing the cuff and allowing circulation to return to normal.

Flexibility of Arteries

The heart is faced with an enormous task on a continual basis: it must pump blood into arteries, arterioles, and ultimately capillaries, where the exchange of materials takes place. The entire circulatory system contains thousands of miles of blood vessels, and the capillaries have diameters of only seven to nine μm (see Figure 36.4). To accomplish this task, arteries must be extremely flexible, allowing blood to continue flowing away from the heart even when the heart is relaxed (diastole). Aging, poor diet, and altered lipid metabolism can cause the arteries to become thick and hard and lose their elasticity, a condition known as arteriosclerosis.

The next activity indicates the importance of flexible arteries. You will be supplied with two hand water pumps (see Figure 36.5). One is fitted with a balloon and a glass tube, the other with just a glass tube. Setup A is analogous to a heart pumping blood into elastic arteries: setup B is analogous to a heart pumping blood into hardened or diseased arteries.

1. Using setup A, squeeze the hand pump often enough to maintain a steady flow of water from one culture dish into another. Count the number of contractions needed to maintain this flow for thirty seconds: _____

2. Pump setup B in a similar manner to maintain a steady flow. Count the number of contractions needed to maintain this flow for thirty seconds: _____. Answer Question 3 in the Evaluation concerning the flow of water in each case.

Capillaries

The heart pumps blood, and the arteries and arterioles distribute the blood to the capillaries. Capillaries are thin-walled tubes composed of epithelial cells (endothelium), usually surrounded by a basement membrane for support. Life-supporting materials are delivered to, and waste materials are removed from, the cells by the same capillaries. This remarkable exchange of materials between the capillaries and cells is facilitated by a complex interaction of osmotic and hydrostatic pressure.

Figure 36.4 Comparison of walls of arteries, veins, and capillaries

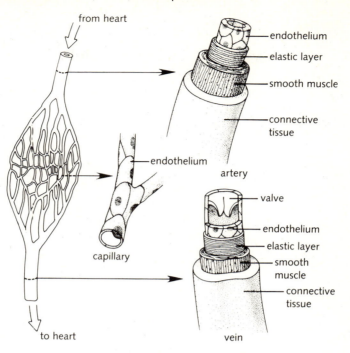

Using the following procedure examine the capillaries present in the tail fin of a goldfish.

1. Remove a fish from the aquarium with a small fish net. Make sure your hands are wet before touching the fish; this will prevent damage to the integument. Carefully place the fish in a petri dish.

2. Soak a small pad of cheesecloth in water and position it over the head and operculum (gill cover) of the fish. This will hold the fish in place and provide water (containing dissolved oxygen) to the capillaries in the gill filaments. Make sure the pad is big enough to keep the fish still in the proper position.

3. Examine the tail fin under low power to observe the capillaries. This must be done quickly. Record your observations in Question 4 in the Evaluation.

4. Return your fish to the aquarium immediately.

Figure 36.5 Apparatus used to demonstrate importance of flexibility of arteries

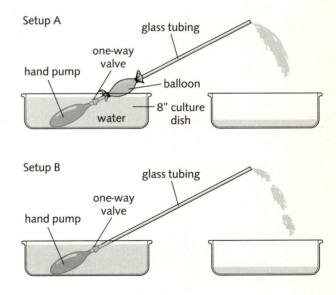

Optional Activity

Your instructor may provide you with some compounds that are **vasodilators** (cause vessels to relax) or **vasoconstrictors** (cause vessels to contract) to add to the water surrounding the tail fin. Obtain another fish from the aquarium and repeat the procedure above, and in addition add one chemical to the water surrounding the fin. After adding one compound be certain to flush out the area with fresh water before using the next compound. After adding each compound observe the blood flow through the capillaries. Record your observations in Question 4(c) in the Evaluation.

Veins

In humans the hydrostatic pressure of blood reaching the capillaries has dropped to approximately 30–40 mm Hg. The pressure of blood flowing through the veins is reduced even further. Follow the directions below to determine your venous blood pressure.

1. Hold your arm at your side and examine your wrist or the back of your hand to locate a prominent vein.

2. With your arm outstretched and palm up hold your arm at heart level for one minute.

3. Have your laboratory partner raise your hand slowly until the vein collapses. Make certain your arm muscles are completely relaxed. You will need to observe this carefully because the veins may not be prominent. Have your partner measure the distance in centimeters from your heart to the bottom of your raised hand.

4. Convert this centimeter measurement to millimeters of mercury by using the following formula:

(distance in cm)(0.74 mm mercury) = venous pressure

5. Record your venous pressure in Question 5 in the Evaluation.

Valve Function in Veins

As you have determined in the previous activity venal blood pressure is quite low. Since the heart does not pump blood back to the heart, the return of the blood to the heart is facilitated by a specialized system of valves. Most veins contain **semilunar valves** that prevent the backflow of blood.

Blood is forced through the veins toward the heart by the movement of skeletal muscles that massage the veins.

With care you can usually locate one of these valves on the back of the hand using the following procedure (see Figure 36.6).

1. Wrap a piece of rubber tubing around your partner's arm above the elbow just tight enough to make the veins in the hand stand out.

2. Select one of the bluish colored veins on the back of the hand. Place a finger on this vein at a point farthest away from the heart, and with another finger push the blood up the vein toward the heart. Lift this finger and observe what happens.

3. Reverse the procedure by lifting the finger that is farthest from the heart after you have pushed the blood out. What did you observe? Record your observations in Question 6 in the Evaluation.

Figure 36.6 Locating a vein valve

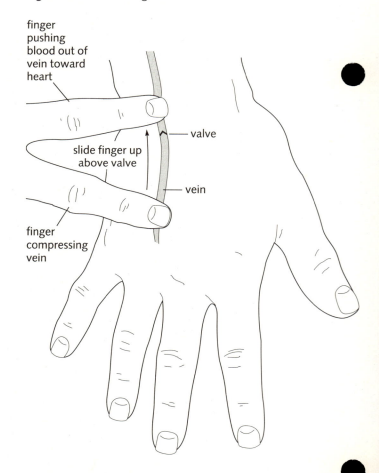

finger pushing blood out of vein toward heart

valve

slide finger up above valve

vein

finger compressing vein

EVALUATION 36

Vertebrate Circulation

1. Complete the following information for your experiment involving pulse rate.

   (a) State your hypothesis.

   (b) Describe the procedure used.

   (c) Use the space below to prepare a data table.

   (d) Interpret the results of your experiment and state your conclusion.

2. Record your blood pressure: _____

3. Review the results of the experiment concerning the flexibility of arteries and answer the following questions.

   (a) What relationship do you think exists between systolic blood pressure and artery flexibility? Explain.

   (b) What is the advantage of flexible arteries?

   (c) What effect would arteriosclerosis have upon the heart?

4. (a) Diagram and label the capillaries you observed in the fin of the fish tail. Indicate the direction of blood flow with arrows.

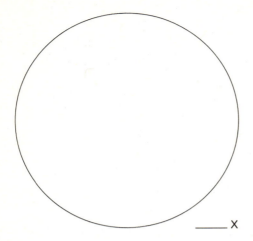

_____ X

(b) Describe the flow of blood in the capillaries.

(c) (optional) List the chemical compounds you added to the tail fin and describe any changes you observed.

Compound	Change observed in capillaries

5. (a) Record your venous pressure: _____

(b) How does this blood pressure compare with the diastolic blood pressure you measured in your arm?

(c) Why is the venous pressure not recorded as a systolic and diastolic pressure?

6. What function did you observe for the vein valve in the hand?

Breathing and the Exchange of Gases

OBJECTIVES

At the end of this laboratory activity, you should be able to:

- identify the major parts of the vertebrate respiratory system.
- calculate and define the major volumes and capacities of the human lung.
- explain the mechanics of breathing.
- demonstrate the effect of carbon dioxide on breathing rate.

INTRODUCTION

The function of a respiratory system is to exchange gases between the organism and its environment. All respiratory systems must meet certain basic requirements if they are to provide the organism with sufficient oxygen to meet its metabolic needs as well as rid the body of excess carbon dioxide. First, the gas exchange surface must be moist, since gases can only diffuse into cells if a layer of moisture is present. Second, the surface must be thin and have adequate surface area for gas exchange to occur. In larger animals, this respiratory surface is closely associated with a blood circulatory system for the transport of gases to the deeper underlying tissues and organs.

In simple one-celled animals the task of gas exchange is easily accomplished since carbon dioxide and oxygen can diffuse directly through the cell membrane from the surrounding water environment. In simple multicellular animals, such as the planarian and *Hydra*, the skin of the body provides the surface for gas exchange. These animals, in spite of being multicellular, require no special respiratory organs (see Figure 37.1). They are thin enough and have sufficient surface area to expose most of the body mass to the environment. Larger animals, such as the earthworm, still rely on their body surface as the site for gas exchange, but have developed an extensive circulatory system to transport these gases to the internal organs.

Terrestrial animals have one additional problem that must be solved if their respiratory systems are to function properly. The move to land meant that the respiratory surface, which must always be moist, was in most environments in danger of drying out.

Figure 37.1 Gas exchange in *Hydra*

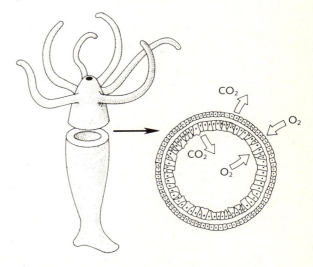

As animals moved onto land, their respiratory systems developed as specialized invaginations with openings for the intake of gases from the environment. By moving the respiratory surface inside the body, this delicate, thin, moist surface was given adequate protection from dehydration and mechanical injury. The vertebrate lung probably evolved in a primitive freshwater fish that gulped air into its pharynx. A small sac developed in the pharynx as a reservoir for air and eventually evolved into a many-branched organ that today we know as the modern vertebrate lung. This ancestral extension of the pharynx also evolved into the swim bladder that you studied in the dissection of the fish (Laboratory Activity 17).

The human lung meets all of the requirements for an efficient respiratory organ. It is moist, has enormous surface area (about 80 m²), is well protected from dehydration, and is intimately associated with a circulatory system for transport of gases to and from the tissues.

In the human respiratory system, air first passes through the nasal passages into the **pharynx** (throat), through the **larynx** (voice box), then into the **trachea,** a long tube leading to the lungs. The trachea branches into two primary **bronchi,** which in turn branch into the smaller **bronchioles.** The bronchioles further divide into small sacs known as **alveoli.** It is in the alveoli that the actual exchange of gases with the blood occurs.

During this laboratory activity you will examine several features of the vertebrate lung. First you will obtain the lungs that were saved from the fetal pig activity and do a more thorough dissection in order to locate the structures named above. You will also examine the microscopic anatomy of lung tissue. You will study the mechanics of breathing using a mechanical model and calculate the different respiratory volumes and capacities of the human lung. Finally, you will evaluate the role of carbon dioxide in the control of breathing.

MATERIALS

Lungs from fetal pig dissection
Dissecting tray
Scissors
Hand lens or dissecting microscope
Forceps
Dissecting needles
Prepared slide of lung tissue
Compound microscope
Bell-jar model
Spirometer
Gauze swab
Isopropyl alcohol (70%)
Breathing apparatus with glass and rubber tubing
Lime water (calcium hydroxide solution)
Straws
Stopwatch
Plastic or paper bags
Beaker (100 mL)
Brom thymol blue solution
Dropping bottle of NaOH solution (dilute)

PROCEDURE

Anatomy of Fetal Pig Respiratory System

Obtain the preserved lungs that were dissected from the fetal pig in Laboratory Activity 34. You may need to refer to that activity again as you begin the dissection of the lungs. Place the lungs in a dissecting tray and proceed with the following:

1. Locate the **trachea** (windpipe) at the top of the lungs and note the rings of **cartilage.** Remove a section of trachea about 5 mm long by cutting with scissors. Examine the cut edge with a hand lens or under a dissecting microscope. Note the shape of the ring of cartilage. Make a diagram of this cross section and answer Question 1 in the Evaluation.

2. The lungs are divided into **lobes,** or sections, which may be located by lifting each lung with your forceps. The right (pig's right) lung is divided into four lobes and the left lung into three.

3. Trace the trachea to the point where it divides into the right and left **bronchi.** Trace these air tubes into each lung and carefully dissect away the lung tissue to expose the finer branches of the bronchi which form the **bronchial tree.** Turn to Question 2 in the Evaluation and label the structures you have just located.

Microscopic Anatomy of Lung Tissue

1. Obtain a slide of a section of lung tissue. Observe under low power of the compound microscope and locate the following structures.

   alveoli: numerous small air sacs with a wall that is one cell thick
   bronchioles: larger, open chambers that carry air to the alveoli

2. In Question 3 in the Evaluation, make a diagram of a section of lung tissue from the prepared slide. Label the structures that you identified.

The Mechanics of Breathing and Lung Capacity

Examine the bell-jar model provided by your instructor (see Figure 37.2). Note that the sides of the jar represent the thoracic wall, the balloons represent the lungs, the glass tubing represents the trachea and bronchi, and the rubber sheet simulates the diaphragm.

1. Move the rubber "diaphragm" up and down and note what happens to the balloons. Describe your results in Question 4 in the Evaluation.

   As you breathe, different quantities of air move in and out of the lungs depending upon the force and depth of your breathing. The total amount of air you can hold in your lungs may be divided into four volumes. Although you will not be able to measure all of these volumes during this activity, with a simple hand-held **spirometer** (a device to measure the volume of air expired from the lungs) you can record several of the volumes listed. In this activity, volumes that cannot be measured directly with the spirometer must be calculated from your data and are indicated with an *.

- **Tidal volume (TV).** Represents the air moved in and out of the lungs during normal relaxed breathing—approximately 500 mL.
- **Expiratory reserve volume (ERV).** The amount of air that can be forcibly exhaled after a normal tidal volume expiration—approximately 1100 mL.
- ***Inspiratory reserve volume (IRV).** The volume of air that can be inspired forcefully in addition to the normal tidal volume—approximately 3000 mL.
- **Residual volume (RV).** The amount of air remaining in the lungs after a maximal expiration. This air cannot be exhaled from the lungs, and you cannot measure or calculate this volume—approximately 1200 mL. The residual volume is necessary to aerate the blood between breaths. If it were not present in the alveoli, the concentrations of oxygen and carbon dioxide would drop markedly with each breath.

When a physiologist considers two or more volumes together, they are referred to as **capacities.**

- **Vital capacity (VC).** The maximum amount of air that can be forcefully expelled from the lungs after a maximum inspiration. This is normally 70–75% of the total lung capacity and represents the sum of the tidal, expiratory reserve, and inspiratory reserve volumes—approximately 4500 mL.
- ***Inspiratory capacity (IC).** The maximum amount of air that can be inhaled after a normal expiration. This capacity is the sum of the inspiratory reserve volume and tidal volume.
- **Total lung capacity (TC).** The sum of all the volumes or the total amount of air in the lungs after a maximum inspiration.

1. Obtain a spirometer and wipe the end with a gauze swab soaked in 70% isopropyl alcohol. Place a disposable mouthpiece over the stem. Your instructor will demonstrate the use of the spirometer.

2. Rest comfortably and breathe normally through the nose. Have your partner determine your respiration rate by watching your chest and counting the number of breaths you take in 60 seconds. Make two additional counts, average the results, and record this data in Question 5 in the Evaluation.

   Reading 1: _____
   Reading 2: _____
   Reading 3: _____
       Total _____   Average _____

Figure 37.2 Bell-jar model of respiratory system
(a) Diaphragm contracted (lowered); lungs fill with air
(b) Diaphragm relaxed (raised); air leaves lungs

(a)

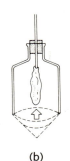

(b)

3. Measuring tidal volume: Set the spirometer on 1000 or 2000. While resting in a sitting position, inhale normally through the nose and then exhale through the spirometer. Do this at least three times without removing the spirometer from your mouth. Do not try to exhale forcibly.

4. Divide the total volume recorded on the dial by the number of times you exhaled. Record your tidal volume in the graph in Question 5 in the Evaluation.

5. Measuring expiratory reserve volume: Set the spirometer dial on zero. Breathe normally at least three times. At the end of a normal expiration, place the spirometer in your mouth and forcibly exhale through the spirometer all the air you can from your lungs. Record your result in the graph in Question 5 in the Evaluation.

6. Measuring vital capacity: Set the spirometer on zero. Take several very deep breaths and then forcefully exhale through the spirometer. Record this result in Question 5 in the Evaluation. Use Table 37.1 to determine your predicted vital capacity. Record this value in the graph in Question 5 in the Evaluation.

7. Calculating inspiratory capacity: Calculate the IC using the following equation.

$$\underset{\text{IC}}{\underline{\hspace{2cm}}} = \underset{\text{VC}}{\underline{\hspace{2cm}}} - \underset{\text{ERV}}{\underline{\hspace{2cm}}}$$

Record your result in the graph in Question 5 in the Evaluation.

8. Calculating inspiratory reserve volume: Calculate the IRV using the following equation.

$$\underset{\text{IRV}}{\underline{\hspace{2cm}}} = \underset{\text{IC}}{\underline{\hspace{2cm}}} - \underset{\text{TV}}{\underline{\hspace{2cm}}}$$

Record your result in the graph in Question 5 in the Evaluation.

Control of Breathing

Although two gases—oxygen and carbon dioxide—are involved in respiration, carbon dioxide is the more important in controlling the rate of breathing under normal circumstances. The following activities will help to demonstrate the role of carbon dioxide in the regulation of breathing.

Table 37.1 Normal vital capacity of adults* (in cubic centimeters)

| | Height in inches | Age in years | | | | | |
		20	30	40	50	60	70
Males	60	3885	3665	3445	3225	3005	2785
	62	4154	3925	3705	3485	3265	3045
	64	4410	4190	3970	3750	3530	3310
	66	4675	4455	4235	4015	3795	3575
	68	4940	4720	4500	4280	4060	3840
	70	5206	4986	4766	4546	4326	4106
	72	5471	5251	5031	4811	4591	4371
	74	5736	5516	5516	5076	4856	4636
Females	58	2989	2809	2629	2449	2269	2089
	60	3198	3018	2838	2658	2478	2298
	62	3403	3223	3043	2863	2683	2503
	64	3612	3432	3252	3072	2892	2710
	66	3822	3642	3462	3282	3102	2922
	68	4031	3851	3671	3491	3311	3131
	70	4270	4090	3910	3730	3550	3370
	72	4449	4269	4089	3909	3729	3549

Adapted with permission from Propper Manufacturing Co., Inc., New York.
*Variations must be at least 20% below predicted normal to be considered subnormal. Variations can also exist depending upon size and body structure.

Comparison of Inhaled and Exhaled Air

1. Obtain a setup of flasks similar to the one shown in Figure 37.3. Examine the arrangement of the glass tubes and note that inhaled and exhaled air must bubble through the water in the flask.

2. To each flask, add approximately 100 mL of lime water (calcium hydroxide solution, which absorbs carbon dioxide).

3. Attach one rubber tube to the long glass tube (A) and the other rubber tube to the short glass tube (B).

4. Attach a straw to the glass Y-tube and breathe continuously through the flask until the solution becomes cloudy. The flask containing the solution with the highest concentration of carbon dioxide will turn cloudy due to the formation of insoluble calcium carbonate. Record your observations in Question 6 in the Evaluation.

Figure 37.3 Apparatus for comparing inhaled and exhaled air

mouthpiece

The effect of carbon dioxide concentration in the blood on the breathing rate can be demonstrated by the following activity.

1. Observe your partner breathing normally for three minutes. Record his/her breathing rate (breaths per minute): _____

2. Hyperventilate (breathe deeply and rapidly) twenty times. Measure the breathing rate immediately after hyperventilating. Record this rate: _____

▶ **CAUTION: Do not continue this activity if a feeling of dizziness develops.**

3. Place a paper or plastic bag over your partner's mouth and have him or her breathe into the bag until breathing becomes difficult.

4. Remove the bag and determine the breathing rate. Record your result: _____ . Answer Question 7 in the Evaluation.

Effect of Exercise

1. Add 100 mL of water to a beaker and then add sufficient brom thymol blue solution (pH indicator) to give a distinct blue color. Add one or two drops of diluted sodium hydroxide solution if the water is not blue. A decrease in pH will indicate an increase in carbon dioxide concentration.

2. One partner should breathe into the beaker through a straw. Record the time in seconds that it takes for a color change to occur. Record your result: _____

3. Exercise by running in place for at least two minutes. Exhale into a second beaker of brom thymol blue solution and record the time it takes for a color change to occur: _____ . Complete Question 8 in the evaluation.

EVALUATION 37

Breathing and the Exchange of Gases

1. (a) Diagram the cross section of the trachea.

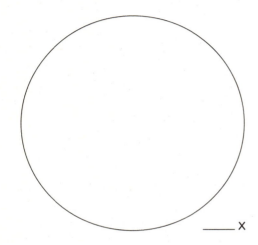

_____ X

(b) Describe the shape of the ring of cartilage and relate this shape to the fact that the trachea lies just ventral to the esophagus.

2. Label the following diagram of the lungs. Include the trachea, lungs, bronchi, bronchioles, and alveolar sac.

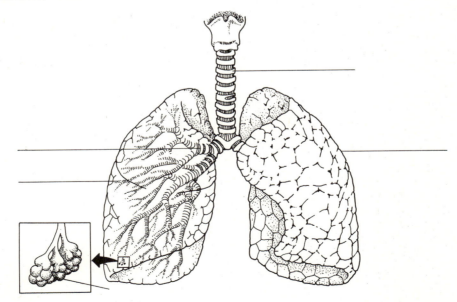

3. (a) Make a diagram of the cross section of the lung tissue and label the alveoli and bronchioles.

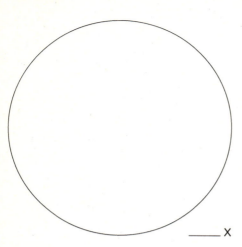

_____ X

(b) The air passages of the lung show continuous branching from the trachea to the alveoli. Why is this branching important to the exchange of gases in the lungs?

4. (a) Explain why the balloons inflated when you lowered the rubber diaphragm. Why did the balloons deflate when the diaphragm was moved up? Consider changes in air pressure between the chest cavity and the environment in your answer.

(b) What would happen if a hole were made in the side of the bell? Explain.

(c) Similarly, explain the consequences of a human chest wound, in which the chest wall is punctured.

5. Record your data from the spirometer activity in the following spirogram. (A graph like this is obtained by breathing through a tube into a closed chamber attached to a marking pen. The pen rests against a rotating drum and, as the subject breathes, the chamber moves up and down, producing a line like the one drawn here.)

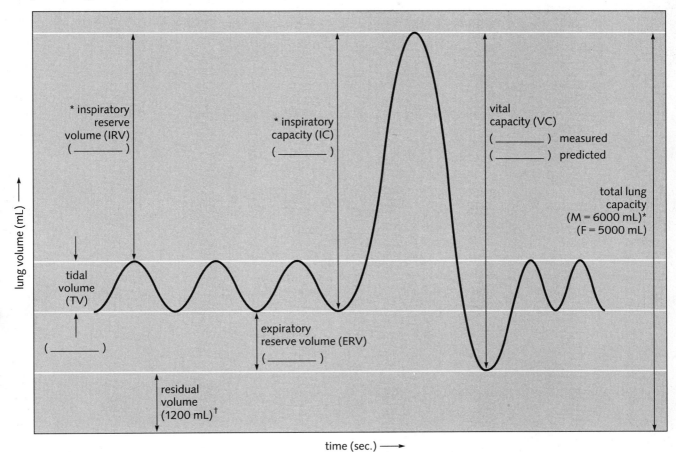

* calculated values
† approximate values

6. Table 37.1 indicates that a person's sex is an important factor in determining vital capacity. Name several other factors that may have an influence on vital capacity.

7. (a) Record your observations from the activity comparing inhaled and exhaled air in the following table. In the parentheses, indicate which type of air (inhaled or exhaled) passed through the flask.

Flask A	Flask B
( )	( )

(b) Based on these observations, which type of air has the highest concentration of carbon dioxide?

8. (a) Record your results from the activity on the effect of carbon dioxide on breathing rate in the following table.

	CO$_2$ level in blood	Breathing rate (breaths/minute)
Normal	Normal	
After hyperventilating	Lower	
After breathing in bag	Higher	

(b) What is the effect of increased carbon dioxide level in the blood on the breathing rate?

(c) Why is this important during exercise?

Digestion of a Carbohydrate

OBJECTIVES

At the end of this laboratory activity, you should be able to:

- state the function of saliva in the process of digestion.
- give the optimum conditions under which the enzyme salivary amylase (ptyalin) functions.
- state the importance of digestion to absorption.

INTRODUCTION

Animals spend a great deal of time acquiring food. Acquiring food is a twofold process—ingestion, the taking in of pieces of food, and digestion, the chemical breakdown of the food. This chemical breakdown is a multistep process accomplished by digestive enzymes. The process is called **hydrolysis** (*hydro* = water, *lysis* = a loosening) and occurs when water is added to the chemical bonds of the large food molecules, breaking them into smaller subunits. In higher animals these enzymes are added to the food as it moves along the digestive tract, breaking the food down into relatively simple molecules that can be absorbed into the body proper. What is food to one animal may not be food to another simply because the specific enzyme needed to digest a particular substance may be present in one animal and not in another. A good example of this is the problem faced by mammals that are herbivores (vegetarians) and ingest cellulose as a major portion of their diet. Because these animals (cattle, deer, giraffes and others) do not have the appropriate enzymes, they cannot digest cellulose. Their problem is solved by harboring certain bacteria and protozoa in their gastrointestinal tract that produce the enzymes necessary to digest cellulose.

The following activity is intended to demonstrate the process of digestion of starch, a complex carbohydrate. Starch digestion begins in the mouth and is completed in the small intestine. Because starch is a very large molecule it cannot be used as a nutrient until it has been broken down into smaller molecules that can be absorbed by the cells lining the intestinal tract. The principles studied in this activity concerning carbohydrate digestion also apply to the digestion of other organic nutrients, such as proteins and lipids (fats and oils).

The two types of starch in foods are amylose and amylopectin, with the former being the more abundant. This activity will demonstrate the action of **salivary amylase** (ptyalin). The final product of salivary amylase digestion is the sugar **maltose.** This reaction may be summarized as follows.

$$\text{Starch} + \text{water} \xrightarrow[\text{hydrolysis}]{\text{salivary amylase}} \text{maltose}$$

MATERIALS

Paraffin
50-mL beakers
Hydrion paper
Test tubes (15 mm × 125 mm)
1% starch solution
Boiling water bath
Dropping bottle of concentrated hydrochloric
   acid (HCl)
37°C water bath
Ice water bath
Test tubes (10 mm × 75 mm)
Dropping bottle of iodine-potassium iodide
   solution (I₂KI)

Dropping bottle of Benedict's solution
Dialysis tubing
Dental floss
Funnel

PROCEDURE

Digestion of Carbohydrate

Use the flowchart in Figure 38.1 as a guide as you
follow these procedures.

1. Chew on a piece of paraffin to stimulate the
   secretion of ptyalin. Collect 15 mL of saliva in
   a 50-mL beaker.

> CAUTION: Saliva should be handled only by
the person donating it for the experiment.

2. Determine the pH of the saliva with a piece of
   Hydrion paper. Record the result: _____

> CAUTION: Hydrochloric acid (HCl) is a skin
and eye irritant. If it contacts your skin or clothing,
flush with water and inform your instructor.

3. Prepare the following tubes.

Tube #1	4 mL starch solution
	4 mL water
Tube #2	4 mL starch solution
	4 mL saliva
Tube #3	4 mL boiled saliva (place tube with saliva in a boiling water bath for 5 minutes)
	4 mL starch solution
Tube #4	4 mL starch solution
	4 mL saliva
	7 drops of concentrated HCl
Tube #5	4 mL starch solution
	4 mL saliva

4. Place tubes 1, 2, 3, and 4 into a 37°C water bath
   and incubate for 1 hour. Place tube #5 in an
   ice water bath for 1 hour.

5. After 1 hour, divide the contents of each tube
   equally into two smaller tubes. Mark each set
   of tubes 1A, 1B; 2A, 2B; and so on.

6. Test each tube labeled "A" for starch by adding
   4–5 drops of I₂KI solution to each tube. The
   I₂KI will turn dark purple-black when the concentration of starch is high. Shades of reddish
   brown indicate lesser amounts of starch. Record your results in Question 1 in the Evaluation.

7. To each tube labeled "B" add 4 mL of Benedict's solution and place in a boiling water bath
   for exactly 2 minutes. Benedict's solution turns
   blue, green, yellow, orange, or red, depending
   on the concentration of sugar. Brick red indicates the highest concentration of sugar. Answer Questions 2–5 in the Evaluation.

A Sampler of Digestion

One purpose of digestion is to chemically break
down food molecules into smaller and simpler ones
to enable these nutrients to be absorbed into the
blood stream. You can demonstrate this function
with the following exercise.

1. Collect approximately 15 mL of saliva in a small
   beaker as you did previously.

2. Cut a section of dialysis tubing 8 cm long. Hold
   the tubing under water until it opens into a
   tube and becomes flexible.

3. Tie one end of the tube securely with a piece
   of dental floss. Make sure that the bag does
   not leak.

Figure 38.1   Flowchart for starch digestion experiment

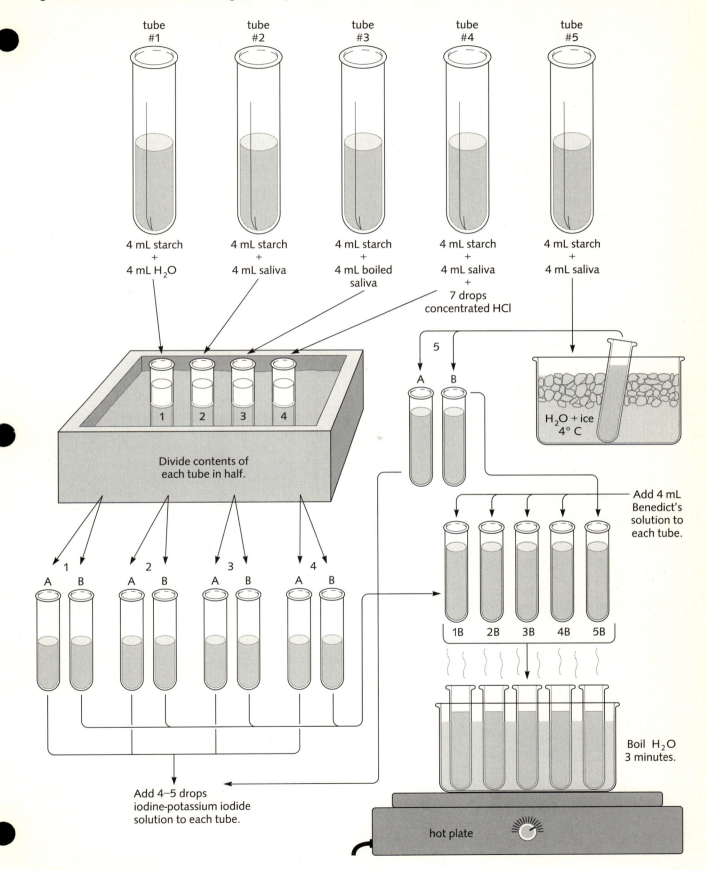

4. Use a funnel to half-fill the bag with saliva.

5. Add starch solution to the bag until it is nearly filled. Leave enough of the bag unfilled to allow for tying off. Tie the open end with dental floss.

6. Place the bag containing the saliva and starch in a 50-mL beaker. Add warm (37°C) water to the beaker until it is nearly full. Add enough I₂KI solution to the beaker water to make it deep yellow in color.

7. Place the beaker containing the bag in a water bath to maintain the temperature at 37°C.

8. After 1.5 hours use Benedict's solution to test the water in the beaker for the presence of reducing sugar. Remove 3 mL of the water from the beaker and add 3 mL of Benedict's solution to a test tube. Place the tube in a boiling water bath for 3 minutes. Test the water again after an additional 30 minutes.

9. Open the bag carefully and place the contents into a test tube. Add an equal amount of Benedict's solution and complete the test for reducing sugar. Complete Questions 6–8 in the Evaluation.

EVALUATION 38

Digestion of a Carbohydrate

1. Complete the following chart for the digestion of starch by recording any color change that occurred.

	Water	Saliva	Boiled saliva	HCl	Ice H_2O
Starch					
Maltose					

2. Describe the ideal conditions for the digestion of starch by amylase.

3. What do you think happens to amylase activity in the stomach? Consult your textbook to determine the pH of the contents of the human stomach.

4. What do you think happened to the amylase when it was boiled to reduce its activity? (Remember that enzymes are proteins!)

5. What is the purpose of tube #1 in the activity on the digestion of starch?

6. Did any of the starch leave the dialysis tubing and enter the beaker water? If it did, how do you know?

7. Did you get a positive test for reducing sugar in the beaker water? If so, explain what occurred.

8. In the experiment employing the dialysis tubing, to what is the tubing analogous in the digestive system? In what ways does it differ? Be as complete as possible in your answer.

Excretion: The Elimination of Waste

OBJECTIVES

At the end of this laboratory activity, you should be able to:

- state the three steps involved in urine formation.
- diagram the main components of the kidney nephron.
- describe the location of the nephron parts within the kidney.
- list the functions of the vertebrate kidney.
- compare normal and abnormal urine.

INTRODUCTION

In vertebrate animals the main excretory organ is the **kidney** (see Figure 39.1a), which excretes nitrogenous wastes (ammonia, urea, and uric acid) from protein and nucleic acid metabolism. In addition the kidney excretes excess water and certain salts from the blood. These functions are important in maintaining homeostasis, the balanced internal environment of the body. The functional unit of the vertebrate kidney is the **nephron** (see Figure 39.1b).

Most fish have relatively simple nephrons to maintain proper water balance and rid the body of wastes. In reptiles, birds, and mammals the nephron is modified by the addition of the **loop of Henle.** This structure enables the nephron to produce a urine more concentrated in salt, an important benefit to land animals that need to conserve water.

The nephron is composed of small arteries that supply blood to a tuft of capillaries, the **glomerulus** (see Figure 39.2). A spherical structure, **Bowman's capsule,** surrounds the glomerulus and is attached

to a convoluted tubule, a portion of which is the loop of Henle. The capillaries leaving the glomerulus are intimately associated with the tubule. The following paragraph is a summary of the basic functions of the nephron; you should, however, read the section on kidney function in your textbook before you complete this activity.

Filtration occurs in the glomerulus. Normal hydrostatic blood pressure forces blood plasma out of the capillaries through pores (fenestrated epithelium). The blood plasma is then filtered through a **basement membrane** (connective tissue layer) before it enters the cavity of Bowman's capsule. This filtrate differs from plasma in that it lacks blood cells and large protein molecules. Most of the useful material in this filtrate is reabsorbed as the filtrate moves through the tubule. This step is called **reabsorption.** The cells of the tubule pass materials out of the lumen (cavity) of the tubule by active transport and passive diffusion. **Elimination** is the last step of the process. Any materials not reabsorbed move into the collecting tubules and ultimately into the ureter, which carries the urine to the bladder. The urine will contain metabolic wastes, excess salts, and some water.

Each human kidney contains approximately one million nephrons. If it were stretched out, each nephron would be 35 mm in length.

The human kidney produces approximately 150 L of filtrate each day but excretes only 1.5 L of this amount. During the process the blood loses most of its nitrogenous wastes, the salt content of the body is balanced, and the pH of the blood is adjusted. This very complex homeostatic mechanism is under hormonal control.

During this activity you will dissect the kidney, examine kidney tissue, and observe the results of nephron activity.

Figure 39.1 The kidney
(a) Kidney section

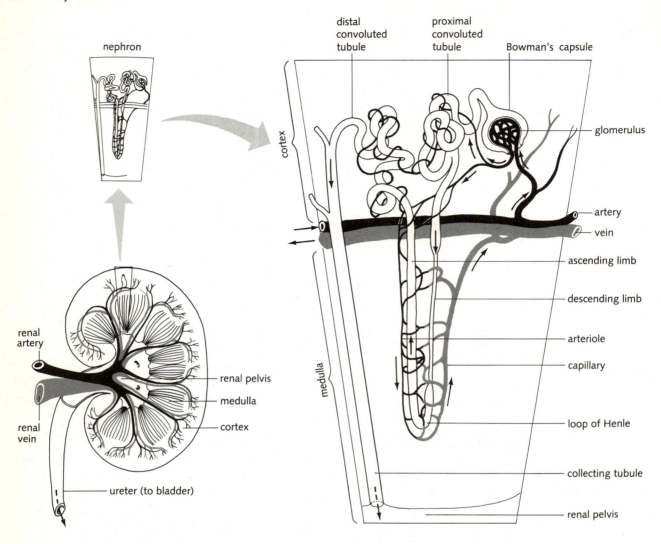

nephron

distal convoluted tubule

proximal convoluted tubule

Bowman's capsule

glomerulus

cortex

artery

vein

ascending limb

descending limb

medulla

arteriole

capillary

loop of Henle

collecting tubule

renal pelvis

renal artery

renal pelvis

medulla

cortex

renal vein

ureter (to bladder)

Figure 39.2 Cross section of kidney tubules

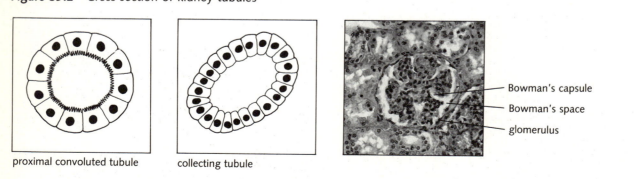

proximal convoluted tubule

collecting tubule

Bowman's capsule

Bowman's space

glomerulus

MATERIALS

Kidney from fetal pig dissection
Razor blade
Paper towel
Dissecting microscope
Prepared slide of mammalian kidney, sagittal
  section
Compound microscope
Living goldfish
Scissors
Watch glass
Ringer's solution
Dissecting needles
Deep-well slide
Chlorophenol red dye
Urine container
Pipette with pump
Microscope slide
Cover slip
Test tube
Benedict's solution
Beaker
Hot plate
Nitric acid
Ferric chloride
Aspirin

PROCEDURE

Anatomy of Fetal Pig Kidney

Obtain the plastic bag containing the fetal pig kidney that you removed during the fetal pig dissection. Follow the instructions below to examine the interior of the kidney.

1. Make a longitudinal section through the kidney using a razor blade. Cut on a paper towel.

2. Place half of the kidney under the dissection microscope and examine the cut side to find the following.
   - Renal capsule: a thin layer of tissue that surrounds the kidney
   - Cortex: a brownish outer layer
   - Medulla: an inner layer of tissue that is smooth and light-colored
   - Renal pelvis: a funnel-shaped area where urine collects and the ureter joins the kidney

3. Make a drawing of the areas of the kidney in Question 1 in the Evaluation.

Microscopic Examination of Mammalian Kidney

The sagittal (longitudinal) section of kidney you are about to study has been made by slicing a kidney into very thin sections. When examining the section, the long meandering nephrons will not appear as a continuous tubule. Figures 39.1 and 39.2 will be helpful in interpreting what you see.

1. Hold the slide of a kidney sagittal section up to the light and view with the unaided eye. The dark outer layer is the cortex containing the glomeruli; the light inner portion is the medulla containing the loops of Henle.

2. Using low power of the compound microscope study this kidney section, starting your observations with the cortex. Notice the large circular areas. These are the glomeruli in cross section. The smaller circles are the tubule sections.

3. Diagram and label a section of the cortex in Question 2(a) in the Evaluation.

4. Examine the medulla of the kidney section slide. You will not find any glomeruli in this area, only portions of the collecting tubule and the loop of Henle. Diagram and label the medulla in Question 2(b) in the Evaluation.

FUNCTIONING KIDNEY TUBULE

Cells of the kidney tubule are capable of active transport; that is, they are able to pump materials against an osmotic gradient. You will be able to demonstrate this remarkable process in the following procedure.

1. Obtain a living goldfish. Decapitate the fish with a single cut, using sharp scissors.

2. Open the abdomen and quickly remove the gut. The kidneys are a mass of brown tissue positioned close to the vertebral column. Refer to the fish dissection included in Laboratory Activity 17 if you need to refresh your memory.

3. Carefully remove the kidneys and place in a watch glass containing enough Ringer's solution (a solution of salts) to cover the tissue.

4. Under the dissection microscope, tease the kidney tissue apart using two dissecting needles until small filaments are seen. These are the tubules of the nephrons.

5. Place several of these tubules onto a deep-well slide containing several drops of Ringer's solution. Examine carefully under low power of the compound microscope.

6. Add several drops of chlorophenol red dye into the deep well.

7. Observe the tubules carefully for the next 15–20 minutes at 2-minutes intervals. Upon completion turn to the Evaluation and answer Question 3.

Urinalysis

If you are going to complete the Optional Activity (Nephron Activity in Humans) at the end of this laboratory activity you can save some time by completing the first two steps now.

► **CAUTION: Take great care in collecting and handling urine samples.**

1. **The urine used in the following urinalysis should be your own. You should not handle or study any urine sample except your own during the following activities.**

2. **Collect your urine sample in an unbreakable leak-proof container. Place the container in a zip-lock bag for transport to the laboratory.**

3. **When you are finished dispose of your urine sample as directed by your instructor.**

A great deal can be determined about the health of an individual from his or her urine sample. A variety of substances can be found normally in the urine but blood, pus, sugar, protein, and bacteria indicate illness or disease. In the following activities you will be examining your urine in a very general way; the results of these tests should not be looked upon as a complete urinalysis.

Microscopic Examination of Urine

1. Upon awakening collect approximately 50 mL of urine in a container. Cap the container and bring it to the laboratory as soon as possible.

2. Allow the urine to remain undisturbed in its container to allow for any solid material to settle to the bottom.

3. Using a pipette with a pump obtain several drops of urine with any sediment from the bottom of the jar.

4. Prepare a wet mount of this urine and observe under low and high power of the compound microscope. You may find red and white blood cells, epithelial cells from the lining of the urinary tract, and even some bacteria if you have a urinary infection. Diagram any contents of the urine that you can identify in Question 5 in the Evaluation.

Testing for Urinary Glucose

Benedict's solution can be used to test for glucose or any reducing sugar. A reducing sugar reduces the copper of the solution and changes the color of the solution from blue to green to yellow to red, depending on how much sugar is present. Normally all the glucose is reabsorbed by the capillaries so that none should appear in the urine.

1. Place 1 mL of urine in a test tube. Add 10 mL of Benedict's solution.

2. Put the test tube and its contents into a beaker of boiling water on a hot plate. Remove after 3 minutes and note any color change in Question 6 in the Evaluation. The greater the change toward red, the greater the quantity of reducing sugar.

Test for the Presence of Urea

In humans much of the nitrogenous waste produced from the breakdown of amino acids and nucleic acids forms urea. The following is a test for this waste material.

1. Place several drops of urine on a clean slide and add an equal number of drops of concentrated nitric acid.

➤ **CAUTION: Nitric acid will burn the skin and damage clothing. If you spill any acid on the skin or clothing flood the area with water and notify your instructor.**

2. Place the slide on a warm hot plate to evaporate about half of the liquid. Do not allow to boil. Allow the slide to cool, causing any urea nitrate present to crystalize.

3. Examine the slide under low power of your microscope. In Question 7 in the Evaluation diagram any crystals found in the urine.

OPTIONAL ACTIVITY

One or more members of your laboratory class may wish to volunteer for this activity. Volunteers should know from past experience that they are *not* allergic to common aspirin.

Nephron Activity in Humans

1. Collect 20 mL of urine in a small beaker. Add 5 mL of ferric chloride. Only a slight change in color should occur.

2. Take one aspirin tablet and drink a glass of water.

3. After waiting 15 minutes collect urine again and add 5 mL of ferric chloride. If aspirin is present in the urine a color change to black should occur.

4. In the event you did not get a positive test for aspirin repeat Step 3 every 15 minutes until you obtain positive results.

5. Answer Question 4 in the Evaluation.

EVALUATION 39

Excretion: The Elimination of Waste

1. Diagram the interior of the fetal pig kidney and label the capsule, cortex, medulla, and pelvis.

2. In the spaces below, diagram and label the microscopic examination of the kidney cross section. In (a) diagram a section of the cortex and label a glomerulus. In (b) diagram a section of the medulla and label several of the tubules.

(a) 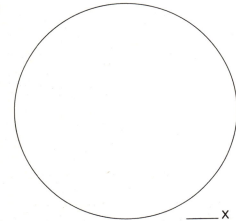 _____ X

(b) 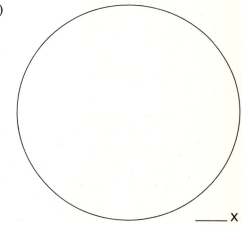 _____ X

3. (a) Describe any change in the tubule that you observed after you added the chlorophenol red.

   (b) Was this an example of active transport? Explain your answer.

4. (a) How long did it take for the aspirin to show up in the urine?

   (b) Why the delay?

   (c) Why did the aspirin appear in the urine?

5. Diagram any contents of the urine that you can identify from your microscopic examination (blood cells, epithelial cells, bacteria, and crystals).

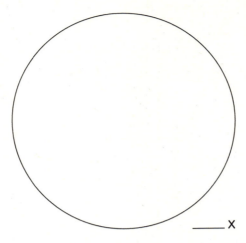

_____ X

6. What were your results from the test for reducing sugar (glucose) in the urine?

7. Draw any urea nitrate crystals you observed.

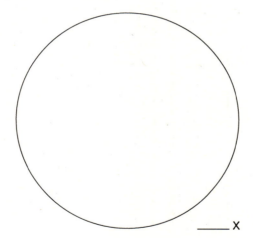

_____ X

Development: The Early Embryology of the Frog

OBJECTIVES

At the end of this laboratory activity, you should be able to:

- diagram and label the egg through the tail bud stage of frog development.
- describe the major events that occur during the early stages of frog development.

INTRODUCTION

The miracle of life has always fascinated humans, and we have had a great deal of curiosity concerning the development of animals: where individuals come from, how they grow, and what controls all these events. The first known treatise concerning animal development was written by the Greek philosopher Aristotle (384–322 B.C.)

Aristotle theorized that developing embryos were either preformed (the theory of preformation) or differentiated from a formless mass (the theory of epigenesis). Although he considered the theory of epigenesis more likely to be correct, the theory of preformation prevailed until the 1800s. Those who accepted the theory of preformation believed that a tiny organism present in the gamete simply grew into a larger form with time. Research conducted during the latter part of the nineteenth century proved epigenesis to be correct. The results of these experiments ushered in the field of embryology.

A multitude of questions arise concerning the progressive changes that occur during the development of an embryo. How do genes control these incredible events? Does the cellular environment play a role? If so, how? What causes a sheet of cells to fold in a specific manner? How do various tissues differentiate from common ancestral cells? The answers to many such questions are not complete, and much exciting research is taking place in embryology.

The study of embryonic development of many animal species has revealed certain common features. Therefore, a study of the progressive development of the embryo in a single species can illustrate the common features found in many animal species. In the following activity you will study frog embryos to uncover some of these important principles of embryology.

MATERIALS

Compound microscope
Dissecting microscope
Plastic metric ruler
Prepared slides of the following:
   development stages, whole mount
   unsegmented egg, typical section
   early cleavage, typical section
   blastula, typical section
   crescent blastopore, typical saggital section
   yolk plug, saggital section
   neural plate, cross section
   neural tube, cross section

PROCEDURE

As you follow the directions given in the procedures you will also be asked to make drawings of the embryos being studied. These labeled drawings should show all the areas of the embryo that appear in **boldface** in the written descriptions. The specific drawings required are listed in the Evaluation. It will be easier if you remove these Evaluation sheets now and have them in front of you as you study these embryos.

You will be using prepared slides of frog embryos in this investigation, including a deep-well slide that contains approximately ten frog embryos at different stages of development. In addition you will be examining a series of prepared slides containing stained sections of similar frog embryos. This will allow you to see the external and internal features of these progressive stages of development. You will view the whole-mount slide with the dissecting microscope and the sections under the compound microscope.

The whole-mount and section drawings shown in Figure 40.1 will be helpful in the identification of the embryological stages of the frog.

1. Obtain a deep-well slide containing the whole embryos and place it on the stage of a dissecting microscope. Light the stage from above and tilt the slide from side to side to roll the embryos over and over. Look at each embryo carefully to gain a general idea of the various stages of development present.

2. Find the egg, which is round in shape and quite smooth. It has a dark pigmented area, the **animal pole,** and a gray portion, the **vegetal pole,** which contains a higher concentration of yolk. Place a small plastic metric ruler under the slide, and specifically under the egg, to measure its diameter (approximately 1–1.5 mm). This egg is truly a remarkable structure. It contains all the information needed to produce an entire frog. In this egg the cytoplasm has become organized to direct future embryonic development.

3. Draw and label the whole mount egg in Question 1a of the Evaluation.

4. Obtain a prepared slide of a typical section of a frog egg. Under the low power of a compound microscope notice that the egg section nearly fills the low-power field of view. The entire inner portion of the egg is filled with **cytoplasm.** Much of the cytoplasm is composed of yolk, which is rich in fats and proteins and serves as nutrient for the developing embryo.

5. Draw and label the egg section in Question 1b in the Evaluation.

The frog egg enters into a period of rapid mitosis soon after fertilization. The first cell division, or **cleavage,** creates two cells of equal size. Each of these cells then divides and produces two cells of equal size.

6. Find a frog embryo composed of two or four cells in your whole-mount slide. Notice that each cell has an equal amount of the animal and vegetal portion of the original egg.

7. Draw and label a two- or four-cell whole-mount embryo in Question 2 in the Evaluation.

8. Rotate the embryos within your deep-well slide and find one in the eight-cell stage. Notice how the third cleavage has cut across or is at right angles to the first two cleavages (radial cleavage). In addition notice that this third cleavage did not divide the four cells at a midpoint. Instead, the third cleavage cut the four cells at a plane closer to the animal pole, thus producing four smaller cells in the animal pole and four larger ones in the vegetal pole.

9. Draw and label the eight-cell whole mount in Question 3 in the Evaluation.

10. Study carefully an early cleavage section under the compound microscope. Notice the large vegetal cells and the smaller animal cells. Draw and label the typical section in Question 3 in the Evaluation.

This process of cleavage produces more cells in a very controlled fashion. Initially the fertilized egg had just one set of chromosomes for the total amount of cytoplasm; but as cleavage continues, many cells are produced, each with its own set of chromosomes. The ratio of genetic information to cytoplasm is greatly increased.

11. Rotate the whole-mount slide and find an early **blastula** embryo (a blastula is a hollow fluid-filled ball of cells, one layer thick). The blastula is not appreciably larger than the fertilized egg, yet it is composed of numerous cells that can be seen clearly. As cleavage progresses the animal cells divide more frequently and are smaller than the vegetal cells. Cleavage ends about 24 hours after fertilization. Prepare a side view of this early blastula stage in Question 4a in the Evaluation.

12. Examine a typical section of an embryo at the blastula stage of development and determine how these cells are arranged internally. Notice that the cells are arranged around the **blastocoel,** a central free space.

Figure 40.1 Embryological development of the frog

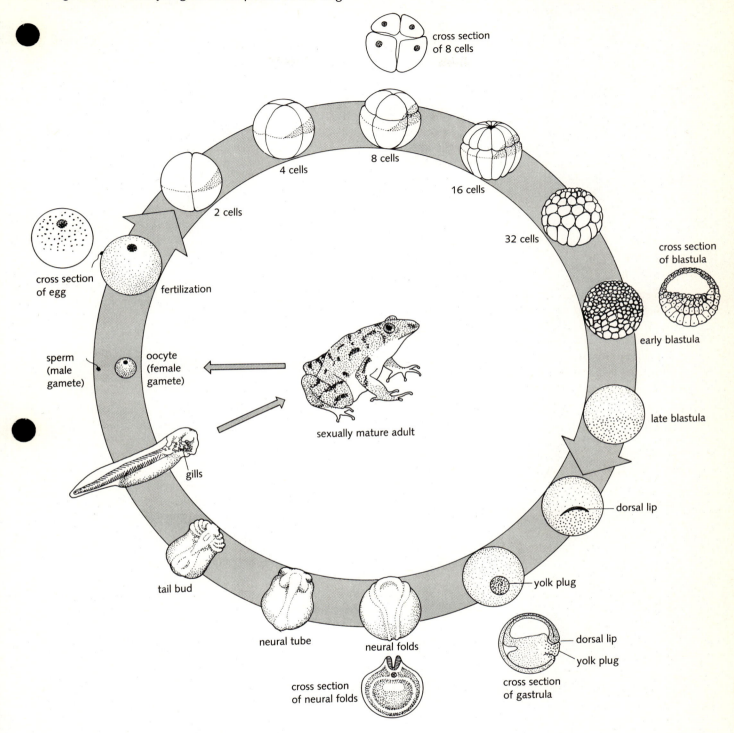

cross section of 8 cells

4 cells

8 cells

2 cells

16 cells

32 cells

cross section of egg

fertilization

cross section of blastula

early blastula

sperm (male gamete)

oocyte (female gamete)

sexually mature adult

late blastula

gills

dorsal lip

tail bud

yolk plug

neural tube

neural folds

dorsal lip

yolk plug

cross section of neural folds

cross section of gastrula

13. Draw and label a typical section of the frog blastula in Question 4b in the Evaluation.

The frog blastulas you have been studying exhibit symmetry and polarity and contain specific areas that will develop into germ layers, precursors of the organ systems of the frog.

14. In the whole-mount slide find a late blastula that has a rough surface like orange peel. You will not be able to discern individual cells, as they are barely visible. (At this stage of development, the study of whole embryos is somewhat difficult because the distinguishing characteristics may not be visible as you view them from above. Since the embryos are bottom-heavy you will need to roll the embryos over to view their ventral (bottom) surface.) Draw and label a whole-mount late bastula in Question 5 of the Evaluation.

The next important process of the progressive development of the frog embryo is that of **gastrulation,** which begins when cells at the animal pole start to push downward due to their rapid rate of mitosis. During this downward movement cells migrate into the blastocoel of the embryo and create a crescent-shaped indentation, the **blastopore,** at the margin between the animal and vegetal poles. As the indentation grows it develops into the shape of a horseshoe and ultimately into the shape of a circle. The upper margin of this indentation is the **dorsal lip.** Internally, a stomachlike structure (gastrula means "little stomach") has begun to form. The entrance to this "stomach" or pouch is the blastopore, while the pouch itself is the beginning of a second internal cavity called the **archenteron** (primitive gut). As the archenteron enlarges, it pushes into the space occupied by the blastocoel, causing the blastocoel to become a small narrow space at the animal pole.

15. Examine a sagittal section of a frog crescent blastopore under the compound microscope. A sagittal section is made longitudinally through the embryo. Notice the large central blastocoel surrounded by smaller cells in the animal pole and larger cells in the vegetal pole. Also note that cells from the animal pole accumulate near the indentation in the spherical embryo.

16. Draw and label a sagittal section of a crescent blastopore in Question 6a in the Evaluation.

17. Find a whole-mount embryo showing the dorsal lip of gastrulation. The blastopore may appear as a dimple and is located below the equator of the embryo in the yellowish-gray vegetal area. The actual shape of the blastopore is determined by the age of the embryo you are observing.

18. Draw and label the whole mount dorsal lip in Question 6b in the Evaluation.

The next stage of gastrulation is the **yolk plug.** Cells high in yolk content have entered the embryo, and at this stage a small portion of these cells are still evident externally, appearing as a small plug.

19. Find a whole-mount embryo showing the yolk plug. Draw and label the whole mount in Question 7(a) in the Evaluation.

20. Examine a frog yolk plug sagittal section slide. Notice the archenteron filled with yolk cells. The blastocoel has become a small narrow space at the animal pole. The yolk plug is formed near the end of gastrulation as the last yolk-filled cells enter through the blastopore.

21. Draw and label the sagittal section of the yolk plug in Question 7(b) in the Evaluation.

The yolk plug stage is a late stage of gastrulation. When gastrulation begins, the embryo has only one tissue or germ layer present, the **ectoderm** (outer layer). At the completion of gastrulation, two additional germ layers, **mesoderm** (middle layer) and **endoderm** (inner layer) are present. These three germ layers give rise to all the tissue areas of the developing frog. In addition to the development of germ layers, during gastrulation the embryo starts to lose its spherical shape and elongate.

Once the three germ layers are present, the elongated embryo enters the process of **neurulation.** During this stage the nervous system begins to develop from the ectodermal germ layer. Ectoderm along the dorsal region of the embryo thickens to create the **neural plate** and later folds in on itself to become the **neural folds.** Ectodermal tissue also produces the skin and the lens of the eye. A region of the mesoderm develops into the **notochord,** the precursor of the spinal column and the muscles of the back. A hollow neural tube forms in the dorsal area as a result of the fusion of the neural fold and ultimately develops into the brain and spinal cord.

The archenteron, which eventually becomes the digestive tract, is lined with endodermal cells. Other endodermal cells produce the linings for such organs as the lungs, liver, and pancreas. The blastopore will develop into the anus, and the mouth will break through at the opposite or anterior end of the embryo at a later time.

22. Find a whole mount of a neural stage. Draw and label the whole-mount neural stage in Question 8(a) of the Evaluation.

23. Examine a neural tube cross-section slide and view under the compound microscope. The slide you have selected probably has three or four sections on it; these will usually differ considerably due to age and section of the embryo from which it came.

24. Locate the hollow neural tube in the upper most dorsal area of the section.

25. Ventral to the neural tube locate the notochord as a smaller circle.

26. The archenteron is the large portion of the section.

27. Draw and label the neural tube cross section in Question 8(b) in the Evaluation.

The final process in the early development of the frog is **morphogenesis** (the beginning of form). The continued precise movement of cells begins to give the embryo definite shape. Much is yet to be learned about this process. The results of morphogenesis are quite apparent when viewing the tail bud stage of development. At this point the animal has elongated and has a head, tail, back, and belly.

28. Find the tail bud stage in your whole-mount slide. Draw and label the tail bud stage in Question 9 in the Evaluation.

The last embryo to be examined is an embryo in the external gill stage. It will be the largest embryo present in the whole-mount slide. At this stage of development the fishlike tail is present, and external gills can be seen. If you examined this embryo internally most of the internal organs would be well developed, but much more specialization is needed before this embryo develops into a mature frog.

29. Draw and label this embryo in Question 10 in the Evaluation.

This tail bud stage will ultimately develop into a free-swimming fishlike animal, the tadpole. The tadpole goes through a metamorphosis to become an adult frog.

EVALUATION 40

Development: The Early Embryology of the Frog

As you follow the directions given in the procedures you will also make drawings of the embryos being studied. Make sure that you label all areas of the embryo listed next to each diagram.

1. (a) Draw a whole mount and label the animal and vegetal pole of the egg.

(b) Draw a whole-mount sagittal section of an egg and label the cytoplasm.

Whole mount

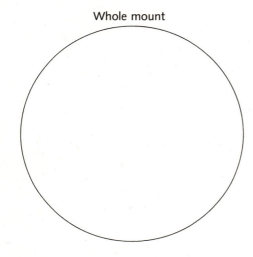

Sagittal

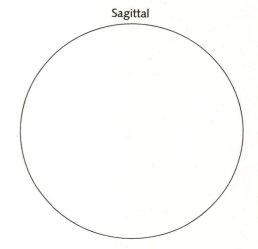

2. Draw a two- or four-cell whole-mount embryo.

Whole mount

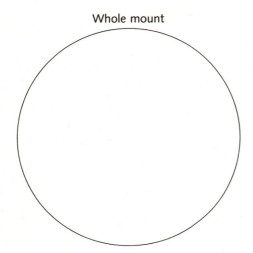

3. (a) Draw an eight-cell whole mount and label: first cleavage, second cleavage, and third cleavage.

Whole mount

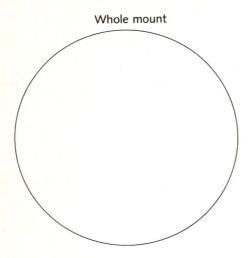

(b) Draw and label an eight-cell section.

Section

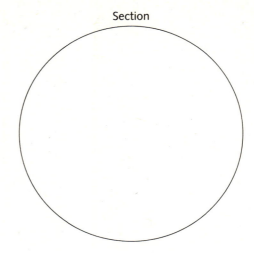

4. (a) Draw a whole-mount blastula and label: animal cells, vegetal cells.

Whole mount

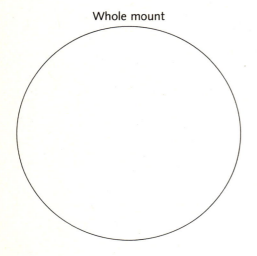

(b) Draw a typical section of an early blastula and label the blastocoel.

Typical section

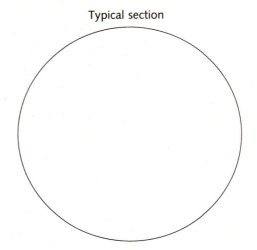

5. Draw a whole mount late blastula.

Whole mount

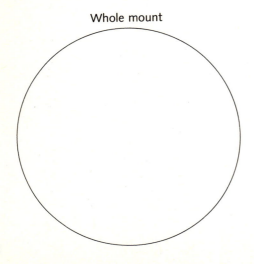

6. (a) Draw a sagittal section of a crescent blastopore and label blastopore and blastocoel.

Sagittal section

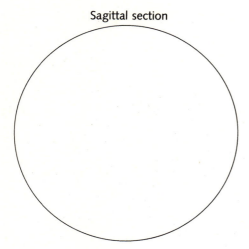

(b) Draw a whole-mount dorsal lip stage and label blastopore and dorsal lip.

Whole mount

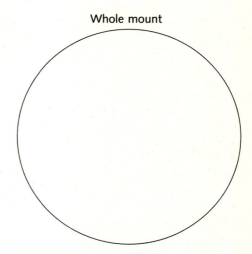

7. (a) Draw a whole-mount yolk plug stage and label the yolk cells.

Whole mount

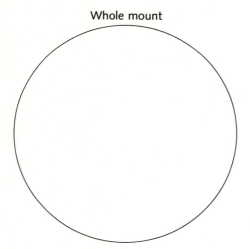

(b) Draw a sagittal section of the yolk plug and label blastocoel, archenteron, and yolk cells.

Sagittal section

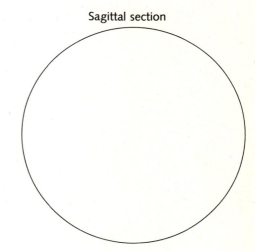

8. (a) Draw a whole-mount neural stage and label neural plate or neural folds.

(b) Draw a neural tube cross section and label the neural tube, notochord, and archenteron.

Whole mount

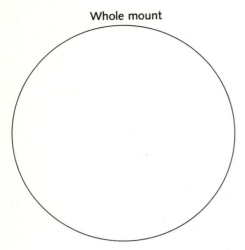

Cross section

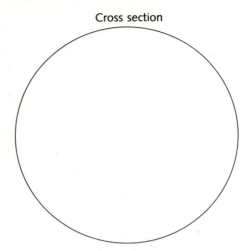

9. Draw and label a tail bud embryo and label the head and tail.

Whole mount

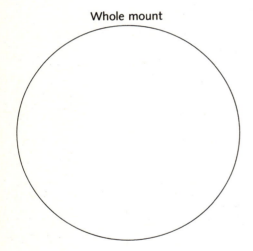

10. Draw an external gill embryo and label the external gills.

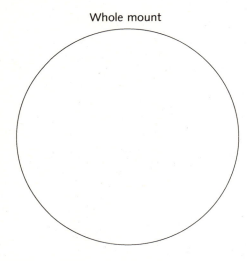

Whole mount

11. During cleavage not only has the number of cells increased, but the amount of DNA has also increased. Can you think of why this would be an advantage to the embryo?

12. Which system begins to develop first in the frog? Do you see any advantage in this?

13. There is considerable cellular migration during the early stages of animal embryology. What cell organelle is responsible for this cellular movement? Consult your textbook for additional information.

Exploring the Human Hand

OBJECTIVES

At the end of this laboratory activity, you should be able to:

- name the major bones, joints, digits, and skin markings of the hand.
- describe the various movements of the fingers and state the purpose of each.
- name the four major tissues of the hand and state the general function of each.
- outline the major steps in the evolution of the human hand.

INTRODUCTION

The forerunner of the human hand emerged between 200–300 million years ago during what paleontologists call the Mesozoic era. At this time the reptiles became firmly established on Earth and with them the beginnings of the pentadactyl (five-digit) hand that we know today. The Cenozoic era began about 65 million years ago and witnessed the establishment of the first mammals. Toward the end of the Cenozoic era, about 6 million years ago, the early hominids (human family) evolved with the five-digit hand virtually as we know it today. In other mammalian groups the bones had become fused into four or fewer digits and specialized for a particular function. The primate hand retained a more general structure and, as a result, more versatility in its function.

The outstanding feature of the human hand is the opposable thumb. With the five-digit opposable-thumb arrangement, we are able to button a shirt, tie our shoelaces or make fine watches. In spite of the remarkable adaptability of this unique structure, there is a tendency for us to take our hands for granted and perhaps even abuse them on occasion. During this laboratory activity you will learn to appreciate the remarkable structure of the human hand and its ability to perform many diverse functions.

MATERIALS

Booklet: *The Human Hand* by J. R. Napier
Piece of plate glass (8″ × 10″)
Tube of fingerprint ink
Roller
Magnifying glass
Skeleton of hand
Watch glass
Hammer
Screwdriver

PROCEDURE

For this laboratory you will use the booklet, *The Human Hand*, by J. R. Napier.[1] As you read each section of the booklet, you will have certain activities to complete in your laboratory manual.

1. Napier, J. R., *The Human Hand*, Carolina Biological Supply Company, Scientific Publications Division (Burlington, North Carolina, 1976).

Introduction

1. Read the introductory material on pages 2–3 of the booklet and answer the following questions.

   Define the term *pentadactyl*.
   Why do bones occasionally fuse during the course of evolution?
   Trace the evolution of the hand in Figure 1, page 3. Identify three changes that occurred in the bony structure of the hand in the past 200 million years.

External Morphology

1. Read the sections on terminology, external morphology, and digital formula as you complete the following questions and activities. Like every other part of the human body, the hand has been described in great detail. Indeed, the fingers have even been named for identification. As you read the section on terminology, complete the following chart.

Modern name	Canute's Latin	Digit number	Comment
Thumb			
Index			
Middle			
Ring			
Little			

Anatomists use a digital projection formula to indicate the length of the fingers in relation to each other.

1. Read the section in your booklet on digital projection and use Figure 41.1 to complete the first column in the following table with the proper notation. Remember that you always number the digits from longest to shortest.

Projection formula table

Hand (Figure 41.1)	Your hand (male/female)	Partner's hand (male/female)
⟩ ⟩ ⟩ ⟩	⟩ ⟩ ⟩ ⟩	⟩ ⟩ ⟩ ⟩

Figure 41.1 Human hand skeleton

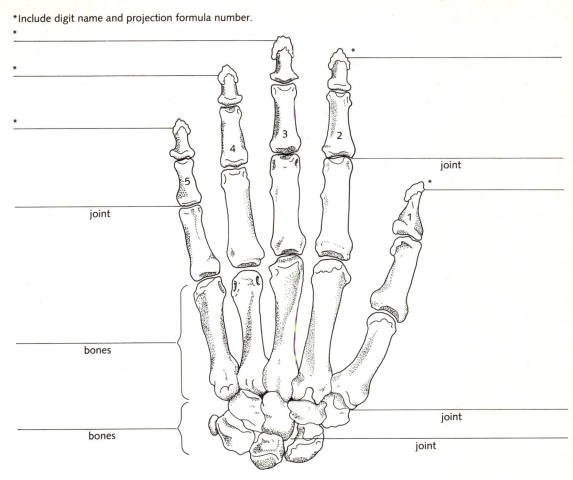

*Include digit name and projection formula number.

*
*
*
joint
bones
bones

*
joint
*
1
joint
joint

Prepare a print of your hand and your partner's hand by doing the following.

2. Place your hand flat against a glass plate that has been rolled with a very thin layer of fingerprint ink.

3. With the fingers held almost parallel, place your hand on the grid in Figure 41.2. The tip of your middle finger (impudicus) should just cover the X on the middle line of the grid and your other fingers (except the thumb) should just about cover the other vertical lines. Roll your hand gently but firmly on the paper to make a print that shows the lines of the hand clearly. Your print should be similar to Figure 6 in the booklet. If your print is too light or too dark, obtain a second grid from your instructor and try again.

4. The grid will help to determine the relative length of the fingers. Record the digital projection formula for your hand and also for your partner's in the Projection Formula Table.

As the text indicates, the third finger in humans is always dominant, but the index and ring finger vie for dominance.

5. Using the information from the booklet, complete the following table and collect data from the class to determine how this particular finger length relates to sex.

	Annularis dominant	Demonstratorius dominant
Male (%): Booklet		
Class data		
Female (%): Booklet		
Class data		

Figure 41.2 Grid for handprint

What does the author of the booklet hypothesize as the reason for the dominance of the index finger in humans as opposed to other primates?

Define the term *sexual dimorphism*.

Palmar Pads

1. Identify the palmar pads on your hand and then circle and label them on the handprint made in the last activity. What are palmar pads?

Why has there been a gradual shrinkage of these pads through the course of evolution?

Skin Markings

1. Locate the flexure lines on your hand and then label the heart and head lines on your handprint. What is the function of the flexure lines of the palm of the hand?

2. Locate Lange's lines on the back of your hand. Why are they most easily seen when the hand is relaxed?

3. Locate the papillary ridges. Label these ridges on your handprint. What are the two functions of these ridges?

What role does moisture play in skin sensitivity?

4. On the fingertips the papillary ridges form the fingerprints. Use your hand and your handprint to see if you can identify the ridge characteristics drawn in Figure 9 of your booklet. Label as many as you can on the handprint. Why is it believed that the conformation of these ridges is not completely determined by genetics?

Internal Morphology

The hand is capable of many complex movements brought about by its intricate musculature and innervation and the various types of joints formed by its 27 bones, including the 8 carpels making up the wrist.

1. Read in the booklet about these different joints, locate them in the hand skeleton provided by your instructor, and then complete the following chart.

Name of joint	Number	Function
Hinge		
Condyloid		
Plane		
Saddle		

2. After completing the chart, turn back to Figure 41.1 and label the joints, the phalanges according to Canute's system, and the metacarpals and carpals.

3. One of the most important movements of the saddle joint is opposition that occurs between the thumb and remaining digits. Describe how this movement occurs.

Why does brachiation (swinging by the arms) relegate the function of opposition to a secondary role in the life of the apes?

4. In addition to the bones and joints, the muscles, nerves, and blood vessels all play an essential role in the many complex functions of the hand. In the chart below, summarize these functions.

Tissue/organ	Function
Connective (fascia)	
Muscle Extrinsic	
Intrinsic	
Nerve Afferent	
Efferent	
Blood vessels	

Movements of the Whole Hand

1. The two main types of movement of the hand are prehensile and nonprehensile. What is the difference between these two types of movement? Give several examples of each.

2. Prehensile movement is more complex than nonprehensile movement. Which type of joint movement is of major importance in this type of movement?

3. On the lab table you will find several objects that will enable you to illustrate the four types of prehensile movement. Complete the chart below as you perform each of these grips.

4. Classify each of the following grips according to the classification scheme in the booklet.
   Picking up a glass slide: _____
   Throwing a ball: _____
   Carrying a book at your side: _____
   Swinging on a bar: _____

Grip	Hand structure used	Examples
Precision		
Power		
Hook		
Scissor		

Evolution of the Hand

1. What has been the general trend in the evolution of the human hand?

2. The evolution of the hand in primates apparently occurred in three stages. What were these stages and how did they differ from each other?

3. What does it mean to become a full-fledged brachiator and have an arboreal way of life?

4. What was the major development that enabled the Hominidae to become tool users, tool modifiers, and finally tool makers?

EVALUATION 41

Exploring the Human Hand

Your evaluation for this laboratory activity will be a quiz. Please see your instructor.

Chronobiology: Human Rhythms

OBJECTIVES

At the end of this laboratory activity, you should be able to:

- plot a graph showing your body temperature, pulse, finger counting speed, and adding speed.
- determine the correlation between temperature/pulse, temperature/finger counting speed, and temperature/adding speed.
- write a conclusion concerning your body rhythms.

INTRODUCTION

Students of nature have long observed rhythms in living organisms as the following biblical passage indicates.

> To everything there is a season,
> and a time to every purpose under the heaven.
> A time to be born, and a time to die;
> A time to plant, and a time to pluck up
> that which has been planted.
> Ecclesiastes 3:1,2

Over the centuries a vast number of such rhythms have been found in living forms and recorded. Evidence indicates that organisms seem to be able to time their major life events; when to migrate, bloom, sprout, actively feed, hibernate, mate, or give birth. Such biorhythms were also found in humans.

In the late 1950s, Dr. Franz Hallberg introduced a new biological discipline, **chronobiology,** which was devoted to the study of biological clocks or **biorhythms.** He found numerous rhythms in humans, including body temperature, heart beat, metabolic rate, and blood cell count.

Where are these clocks? Are they specific organs? Are they within the organism, or is the organism sensitive to a timing device located externally? Although much research has been done in this discipline (see Figure 42.1), during the last thirty years a unified model concerning the nature of the biological clock is still wanting. Some principles, however, have emerged. Many of these rhythms

Figure 42.1 Person involved in biorhythm research

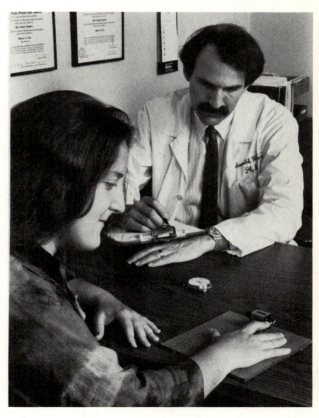

cycle around a 24-hour period, the time required for the earth to complete a single rotation on its axis. Energy enters the environment during the day and provides light, produces heat, causes variation in humidity, and is essential to the process of photosynthesis. Thus it is not surprising that animals and plants have developed patterns repeated in approximately 24 hours, cycles called **circadian rhythms** (*circa* = about, *dies* = day).

The rotation of the earth is only one regularly repeated cycle that affects organisms. Many organisms adapt to changes caused by the phases of the moon (**circalunadian rhythms,** approximately 28 days) and the cycle of the tides (**circatidal rhythms,** 12.4 hours). The orbit of the earth around the sun causes the changes in day length and temperature and puts great evolutionary pressure on plants and animals to cope with seasonal variation in climatic conditions. The adjustments that organisms make to such variation are known as **circannual cycles.**

Although scientists have devoted much effort to locating the biological clock, no one has determined where it is located and how it works in all organisms. Certain pacemaker organs involved in some biological clocks have been identified. For example the hypothalamus in the rat is involved in rhythms of running, temperature, estrus, and drinking. The pineal body in birds is involved in the rhythm of perch-hopping. Scientists have even discovered clocks in life forms composed of a single cell, implying that the clock may be a molecule or a group of molecules. Some researchers believe that the clock is located in the cell membrane, while others believe it is a molecule involved in protein synthesis.

Dr. Frank Brown of Northwestern University sees this entire matter differently. He supports the cosmic theory that holds that organisms need no timing device since the environment generates many of its own rhythms. The cosmic theory states that organisms need only to be sensitive to the cycles that are produced in the environment. These rhythmic signals would include variations in gravity, electrical fields, and magnetism.

Brown found that oxygen consumption in shellfish, rats, and potatoes varies with cosmic changes. In one classic experiment concerning oysters, Brown studied the shell-opening activities of oysters transplanted from New Haven, Connecticut to Evanston, Illinois. For two weeks, the oysters opened in synchrony with the high tides in Connecticut. Then the oysters rapidly shifted their cycle to match one that would occur in Evanston, as if the ocean spread that far west!

Regardless of why biorhythms occur, their mechanisms are deeply rooted within the general metabolism of living organisms. Below is a list of human rhythms for your consideration. Think of the many implications they may have in your daily life.

- Most people show their best mental performance and physical fitness between noon and 7 P.M.
- In 1000 tested workers, 45% could not adjust to a 7-day rotation in shift work.
- Children 0–3 years of age have not developed some circadian rhythms concerning eating and sleeping.
- Depression may be caused by desynchronization of temperature, adrenal hormones, and potassium excretion.
- A specific drug may be therapeutic if taken at one time of day and poisonous if taken at another.
- Humans have a greater tolerance for alcohol during the evening hours.
- Human clotting time is longest at midday and shortest at night.
- Humans have greatest immunity to disease at the end of the day and are most susceptible to disease during the early hours of the morning.

Research in this field has demonstrated that many functions of the human body are rhythmic. Indeed most physiological activities are probably cyclic, making it very important that you be aware of your personal biorhythms. When do you work well and feel the best? Are you a night person (an owl) or are you a day person (a lark)? What kind of a life schedule should you establish to feel the best and accomplish the most? This laboratory activity will help you to answer some of these questions. Before you begin this study complete Question 1 in the Evaluation.

MATERIALS

Thermometer
Watch with second hand

PROCEDURE

The functions that were chosen for you to monitor during this activity were selected because they could be easily measured without expensive equipment and will give you a good idea of your rhythms.

Each time you collect data you should follow exactly the same procedure given in the instructions below. Collect data 7 times a day for 5 days. Take your first readings as soon as you awake in the morning. All of your equipment should be at bedside, and these initial readings should be taken before you begin any morning activities. Any activity will cause the pulse to increase and not give a true indication of your body temperature when you wake up. Other readings should be taken at 9:00 A.M., 11:00 A.M., 2:00 P.M., 5:00 P.M., 8:00 P.M., and just before going to bed.

The suggested times above can be adjusted to meet your schedule if need be. There is nothing sacred about 9:00 A.M. (9:17 A.M. would be fine), but once you have selected 7 reading times you must take your readings precisely at these times for all 5 days of your study. Start your readings by taking your temperature. Use the same procedures and sequence each time.

Remember that the data you collect may provide you with some valuable information about your own cycles. Be as precise as possible.

Temperature

1. Shake down your thermometer so the mercury column reads below 96°F. Place the thermometer as far back as comfortable under the tongue. The thermometer should remain in the closed mouth in this position for at least 3 minutes. You can remain seated and continue to take your other readings while the thermometer is positioned under the tongue.

2. Record your temperature in the daily record chart found in the Evaluation.

Pulse

1. Place your index and your second finger on the inside of your wrist at the base of the thumb to feel a pulse. Count the pulse for a full minute. If your pulse is weak you may be able to find it more easily in the carotid artery of the neck.

2. Record your pulse in the daily record chart found in the Evaluation.

Finger Counting: Eye-Hand Coordination

1. Glance at your watch and immediately touch the index finger with the thumb and count "1," then touch the thumb to the middle finger and count "2," continue by touching the thumb to the ring finger and count "3." Next place the thumb to the little finger at the count of "4." At this point return the thumb back over the fingers by first moving the thumb to the ring finger at the count of "5," the middle finger at the count of "6," and to the index finger at the count of "7." Keep repeating the procedure again and again until the thumb rests on the index finger at a count of "25." Practice this procedure moving as fast as possible until you are comfortable with the movements.

2. Record the time required to complete the count to "25" in the daily record in the Evaluation. If you make a mistake, start over and continue until you make a complete correct count. Record the total time needed to reach a complete correct count.

Adding Speed

In this activity you will use Table 42.1. Time yourself accurately as you add a single column of numbers according to the following procedure.

1. Place a strip of paper at the bottom of the second number in the first column of Table 42.1 and add the two numbers ($8 + 0 = 8$). Now move the paper down one number and add the second and third numbers of the column ($0 + 1 = 1$). Continue down the column, adding the third and fourth numbers ($1 + 8 = 9$), and so on.

2. Add an entire column each time you take a reading. Use another column for the next reading. After 16 readings use the first column again.

3. Record the time required to add a single column of numbers in the daily record in the Evaluation.

4. Turn to Question 3 in the Evaluation and complete the chart (Record of Average Readings).

5. Complete Questions 4–6 in the Evaluation.

Table 42.1 Random numbers

8	3	4	7	4	2	9	2	2	6	2	8	3	1	3	5
0	3	9	2	3	1	2	2	8	2	9	3	8	1	9	5
1	4	5	1	3	5	1	7	4	5	2	5	9	8	7	4
8	9	8	1	3	7	5	3	6	2	1	3	8	0	1	7
3	2	1	6	7	0	8	6	5	2	4	9	5	2	4	2
4	1	4	1	9	1	0	3	9	5	9	0	3	8	9	9
9	2	5	9	8	0	5	4	8	5	4	7	3	5	8	3
4	4	3	7	0	6	1	7	4	1	8	0	2	5	9	6
6	0	7	7	0	7	4	3	4	5	4	5	1	8	2	1
1	8	8	8	5	5	5	6	1	5	5	9	8	6	3	0
9	5	9	7	7	9	9	9	4	3	2	7	8	7	4	6
1	3	9	4	8	2	4	8	9	3	9	0	7	2	7	7
3	9	8	2	8	8	7	0	2	1	3	2	7	2	6	4
3	2	7	3	6	2	9	7	2	2	4	3	5	4	5	6
2	2	3	4	4	7	7	7	4	2	7	1	5	7	2	0
6	9	6	9	4	1	1	0	8	0	1	5	7	5	9	0
3	4	1	7	6	4	2	2	7	1	7	0	1	7	6	1
9	7	1	4	7	6	9	7	3	4	6	8	0	4	7	3
5	1	3	9	6	5	3	5	0	0	2	6	9	2	6	3
6	1	0	7	2	3	8	8	7	3	2	0	6	2	7	5
7	4	1	5	3	7	2	2	3	2	5	9	4	9	1	5
0	6	0	8	1	0	3	0	4	9	9	5	9	6	4	5
8	6	5	8	4	0	9	8	3	6	8	2	7	0	3	6
7	3	1	5	8	5	9	9	4	8	8	2	1	2	9	2
6	5	9	2	9	5	4	4	1	1	4	7	0	8	7	2
9	5	8	4	7	6	2	6	1	2	1	9	9	6	9	5
0	4	6	7	6	6	9	3	6	1	0	0	0	6	5	6
0	3	4	6	2	1	2	2	7	3	8	5	7	2	3	8
8	4	9	9	4	4	2	3	3	4	7	2	6	6	9	9
7	9	3	6	9	4	2	5	7	5	4	2	2	6	0	2
4	7	6	1	9	6	3	2	7	2	1	3	0	6	0	4
9	1	9	3	6	4	6	6	0	0	4	6	2	4	1	8
3	1	2	4	3	7	0	1	7	2	8	0	2	3	7	0
9	6	0	4	6	4	8	6	8	0	1	8	6	5	2	6
9	7	9	9	2	8	1	3	9	5	1	1	0	8	1	4
5	9	1	6	0	8	5	1	8	7	6	5	8	0	7	4
8	5	9	1	9	3	8	2	6	3	7	8	6	2	7	3
1	0	5	2	9	1	7	9	0	3	0	7	4	7	4	3
5	6	3	6	7	9	3	5	2	7	2	3	7	8	0	3
2	4	7	1	5	2	1	3	2	4	4	1	0	5	7	3
5	8	5	4	3	8	7	3	4	1	2	7	9	3	3	2
0	1	9	7	5	4	5	1	8	1	4	5	7	0	0	3
7	6	4	1	1	4	9	0	3	3	8	9	1	4	3	6
9	1	9	7	5	0	0	2	7	7	4	1	5	9	4	3
6	8	3	6	5	0	7	3	5	5	3	7	2	6	1	1
7	0	2	3	3	7	8	6	2	0	1	8	5	5	8	3
1	0	3	9	6	8	1	0	4	6	0	1	5	9	0	3
3	3	0	3	5	8	6	1	0	8	4	6	7	3	1	9
0	1	4	8	8	5	4	1	0	5	5	6	7	1	0	3

EVALUATION 42

Chronobiology: Human Rhythms

1. Before you begin this study state when you think you function best during the day. Explain the basis of your thinking.

2. Daily record of readings

Date	Time	Temp.	Pulse	Finger counting	Adding speed

Daily record of readings (*continued*)

Date	Time	Temp.	Pulse	Finger counting	Adding speed

3. Determine a mean of each measure for each reading time. For example, you will determine a mean for your 5 temperature readings at the start of the day, a mean for your 9:00 A.M. temperature readings, and a mean for each temperature reading throughout the day. Repeat this process for pulse, finger counting, and adding speed. If a particular reading deviates greatly from all the others taken at a particular time it may be best to eliminate this particular piece of data and compute your average with the remaining data. If you do eliminate a piece of data, make sure you mention it and indicate why you think the data deviated from the other readings. Record your means in the following chart.

Record of average readings

Time of day reading recorded	Awakening	————	————	————	————	————	Retiring
Temperature
Pulse
Finger counting
Adding speed

4. On the graph paper provided, prepare a graph plotting your 7 mean temperature readings against time. Record time on the *x*-axis and temperature on the *y*-axis. Repeat this graphing procedure for pulse, finger counting, and adding speed. Your finished graph will have a temperature line, a pulse line, a finger counting line, and an adding speed line. Examine your graph carefully. Do the lines indicate that you have a time of day that is best for you? (Are you a night owl or a morning lark?) State any other conclusions you can make concerning your body rhythms.

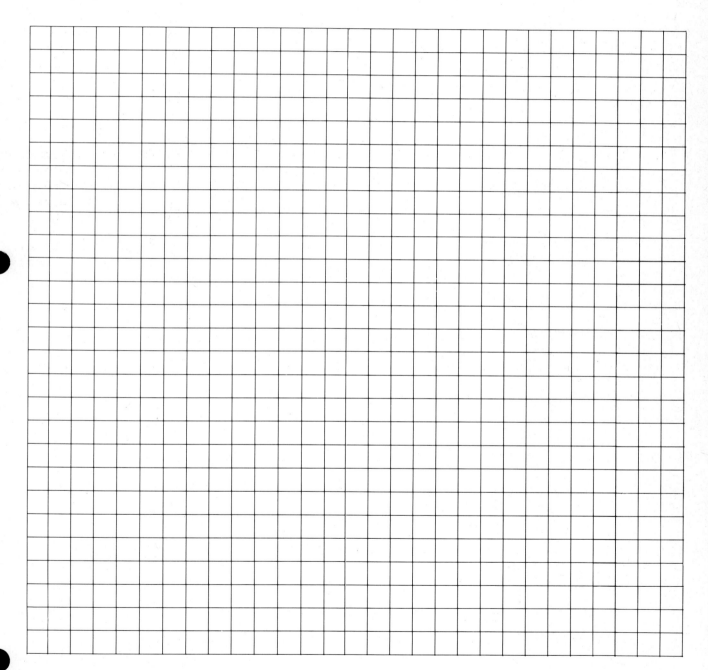

5. Body temperature is an important factor in the life of a warm-blooded animal. Is it possible that as temperature fluctuates the physiological characteristics also change? To determine if this is the case calculate the coefficient of correlation between your mean temperature readings and your mean pulse readings. Also calculate the coefficient of correlation between your mean temperature readings and your mean finger counting times; and between your mean temperature readings and your mean adding speeds. Information concerning the calculation of the Pearson r correlation is found in Laboratory Activity 1. Show your calculations. Does your data indicate that temperature and these other rhythms are related? If so how?

6. Attempt to relate all of your data to what you know about yourself and how you function throughout the day and night.

A Second Look at Biology

OBJECTIVES

At the end of this laboratory activity, you should be able to:

- show a photograph you have taken of a biological structure, an organism, or activity relating to biology.
- write several paragraphs explaining the topic in your photograph.

INTRODUCTION

It is our hope that this laboratory guide and your course in biology have caused you to look and think about the living world a little differently than you did before. Because of what you have learned, what may have been ordinary or commonplace to you before now may stir your curiosity and cause you to take a second look at the world around you.

This activity will enable you to record one such second look. You will photograph a structure, organism, or activity relating to biology. The important point is that you saw something that stirred your curiosity and you wished to make a record of it.

MATERIALS

Camera
Film

PROCEDURE

1. Take a color photograph using any type of camera available. Since you will be handing this in to your instructor, you should produce a print, not a slide.

2. Mount your photograph on the evaluation sheet. Give your topic a title.

3. Explain why your topic caused you to take a second look. That is, how does your topic relate to what you have learned in biology? This may require you to do some research on your topic. Record this information under the photograph on the evaluation sheet.

EVALUATION 43

A Second Look at Biology

Title

Research

Graph Construction

When you wish to show a relationship between two phenomena or events, it is often useful to plot this information on a graph. The reason is simply that it is often difficult to recognize relationships between variables by looking at a table of data; these relationships, however, become quite apparent when the data is presented on a graph.

Many phenomena in biology involve the relationship between two variables, such as amount of sweating and heat loss from the body, rate of heartbeat and temperature in a cold-blooded animal, or the amount of light available and the rate of photosynthesis.

In order to graph the manner in which one variable of a pair changes with respect to another, you must decide which is the independent variable and which is the dependent variable. In discussing independent and dependent variables and other terms relating to graphs in general, refer to the model graph in Figure A.1. After looking at the graph it becomes quite apparent that the number of stomata/mm^2 (openings in the leaves per square millimeter) is the independent variable, and the loss of water is the dependent variable. The reasoning is: The loss of water depends upon the number of stomata; the number of stomata does not depend upon the loss of water. Thus the number of stomata is the independent variable.

In the construction of a graph, the independent variable is plotted on the X-axis (horizontal) and the dependent variable on the Y-axis (vertical). The coordinates on the X-axis are called **abscissas;** those on the Y-axis are **ordinates.** In order to describe a specific point on a graph, first locate the abscissa on the X-axis, next locate the ordinate on the Y-axis, and then plot a point where these two intersect. (Keep in mind that in some cases you will not

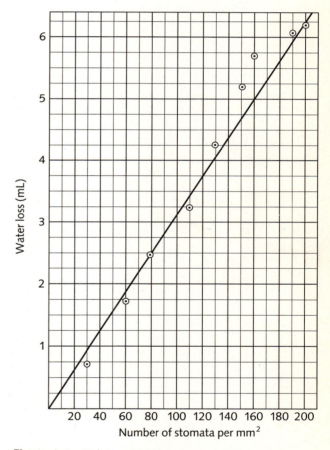

Figure A.1   Relationship between water loss and number of stomata

be concerned with variables that are dependent and independent, but when you can discern them as such they should be placed in the manner just described.)

One of the major problems concerning graph construction is selecting the proper scale. Not only should you choose a scale that will allow the graph to fill the appropriate sheet of graph paper, but the graph must also be large enough to show the proper relationships between the variables. You can achieve this in the following manner: Select either variable and find the difference between the lowest and the highest value. For example, in the model graph the range of values for water loss is $6.1 - 0.7 = 5.4$. If we set each block of the graph equal to one point, our scale will be only 5.4 blocks long. A scale this small cannot show the proper relationship. However, if we assign a value of one-quarter unit to each block, our scale will expand to 20 blocks and will better accommodate our data.

This same procedure can be applied to the scale on the X-axis. Our range is 170, the difference between 30 and 200 stomata; obviously one block cannot represent one unit. However, if we allow one block to equal 10 units we will need 17 blocks. This is obviously more appropriate since it fills more of the graph paper.

After determining the scale you must designate what is being plotted on each axis. The labels along each axis are usually written in complete words, with the exception of standard abbreviations for measurements, such as "mL" for milliliter, "sq" for square, and "mm" for millimeter.

After plotting the points on the graph, draw a small circle around each point. If more than one line is needed on a single graph, you can use triangles, squares, or diamonds to enclose the dot. Doing so will enable you to identify each set of points more easily. When constructing a graph with more than one line, you should also provide a key or legend to identify each line on the graph.

In the graph presented here it is obvious that the points form an almost straight line. In such cases, draw a straight line that leaves a balance of points above and below the line. A line plotted in this way is called a **line of best fit.** This line should not go through the circles around the dots—it should just touch the side so that future reference to the points will not be obscured by the line. If data produces a curved line, it should be drawn in a similar manner.

The last major consideration concerning the construction of a graph is the title. It should be placed at the top of the graph and written in the most convenient open space. It should be said in words, rather than abbreviations, that the dependent variable is a function of the independent variable.

Sample Calculation for the Shannon-Weaver Diversity Index

The formula for the Shannon-Weaver index is

$$N = (P_i)(-\ln P_i)$$

where N = Shannon-Weaver index

$$P_i = \frac{\text{number of organisms}}{\text{in a given species}} {\text{total number of}} {\text{all organisms}}$$

and $-\ln P_i$ = natural log of P_i

Sample Calculation

A population of organisms collected from a stream bed has the following composition:

Organism A = 25
B = 17
C = 37
D = 5
Total = 84

Organism	Frequency (P_i)	Natural log $(-\ln P_i)$	$(P_i)(-\ln P_i)$
A	$\frac{25}{84} = 0.298$	1.211	0.361
B	$\frac{17}{84} = 0.202$	1.598	0.323
C	$\frac{37}{84} = 0.440$	0.820	0.362
D	$\frac{5}{84} = 0.060$	2.821	0.169
			$N = 1.22$

The Shannon-Weaver index is the sum of the products in the last column. Thus, $N = 1.22$, indicating a stream of intermediate water quality.

The Chi Square Test

The chi square (x^2) test is a statistical tool used by researchers to determine if obtained data constitutes a good fit with what is expected to occur. In genetics, this goodness of fit means: Does the observed ratio between different types of offspring from a cross agree with the theoretically expected ratio? The chi square test enables one to determine whether it is reasonable to attribute deviations from the expected ratios to chance. When the deviation from what is expected is small, you would ordinarily not be disturbed—it is normal to expect some deviation to occur by chance. However, if the deviation is large, you should begin to ask questions about the way in which the experiment was run. Was the cross made properly? Was my sample large enough? Did I count accurately? Is the ratio I expected to obtain a reasonable one or should I review my hypothesis and offer an alternative ratio? The problem is knowing how to determine if the deviation is large enough to suspect that it was not due to chance alone. The chi square test is a tool that helps you to make this decision.

The chi square (x^2) formula is:

$$x^2 = \frac{(O - E)^2}{E}$$

where x^2 = chi square
O = observed value
E = expected value

The expression $(O - E)^2/E$ gives the chi square value for each class of data in which a deviation can be expected to occur.

To illustrate the use of this statistical tool, let's consider a situation in which a coin is tossed 1000 times. Theoretically we would expect such a series of tosses to produce 500 heads and 500 tails, or a ratio of 1:1. However, our results are 494 heads and 506 tails. Is this deviation due to chance or might some other factors be causing our observed results to differ from the expected? The chi square test will help to answer this question.

The data are summarized in Table C.1.

Table C.1

Classes	Observed number O	Expected number E	Deviation	Deviation squared $(O - E)^2$	$\dfrac{(O - E)^2}{E}$
Heads	494	500	6	36	0.072
Tails	506	500	6	36	0.072
Totals				$\dfrac{(O - E)^2}{E} =$	0.144
				$x^2 =$	0.144

To interpret our χ^2 value of 0.144 we must turn to a table of chi square values (see Table C.2). It is important to remember that a chi square of zero means that no deviation has occurred, and we have a perfect fit of our expected (hypothetical) ratio with what we actually observed. As chi square values increase, they indicate increasing deviation from the expected results.

The table of chi square tells us whether or not these deviations were due to chance. First you must determine what is called the **degrees of freedom.** This value is always one less than the number of classes of data under consideration. For this experiment, the degrees of freedom is 1.

The values across the top of the table (P) express the probability that deviations within that column are due to chance. In this situation, we find the number 1 in the degrees of freedom column (n) and move across until we locate the numbers on either side of our chi square value. These numbers are 0.0642 and 0.148. The numbers at the top (P) tell us the probability that a chi square value of this magnitude would occur by chance alone is between 80% and 70%. This means that if we were to repeat this experiment 100 times we could expect a deviation this large to occur by chance alone between 70% and 80% of the time. We would, therefore, accept the deviation of our observed ratio from the expected ratio as being due to chance and not question the validity of our results.

Geneticists have arbitrarily established a P value of 0.05 (5% or 1 in 20) as the cutoff point for the acceptance or rejection of the results of an experiment. This means that chi square values below 5% are too large to be attributed to chance alone and that factors other than chance may be at work to produce these results.

Table C.2 Chi Square Values

n	P-.99	.98	.95	.90	.80	.70	.50	.30	.20	.10	.05	.02	.01
1	.000157	.000628	.00393	.0158	.0642	.148	.455	1.074	1.642	2.706	3.841	5.412	6.635
2	.0201	.0404	.103	.211	.446	.713	1.386	2.408	3.219	4.605	5.991	7.824	9.210
3	.115	.185	.352	.584	1.005	1.424	2.366	3.665	4.642	6.251	7.816	9.837	11.345
4	.297	.429	.711	1.064	1.649	2.195	3.357	4.878	5.989	7.779	9.488	11.668	13.277
5	.554	.752	1.145	1.610	2.343	3.000	4.351	6.064	7.289	9.236	11.070	13.388	15.086
6	.872	1.134	1.635	2.204	3.070	3.828	5.348	7.231	8.558	10.645	12.592	15.033	16.812
7	1.239	1.564	2.167	2.833	3.822	4.671	6.346	8.383	9.803	12.017	14.067	16.622	18.475
8	1.646	2.032	2.733	3.490	4.594	5.527	7.344	9.524	11.030	13.362	15.507	18.168	20.090
9	2.088	2.532	3.325	4.168	5.380	6.393	8.343	10.656	12.242	14.684	16.919	19.679	21.666
10	2.558	3.059	3.940	4.865	6.179	7.267	9.342	11.781	13.442	15.987	18.307	21.161	23.209

Reprinted with permission of Hafner Press, a division of Macmillan Publishing Company, from *Statistical Methods for Research Workers* (14th ed.) by R. A. Fisher. Copyright © 1970 by University of Adelaide.

Use of the pH Meter*

Typically, several models of pH meters are available in the laboratory (see, for example, Figure D.1). Your instructor will provide specific instructions for the operation of the meter you will use.

In all cases, the meter must be calibrated before being used for pH measurements. Samples of pH reference buffer solutions are available for calibrating the meters. Generally, the combination electrode is dipped into one of the reference buffer solutions and allowed to stand for several minutes; this permits the electrode to come to equilibrium with the buffer. The "set" or "calibrate" knob on the face of the meter is then adjusted until the meter display reads the correct pH for the reference buffer. For precise work, a two-point calibration, in which the meter is checked in two different reference buffers, may be necessary. Your instructor will explain the two-point calibration method if you will be using it.

Remember that the pH-sensing electrode is made of glass and is therefore very fragile. (Combination pH electrodes are very expensive, so you may be asked to pay for the electrode if you break it.) Handle the electrode gently, do not stir solutions with the electrode, and keep the electrode in a beaker of distilled water when not in use to keep it from drying out. Rinse the electrode with distilled water when transferring it from one solution to another.

*Adapted from James F. Hall, *Experimental Chemistry*, pp. 367–368. Copyright © 1989 by D. C. Heath and Company, Lexington, MA.

Figure D.1 (a) A typical pH meter/electrode setup. The two electrodes may be combined in one container to form a "combination" pH-sensing electrode. (b) pH-sensing electrode. The glass membrane measures difference in [H$^+$] between interior of electrode and the solution being tested. (c) Calomel reference electrode. This completes the electrical circuit in the solution and is very stable and reproducible.

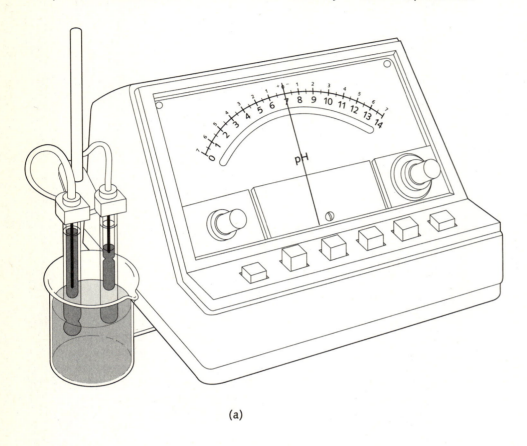

(a)

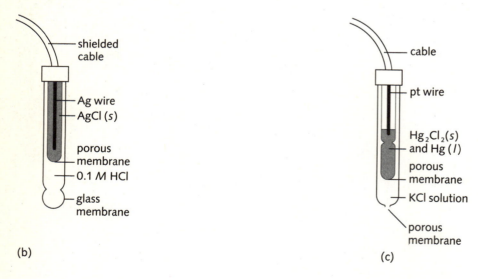

shielded cable
Ag wire
AgCl (s)
porous membrane
0.1 M HCl
glass membrane

(b)

cable
pt wire
Hg$_2$Cl$_2$(s) and Hg (l)
porous membrane
KCl solution
porous membrane

(c)

Use of the Milton Roy Spectrophotometer

Several of the activities in this laboratory text require the use of an instrument that is able to measure the transmission of light through a solution. The one illustrated in Figure E.1 is the SPECTRONIC® 20 manufactured by Milton Roy.* This versatile instrument enables you to measure the amount of light that passes through (percentage of transmission) or is absorbed by (optical density) solutions of varying densities. Furthermore, the wavelength of light can be changed to suit a particular experimental situation.

The principal of operation of the SPECTRONIC 20 is based on the passage of light from a source through a series of lenses and then through a diffraction grating. The diffraction grating disperses the light and produces a band of light of one wavelength (monochromatic), which then passes through the solution in the test tube (cuvette). The amount of light that passes through the solution depends on the density of the solution; that is, less light passes through a denser solution. The light that exits from the tube is then picked up by a phototube, which enables the intensity of light to be read on a meter as percentage of light transmittance or optical density. This value is compared to a standard solution that is used to set the machine before readings of the unknown sample are taken. The wavelength of light used is determined experimentally (see Laboratory Activity 27) or by consulting a reference.

The special optical quality of test tubes used with the SPECTRONIC 20 allow them to reduce the amount of interference the glass might have with the light as it passes through the solution. These tubes are usually marked, and this mark should be toward the front of the sample holder whenever you insert the tube into the well. You should also be certain that the tube is filled sufficiently so that the light will pass through the solution when a reading is taken. Either use the mark on the tube or fill approximately three-quarters full.

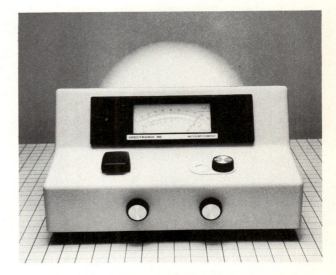

Figure E.1 The SPECTRONIC 20 spectrophotometer.

*SPECTRONIC is a registered trademark of the Milton Roy Company.

Specific directions for using the SPECTRONIC 20 follow.

1. Set the **wavelength control** knob to the desired wavelength. Wavelengths are measured in nanometers (nm).

2. Turn the instrument on using the **power switch/knob** and allow it to warm up for five minutes. Next, ensure that the cover of the **sample holder** (well) is closed, and adjust the **power switch** until the needle reads 0% transmittance on the scale. If the needle fluctuates, the instrument is not warmed up sufficiently.

3. Place the cuvette containing the blank solution—water or other liquid—in the sample holder and close the cover. Then rotate the **light control knob** until the needle of the meter reads 100% transmittance, or 0.0 absorbance. The instrument should now be set for reading samples. It is important to remember that the instrument must always be set with a blank solution before measuring unknown samples. Even distilled water will absorb some light, and by setting the meter to 100% transmittance with a blank, light absorbance due to the solvent is "subtracted" out of the reading for the various unknown solutions.

4. You may now read any unknown solutions by placing them in the sample holder. Always make sure that the cover is closed. Occasionally check the meter setting with the blank to make certain that the unit is calibrated. Record your readings as percentage of transmittance or optical density.

An Introduction to Bacteriology

OBJECTIVES

At the end of this laboratory activity, you should be able to:

- identify the three basic shapes of bacteria.
- prepare a bacterial smear, a simple stain, and a Gram stain from a bacterial culture.
- aseptically transfer bacteria from one culture to another.
- prepare a pour plate and a streak plate.

INTRODUCTION

Microbiology is the study of microscopic organisms such as bacteria, algae, fungi, viruses, and protozoa. This laboratory activity is concerned just with bacteriology, the study of bacteria.

Bacteria are relatively simple in their structure and metabolism and so are uniquely designed to allow biologists to study many of the fundamental life processes that occur in all organisms. In addition, bacteria are easy to grow in culture since they require little space, and although they are at times particular in their nutritional requirements, their demands can usually be met without much difficulty. Bacteria also reproduce at a rapid rate. Under ideal growth conditions, many species will easily produce more than 50 generations within 24 hours. The beginning student in bacteriology is usually surprised when a mere streak on a freshly inoculated agar growth medium produces a blanket of bacteria in only 24 hours.

In the practical world, a knowledge of microorganisms in general and bacteria in particular is valuable because so many of their activities either directly or indirectly influence our lives. For example, microorganisms are one of the major causes of disease in both animals and plants and are a major cause of food spoilage. As decomposers, bacteria help cycle elements in the soil, water, and atmosphere. In industry, they produce many useful organic compounds and are responsible for the production of foods such as yogurt and cheese.

The development of microbiology in the early years parallels the development of the microscope. Although microbes were used by humans for hundreds of years before they were first observed, it was Anton van Leeuwenhoek (1632–1723) who, using his primitive microscope, first accurately described and recorded his observations of microbes in water and other media. The scientists who followed van Leeuwenhoek made major contributions to the health and improvement of humanity through their work in microbiology. Louis Pasteur (1822–1895) proved that microorganisms caused fermentation. Through his studies of microorganisms and heat, Pasteur developed a method for protecting food from bacterial contamination, a process known as pasteurization. Pasteur was also responsible for developing the immunization process and the rabies vaccine. In 1798, Edward Jenner (1749–1823) successfully immunized a child against smallpox, and Robert Koch (1843–1910) discovered the bacteria that causes anthrax, the dreaded disease of cattle. Finally, our understanding of the structure and function of nucleic acids (DNA and RNA) and the burgeoning field of recombinant DNA technology is based on knowledge accumulated over the last several decades from studying bacteria and viruses. The history of microbiology is replete with discoveries that not only have improved our lives but are in their own right intellectually exciting.

The purpose of the exercises in this laboratory activity is to introduce you to the exciting field of bacteriology. The activities that follow are not exhaustive but will help you develop skills for culturing bacteria just as a bacteriologist would in a research laboratory. In addition, you will learn several techniques for identifying unknown bacteria. The techniques you will use have been developed and practiced by bacteriologists over the years and should not prove difficult to learn. Before you work with bacteria, there are two areas that need to be discussed to make your work go smoothly and safely.

SAFETY RULES IN MICROBIOLOGY LABORATORY

The bacteria you are working with are relatively harmless and are normally associated with our bodies and the environment in which we live. However, most microbes are opportunistic pathogens and may cause disease under certain circumstances. It is your responsibility to see that these organisms are handled safely so that you do not contaminate your work area or yourself. You must adhere to the following rules as you work in the laboratory.

- Wear an apron or laboratory coat. All outer garments must be hung in the designated area of the laboratory, not on the table or back of chairs. Keep the tabletop clear of all books except your laboratory manual.
- Never eat or store food in the laboratory. Do not smoke.
- Never place pencils, labels, or hands in your mouth. When necessary, use a wet sponge to moisten any labels that are to be applied to cultures.
- If a culture is spilled, immediately saturate the area with disinfectant and then cover with a paper towel. Inform your instructor immediately.
- Begin each laboratory session by disinfecting your work area with the disinfectant provided by your instructor. Spread the disinfectant liberally over the area with a paper towel and allow to dry. Repeat this procedure before you leave the laboratory at the end of the session.
- Always wash your hands before you leave the laboratory. This applies to coffee and restroom breaks as well as when you have finished your work for the day.

- Label all culture material with your name, class, date, and experiment.
- Dispose of all used cultures, pipettes, glassware, and any material that has become contaminated with bacteria (including paper towels) in the proper receptacle. This material must be sterilized before disposal or reuse.
- Tie back long hair to keep it away from open flames. Long, dangling sleeves also present a fire hazard and must not be worn in the laboratory.
- As indicated earlier in this laboratory manual, report all accidents no matter how minor, to your instructor.

CULTURING AND OBSERVING BACTERIA

When bacteria are cultured in the laboratory they must be supplied with certain basic nutrients. These nutrients include organic compounds, water, and a variety of mineral nutrients such as nitrogen, phosphate, iron, and magnesium. When these nutrients are added to agar (a gelatinlike material) a solid medium results allowing bacteria to grow on its surface. Broth cultures (without agar) may also be used. Bacteria are widely distributed in our environment, and when a culture medium is prepared, it must be sterilized to destroy these unwanted organisms. Sterilization is usually done under high temperature (121°C) and pressure (15 lb./in.^{2}) in a large pressure cooker–type machine called an autoclave. After sterilization, cultures are free of all potentially contaminating organisms and are suitable for inoculation with the bacteria that you wish to grow. It is vitally important that you keep your cultures free from contamination so that only the single species you are concerned with grows in or on your medium.

Bacteriologists have developed techniques for transferring bacteria from one culture to another to minimize the risk of contamination. These techniques are referred to as **aseptic techniques,** and learning them will help ensure that your cultures remain free from contamination.

Bacteria, like other organisms, come in a range of sizes and present a variety of shapes as they grow. The three distinct forms that are most commonly recognized are: cocci, bacilli, and spirilla.

The **cocci** (the singular *coccus* means "berry") have a spherical shape. Cells may grow singularly, in pairs, cubes, or chains. The **bacilli** (the singular *bacillus* means "little staff") are shaped like rods or cylinders. There is often great variation in the length of these bacterial cells. Many are long and slender, while others are so short that they may resemble cocci. Sometimes the bacilli will remain attached to each other after they divide and form a long chain. The **spirilla** (the singular *spirillum* means ''coil'') are primarily unattached individual cells. Spirilla resemble a corkscrew, although the number of spirals may vary considerably from species to species.

Bacteria are extremely small and must be observed under the oil immersion lens. They are measured in micrometers (refer to Laboratory Activity 2) with the cocci varying in diameter from 0.5–1.0 μm. The bacilli vary from 0.5–1.0 μm wide by 2.0–5.0 μm long. As you observe each slide of bacteria you should note their individual shapes and the patterns of cell arrangement, such as chains, clusters, filaments. In order to observe bacteria it is necessary to stain the cells with a dye to enhance their visibility in the microscopic field.

MATERIALS

Apron or lab coat
Disinfectant
Nutrient agar slants of *Bacillus subtilis* and
  Micrococcus luteus
Microscope slides
Inoculating loop
Wash bottle of distilled water
Bunsen burner
Staining tray
Crystal violet stain
Paper towels
Compound microscope
Prepared slide of a spiral-shaped bacterium
Mixed broth culture of *Micrococcus luteus* and
  Escherichia coli
Tubes of melted nutrient agar in a 50°C water
  bath
Sterile petri dishes
Gummed labels
Marking pencil or pen
Gram's iodine
Ethanol (95%)
Safranin stain

PROCEDURE

Your instructor will give you two nutrient agar slants, one containing *Bacillus subtilis* and the other *Micrococcus luteus*. The agar is slanted to provide adequate surface for bacterial growth. Bacteria are normally transferred with an instrument called an inoculating loop. This loop is 3–4 mm in diameter and is attached to a metal handle. To prepare a slide of bacteria carry out the following procedure.

1. Obtain a glass microscope slide that is completely free of lint and grease. Using an inoculating loop, add one or two loops of distilled water to the center of the slide. This will provide a medium in which to spread the bacteria.

2. Sterilize the inoculating loop by holding it over a Bunsen burner and flaming the wire where it joins the handle and then moving out toward the loop end until the wire glows red. Do not set the loop down once it is sterilized. Allow it to cool for 45 seconds.

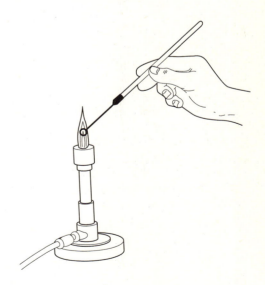

3. Hold the culture tube of *B. subtilis* in the opposite hand and remove the plastic cap or cotton plug with the last two fingers of the hand holding the loop. Flame the neck of the test tube by passing it through the flame twice. While it is open, hold the tube nearly horizontal, not vertical, to reduce the possibility of bacteria entering the culture.

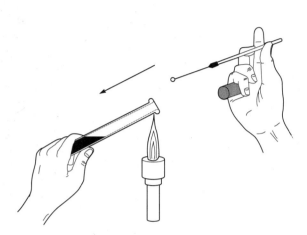

4. While holding the cap, insert the loop into the culture tube and gently remove a small amount (about the size of a pinhead) of the bacterial growth. Be careful not to break the surface of the agar. Reflame the test tube neck, replace the cap, and return the tube to the rack.

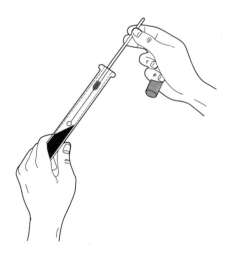

5. Place the loop into the drop of water on the slide and spread the mixture out to about the size of a quarter. Allow to air dry. When finished, flame the loop to prevent contamination of your work area.

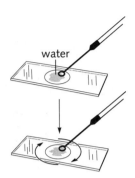

6. Fix the smear to the slide by passing the slide through a Bunsen burner flame several times, *smear side up.* Count to yourself "one and out, two and out, three and out" each time you flame the slide to prevent it from getting too hot.

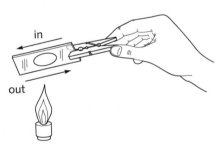

7. Repeat Steps 1–6 to prepare a slide of *M. luteus.*

8. Place both slides on a staining tray and flood with crystal violet stain for 1 minute.

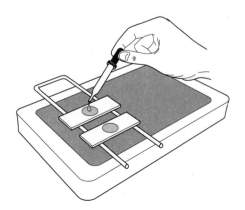

9. Gently wash with distilled water from a wash bottle to remove excess stain.

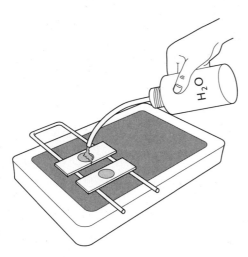

10. Blot with a paper towel to remove excess water and allow to air dry. Examine both slides under the oil immersion lens of your microscope. Also examine a commercially prepared slide of spiral-shaped bacteria. Record your observations of the three bacteria in Question 1 in the Evaluation.

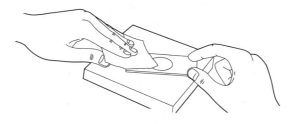

Your next task in learning to work with bacteria is to transfer aseptically a mixture of bacteria to a new culture, isolate each species of bacteria from the new culture, and do a Gram stain on each isolated species. Although this may seem a formidable task at first, bacteriologists routinely isolate individual species of bacteria from a mixture by means of the streak plate culture technique, and you can carry out this same procedure with an excellent chance of success. The streak plate technique is used to dilute a concentrated culture of bacteria and to allow the bacteria to form individual colonies that can be studied for their growth characteristics. Each colony of bacteria that grows will be

from one bacterium that was isolated from the others in the streaking process. Thus streaking provides a means of isolating individual bacteria from a mixture to obtain a pure culture. Bacteriologists frequently use this method to aid in the identification of unknown bacteria contained in a culture.

1. Obtain a broth culture that contains a mixture of the bacteria *Micrococcus luteus* and *Escherichia coli*.

2. Obtain a tube of melted nutrient agar from a 50° C water bath. Allow the tube to cool for a few minutes.

3. Remove the cap from the tube and flame the neck of the culture tube. Raise the lid of a sterile petri dish just enough to allow you to pour the melted agar into the dish. Replace the lid quickly and, if the agar has not spread over the bottom, gently rotate the dish on the tabletop until it spreads evenly. Allow to harden (about 10 minutes) and then label the bottom of your pour plate with your name, date, and the words "Isolation experiment."

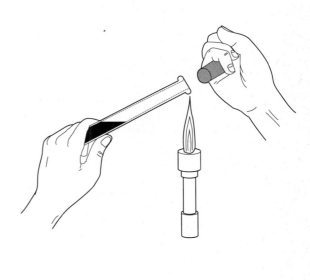

4. Roll the broth culture of mixed bacteria gently between your hands or use a vortex stirrer to distribute the organisms.

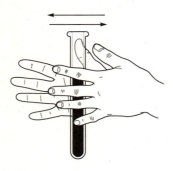

5. Flame an inoculating loop to redness. Remove the cap from the culture tube and flame the neck.

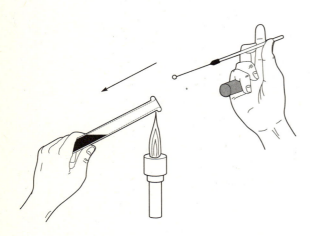

6. Allow the loop to cool for 45 seconds and then remove a loop of culture medium containing bacteria from the tube. Lift the lid of the petri dish and place the loop at the edge farthest from you. Carefully move the loop back and forth, making a series of streaks covering about one-quarter of the dish. Make certain that the loop handle is almost horizontal to the agar surface and that the loop itself is parallel to the surface. *Do not dig the agar*. Close the lid and flame the loop.

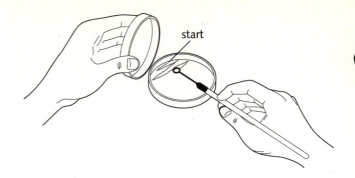

7. Rotate the petri dish one-quarter turn, flame the loop, and then skim the loop over the surface of the agar, touching the original streaked area once. This second set of streaks should cover about one-quarter of the dish. Rotate the dish another quarter turn and repeat the streaking procedure to cover another quarter of the dish.

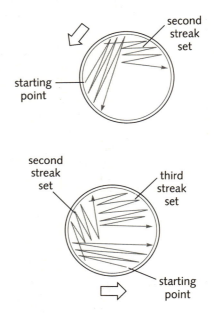

8. Grow and examine the culture in your streak plate according to your instructor's directions.

After the required period of growth, examination of the bacterial colonies should reveal two types. Colonies of *E. coli* are grayish white, and those of *M. luteus* are yellow. Other characteristics of colony growth are the shape of the colony, its texture, how it is elevated from the medium and the contour of the margin. Table F.1 summarizes several of the important colony characteristics. Consult it as you examine your cultures and record the word that best describes the colonies growing in your petri dish in Question 2 of the Evaluation.

Table F.1 Selected characteristics of bacterial colonies

Shape of entire colony	circular	irregular	spindle	filamentous	rhizoid
Elevation (seen from side)	raised	convex			
Shape of edge or margin	entire	undulate	lobate	filamentous	curled
Surface texture	smooth shiny	wrinkled	rough	dry	

From Michael J. Pelczar, Jr., *Laboratory Exercise in Microbiology*, 2nd ed., McGraw-Hill Book Co.

The next activity involves the **Gram stain,** one of the most useful staining procedures in microbiology. The Gram stain is a **differential stain** because it divides bacteria into two groups: Gram-positive and Gram-negative. This staining technique was developed in 1884 by the Danish bacteriologist Christian Gram as he was studying the etiology (origin) of respiratory diseases. While attempting to differentiate the microorganisms from surrounding lung tissue, he discovered that the stains he was using stained some bacteria and not others. The Gram stain has since become one of the most important and widely used tools in the identification of unknown bacteria.

The procedure requires four different solutions: a basic dye, a mordant, a decolorizing agent, and a counterstain. A **basic dye** (crystal violet) is one in which the ion that produces the color is positively charged. A **mordant** (iodine) is a substance that increases the affinity of the cell for the dye. When a mordant is used, it is more difficult to wash out the dye from the cell. A **decolorizing agent,** such as ethanol, removes stains from cells. This is of interest since bacteria differ in the rate at which they decolorize. The **counterstain,** safranin, is a basic dye that is a different color from the crystal violet and enables you to compare those cells that decolorize with ethanol to those that do not.

In Gram-positive bacteria the cell wall is able to retain the crystal violet stain after washing with ethanol and therefore appears deep violet. Bacteria that are Gram-negative lose the crystal violet stain after washing with ethanol, are counterstained by the safranin, and appear light red.

1. Place a slide in the staining tray and add a loopful of water. Aseptically remove a pinhead-size sample of bacteria from one colony in the petri dish and stir in the water drop until dry. Sterilize the loop in the flame when you are finished.

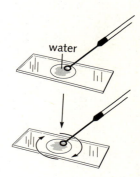

2. Flame the slide to fix the smear, using the same procedure as in the previous activity.

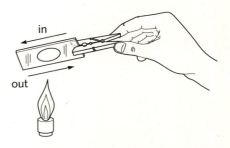

3. Flood the smear with crystal violet stain for 1 minute and then tilt the slide to drain off the excess stain.

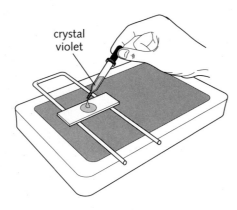

crystal violet

4. While tilting the slide, gently rinse with distilled water from the wash bottle.

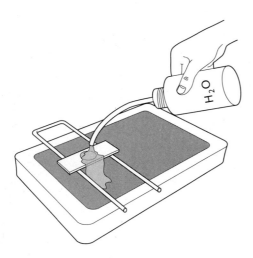

H₂O

5. Replace the slide on the rack and flood with Gram's iodine for 1 minute.

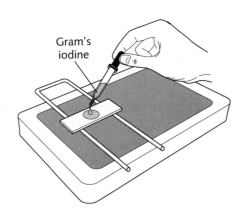

Gram's iodine

6. Drain excess stain from slide and wash with distilled water from a wash bottle.

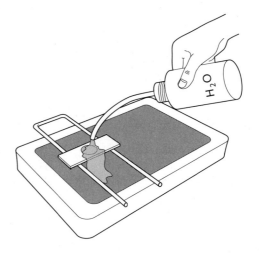

H₂O

7. While holding the slide at an angle, decolorize the smear quickly by allowing alcohol to run over the smear until no more dye runs off. Rinse *immediately* with distilled water to stop the decolorizing process.

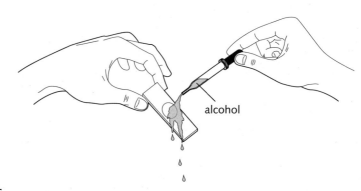

alcohol

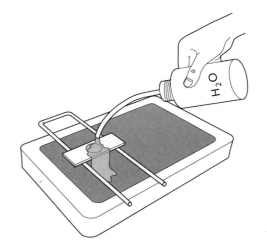

H₂O

8. Replace slide on rack and flood with safranin stain for 30–60 seconds.

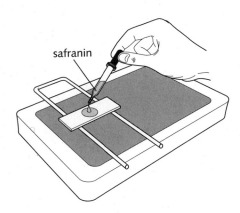

safranin

9. Drain excess stain and rinse with distilled water from a wash bottle. Blot carefully with a paper towel and allow to air dry.

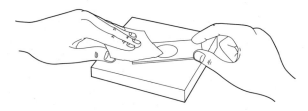

10. Repeat this process with the other bacterial colony. When this second slide is finished, observe under the oil immersion lens and complete Question 3 in the Evaluation.

E V A L U A T I O N **F**

An Introduction to Bacteriology

1. Diagram each of the bacteria studied and label with the appropriate name and magnification. Estimate their sizes using the method learned in Laboratory Activity 2. Describe the shape of each bacterium and the pattern of growth of the cells; e.g., in chains, clusters, or isolated.

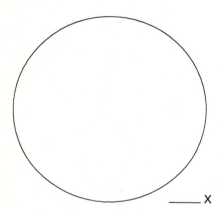

Name:
Mag.
Size
Growth pattern

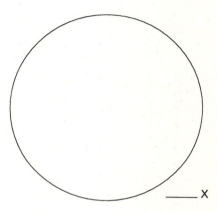

Name:
Mag.
Size
Growth pattern

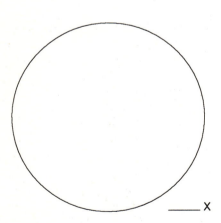

Name:
Mag.
Size
Growth pattern

	E. coli	*M. luteus*
Diagram of colony		
Pigmentation		
Colony shape		
Elevation		
Shape of margin		
Surface texture		

3. Draw and label each of the Gram-stained bacteria below. Record the magnification used, the name of the bacterium, and whether it was Gram-positive or negative.

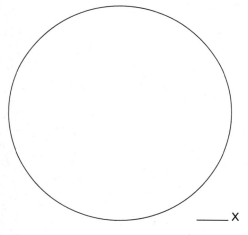

_____ X

Name _____

Gram stain _____

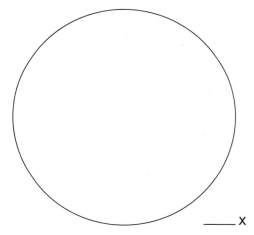

_____ X

Name _____

Gram stain _____

The Identification of Unknown Bacteria

OBJECTIVES

At the end of this laboratory activity, you should be able to:

- perform the various tests used to identify unknown bacteria.
- describe the general procedures used to identify unknown bacteria.

INTRODUCTION

One of the major tasks of microbiologists is to identify unknown microorganisms. This procedure requires the investigator to culture the unknown organism in a variety of ways in order to accumulate information about its **morphology** (structure) and **physiology** (function). Although bacteria have many distinctive morphological features, they are very difficult to identify precisely unless their biochemical characteristics are also examined. Bacteria biochemically differentiate themselves in a number of ways, such as the ability to grow only on certain nutrients, the production of various waste products, the production of pigments, and, as you have already learned, their reaction to certain staining procedures.

The biochemical tests that you will use in this activity have been selected because they are relatively easy for the beginning student to perform and interpret. You will use the results of these tests along with the other characteristics mentioned above to identify your unknown bacteria.

MATERIALS

Apron or lab coat
Disinfectant
Sterile petri dishes
Sterile tube of starch agar
Bunsen burner
Inoculating loop
Cultures of unknown bacteria
Gram's iodine
Gummed labels
Marking pencil or pen
Sterile tube of SIM agar
Sterile tube of tryptone broth
Kovac's reagent
Sterile tube of nitrate broth
Sulfanilic acid
Dimethyl alpha naphthalamine

PROCEDURE

The key in Figure G.1 is designed to enable you to identify eight unknown bacteria. Your instructor will give you one or more unknown cultures labeled with a code number. By careful culturing and observation, you should be able to identify them. Note that at each step in the key you are given two choices based on the morphology and chemical characteristics of the unknown bacteria. This dichotamous key is similar to the one used in Laboratory Activity 3.

Figure G.1 Dichotomous key for the identification of unknown bacteria

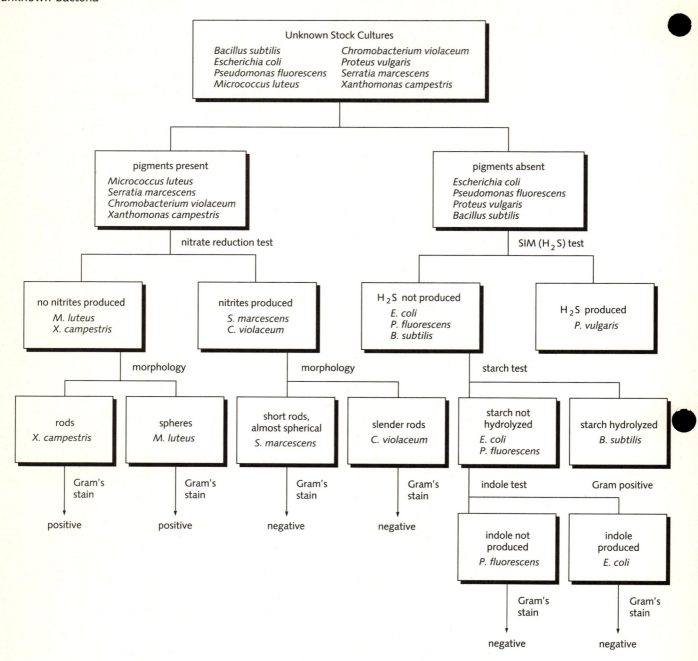

The following pages contain the procedures for doing the chemical tests you will use in the identification of your unknown bacteria. Always read the procedures carefully before testing your culture. As you complete each test, record your observations in the Summary Chart for Unknown Bacteria in the Evaluation.

Starch Hydrolysis

Bacteria produce enzymes known as exoenzymes because they are secreted by the cell into the environment. One class of exoenzymes, the **hydrolyses,** facilitate the breakdown of large molecules into smaller subunits that are able to cross the cell membrane and be used by the bacterial cell during metabolism. Starch is one of the main foods encountered by bacteria, and many are able to digest it into smaller molecules such as glucose and maltose. The following test demonstrates the ability of a bacterium to digest starch.

1. Obtain a sterile petri dish and prepare a pour plate of starch agar using the technique you learned in Appendix F.

2. After the agar solidifies, make a single streak of the unknown bacteria on the agar surface (see Figure G.2). Label the bottom of the petri dish with your name, date, section, culture number, and the test performed. Store the petri dish upside-down.

3. Incubate according to your instructor's directions. If the bacteria have produced exoenzymes, the starch will be hydrolyzed to sugar. You can determine this by flooding the plate with a thin layer of iodine solution. If starch is present, it will turn blue-black. A clear area around the streak of bacteria indicates hydrolysis of starch. Record your observations in the Evaluation.

Figure G.2 Starch test

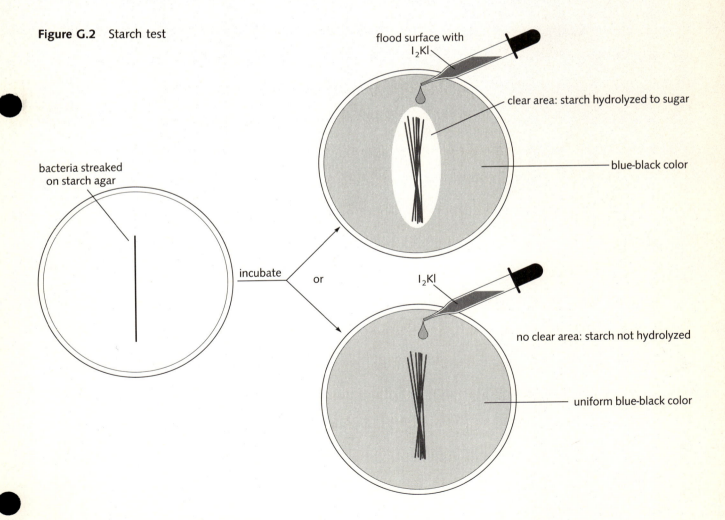

bacteria streaked on starch agar

incubate or

flood surface with I$_2$KI

clear area: starch hydrolyzed to sugar

blue-black color

I$_2$KI

no clear area: starch not hydrolyzed

uniform blue-black color

Hydrogen Sulfide Production

Certain bacteria, when grown in a medium made with a sulfur-containing amino acid such as cysteine, will produce hydrogen sulfide gas. The hydrogen sulfide gas is produced as the sulfur is released from the amino acid and combines with hydrogen. Hydrogen sulfide gas has the odor of rotten eggs and is often detected when certain foods become contaminated with bacteria and spoil.

All bacteria are not able to decompose cysteine and release sulfur. The SIM test is one of several tests developed to detect the production of hydrogen sulfide by bacteria. The SIM medium contains the amino acid cysteine and iron. The iron will combine with the hydrogen sulfide and produce a black color. Perform this test using the following procedure.

1. Obtain a tube of sterile SIM agar. Using a straight inoculating wire, aseptically remove a small portion of bacteria from the stock culture and inoculate the SIM agar by stabbing the wire straight into the agar (Figure G.3). Label your tube with your name, date, section, culture number, and test.

2. Incubate the culture according to your instructor's directions. After incubation, examine your culture for a black streak, which indicates the production of H_2S.

Indole Production

Indole is a nitrogen-containing compound formed by certain bacteria as they metabolize the amino acid tryptophan. Certain intestinal bacteria have this ability, and it is useful in distinguishing them from other bacteria. The medium used is a 1% tryptone broth.

The indole test uses a solution called Kovac's reagent to detect the presence of indole. This reagent will turn a deep red color if indole is present.

1. Obtain a sterile tube of tryptone broth containing the amino acid tryptophan.

2. Using aseptic techniques, inoculate the broth culture with a loop of the unknown culture. Label your tube with your name, date, section, culture number, and test.

3. Incubate according to your instructor's directions.

4. Test for indole by adding about 10 drops of Kovac's reagent to the broth culture. Recap the tube and mix with a vortex stirrer or by gently rolling the tube between your hands. Observe for 10–15 minutes for a red color to develop in the surface layer.

➤ **CAUTION: Kovac's reagent contains hydrochloric acid and will irritate the skin. If you spill any on your skin or clothes, wash immediately with copious amounts of water and notify your instructor.**

Figure G.3 Hydrogen sulfide test

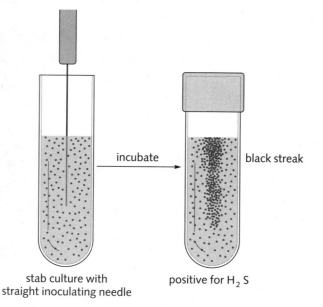

incubate → black streak

stab culture with
straight inoculating needle

positive for H_2S

Nitrate Reduction

Bacteria, like other organisms, use certain organic compounds as an energy source. As these compounds are broken down, electrons are produced and must be disposed of in some manner. In aerobic organisms this is accomplished by combining the electrons with molecular oxygen and hydrogen ions to make water (Laboratory Activity 26). Some anaerobic organisms use other electron acceptors such as nitrate (NO_3^-), reducing it to nitrite (NO_2^-). The fact that only certain bacteria are able to reduce nitrate to nitrite is a useful characteristic in their identification.

The test is done by growing the bacteria in a nitrate broth medium. If nitrites are produced, they can be detected by adding two test solutions to the culture. A red color indicates the presence of nitrites.

1. Obtain a sterile culture tube of nitrate broth.

2. Aseptically inoculate the broth tube with a loop of culture from the unknown culture. Label with your name, date, section, culture number, and test.

3. Incubate according to your instructor's directions.

4. After incubation, add 2–3 drops each of sulfanilic acid and dimethyl alpha naphthalamine to a broth culture. If nitrites are present, a red color will appear immediately.

➤ CAUTION: Dimethyl alpha naphthalamine is a potential carcinogen. Follow your instructor's directions carefully and wash your hands after using this reagent.

EVALUATION G

Identification of Unknown Bacteria

In the following chart record the results of any of the biochemical tests that
you did. Indicate a plus (+) if the reaction occurred and a minus (−) if the
reaction did not occur. In addition, make a brief statement that describes
what happened in each test.

| Test | Unknown bacteria | | |
	# _____	# _____	# _____
Morphology (include diagram and description)			
Pigmentation			
SIM			
Nitrite			
Indole			
Starch			
Gram stain			
Name of unknown culture			

Illustrations

Charles Boyter   p. 350

Cecile Duray-Bito   pp. 6, 44, 54, 62–64, 72, 81, 82, 96, 106, 129, 130 bottom, 131, 139–141, 143, 145, 146, 148, 149, 153, 155, 163, 174, 227, 288, 293, 299, 310, 318, 320, 328, 331–335, 363, 374, 383, 395

Arleen Frasca   pp. 340, 353 top

Marlene DenHouter   pp. 151, 357

Illustrious Inc.   pp. 3, 4, 16, 17, 20, 21, 24, 35, 45, 48, 52, 53, 58, 71, 73, 75, 90, 107, 108, 111, 112, 113, 115, 116, 117, 126, 147, 158, 159, 162, 179–182, 184, 185, 195, 205–211, 215, 216, 219, 234, 236, 243, 244, 250, 251, 255, 259, 265, 272, 284, 290, 296, 300, 305–307, 313, 351, 353 bottom, 354, 359, 361, 365, 369, A1–A26

Patrice Rossi   2, 12, 130 top, 150, 193, 311

Marcia Smith   302

Michael Woods   42, 43

Photographs

All photos researched by Sharon Donahue.

p. 19 (top), Ray Simons/Photo Researchers, Inc; p. 19 (bottom), Arthur M. Siegelman; p. 23, Arthur M. Siegelman; p. 47, Janice M. Sheldon; p. 48, Janice M. Sheldon; p. 73, Janice M. Sheldon; p. 80, South Tahoe Public Utility District/Biological Photo Service; p. 88 (all) Andrew Bajer; p. 89 (counterclockwise from top), Arthur M. Siegelman; Carolina Biological Supply Company; Arthur M. Siegelman; Arthur M. Siegelman; Bruce Iverson, BSc; p. 96 (top two), Carolina Biological Supply Company; p. 96 (bottom), D. A. Glawe, University of Illinois/Biological Photo Service; p. 97, Carolina Biological Supply Company; p. 120 (both), Biophoto Association/Photo Researchers, Inc; p. 122 (all), Janice M. Sheldon; p. 128 (from top to bottom), David Woodward/

Taurus Photos, Inc; C. R. Wyttenbach, University of Kansas/Biological Photo Service; Arthur M. Siegelman; C. E. Mills/University of Washington/Biological Photo Service; Bruce Iverson, BSc; Harold W. Pratt/Biological Photo Service; p. 131, Omikron/Photo Researchers, Inc; p. 138 (from top to bottom), Carolina Biological Supply Company; Arthur M. Siegelman; Bruce Iverson, BSc; Scott Camazine; Eric Grave/Photo Researchers, Inc; Jack Dermid/Photo Researchers, Inc; Carolina Biological Supply Company; p. 150, Carolina Biological Supply Company; p. 159, Samuel W. Woo; p. 174 (both), Paul Silverman/Fundamental Photographs; p. 175, Michael Dalton/Fundamental Photographs; p. 180, Herman Eisenbeiss/Photo Researchers, Inc; p. 203, N. L. Max/University of California/Biological Photo Service; p. 215, Science Source/Photo Researchers, Inc; p. 218, Samuel W. Woo; p. 226, R. Rodewald/University of Virginia/Biological Photo Service; p. 232, Dennis D. Kunkel/Biological Photo Service; p. 235, Don Fawcett, M.D., and Keith Porter/Photo Researchers, Inc; p. 271, E. H. Newcomb and W. P. Wergin/University of Wisconsin/Biological Photo Service; p. 283 (a), Dr. E. R. Degginger/FPSA; p. 283 (b), Jerome Wexler/Photo Researchers, Inc; p. 289 (a and c), Michael S. Thompson/Comstock Editorial Photography; p. 289 (b), Dr. E. R. Degginger/FPSA; p. 290, Farrell Grehan/Photo Researchers, Inc; p. 291, Biophoto Association/Photo Researchers, Inc; p. 303 (top), Scott Camazine/Photo Researchers, Inc; p. 303 (bottom), Carolina Biological Supply Company; p. 304, Carolina Biological Supply Company; p. 339 (left), G. W. Willis, M.D./Ochsner Medical Institution/Biological Photo Service; p. 339 (right), Arthur M. Siegelman; p. 340, Bruce Iverson, BSc; p. 341 (top left), Chuck Brown/Photo Researchers, Inc; p. 341 (top right), Carolina Biological Supply Company; p. 341 (bottom), L. Winograd/Stanford University/Biological Photo Service; p. 342 (left), Carolina Biological Supply Company; p. 342 (right), G. W. Willis, M.D./Ochsner Medical Institution/Biological Photo Service; p. 343 (top two), Carolina Biological Supply Company; p. 343 (bottom); G. W. Willis, M.D./Ochsner Medical Institution/Biological Photo Service; p. 344 (left), Robert Knauff/Photo Researchers, Inc; p. 344 (right), M. I. Walker/Photo Researchers, Inc; p. 350, Taurus Photos, Inc; p. 351, David Powers/Stock, Boston; p. 374, Bruce Iverson, BSc; p. 403, Janice M. Sheldon; p. A9, courtesy Milton Roy/Analytical Products Division.